Michael Falk
Frank Marohn
Bernward Tewes

# Foundations of
# Statistical Analyses and
# Applications with SAS

Springer Basel AG

Michael Falk
Institute of Applied Mathematics und Statistics
University of Würzburg
Am Hubland
97074 Würzburg
Germany
falk@mathematik.uni-wuerzburg.de

Frank Marohn
Faculty of Mathematics and Geography
Catholic University of Eichstätt-Ingolstadt
85071 Eichstätt
Germany
frank.marohn@ku-eichstaett.de

Bernward Tewes
Computing Center
Catholic University of Eichstätt-Ingolstadt
85071 Eichstätt
Germany
bernward.tewes@ku-eichstaett.de

Originally published in German under the title «Angewandte Statistik mit SAS. Eine Einführung»
by Springer Verlag.
© 1995 Springer Verlag

2000 Mathematics Subject Classification 62-01, 62-04, 62-07, 62-09


A CIP catalogue record for this book is available from the Library of Congress, Washington D.C.,
USA

Deutsche Bibliothek Cataloging-in-Publication Data
0101 deutsche buecherei

Falk, Michael:
Foundations of statistical analyses and applications with SAS / Michael Falk ; Frank Marohn ;
Bernward Tewes. - Basel ; Bosten ; Berlin : Birkhäuser, 2002
    ISBN 978-3-7643-6893-7     ISBN 978-3-0348-8195-1 (eBook)
    DOI 10.1007/978-3-0348-8195-1


ISBN 978-3-7643-6893-7

© 2002 Springer Basel AG
Originally published by Birkhäuser Verlag in 2002


Cover design: gröflin. graphic design, www.groeflin.ch
Printed on acid-free paper produced of chlorine-free pulp. TCF ∞

ISBN 978-3-7643-6893-7

9 8 7 6 5 4 3 2 1                              www.birkhasuer-science.com

# Contents

# Preface

The analysis of real data by means of statistical methods with the aid of a software package common in industry and administration, usually is not an integral part of mathematics or mathematical statistics studies, but it will certainly be part of a future professional work of each student.

Nevertheless, there is commonly no natural place in a traditional curriculum for mathematics or statistics, where a bridge between theory and practice fits into. On the other hand, the demand for an education designed to supplement theoretical training by practical experience has been rapidly increasing.

There exists, consequently, a bit of a dichotomy between theoretical and applied statistics, and this book tries to straddle that gap. It links up the theory of a selection of statistical procedures used in general practice with their application to real world data sets using the statistical software package SAS (Statistical Analysis System). These applications are intended to illustrate the theory and to provide, simultaneously, the ability to use the knowledge effectively and readily in execution.

### Targeted Audience
This book addresses students of statistics and mathematics in the first place. But students of other branches such as economics or biostatistics, where statistics has a strong impact, and related lectures belong to the academic training, should benefit from it as well. It is also intended for the practitioner, who, beyond the use of statistical tools, is interested in their mathematical background.

### Contents
The book fits in a statistics program, which has a strong theory component at the beginning and an applied course at the end. It is meant for a two semester course (lecture, seminar or practical training), where the first four chapters can be dealt with in the first semester. They provide a basic knowledge of exploratory data analysis, theory of normal data, regression analysis and of the analysis of categorical data. The second semester deals with multivariate techniques, where a selection can be made from chapters on the analysis of variance, discriminant analysis, cluster analysis including multidimensional scaling, principal components and factor analysis. Numerous problems illustrate the applicability of the presented statistical procedures, where SAS computes the solution. The programs used are explicitly listed and commented. Each chapter includes exercises and an exhaustive treatment is recommended. An appendix offers a brief introduction to SAS, its program structure syntax and its display manager system under a windows shell.

### Level of the Book

The book requires a higher level of mathematics than is assumed for most applied statistics books. This permits us to not only present the methods but to also provide an understanding of the theoretical foundation underpinning them. The book is, consequently, put at a level, where the student has already had an introduction to probability theory and mathematical statistics. For the targeted audience the inclusion of proofs in the text is mandatory and excercises requiring proofs are necessary for their training.

### Level of SAS Programming

No previous experience in SAS is required and the level of the SAS programming should be no serious problem for a student of mathematics or statistics, who has some practical knowledge of an arbitrary software package.

The SAS version used in most of the programs is version 6.12. In cases, where the current version 8 offers crucial new features such as PROC KDE for kernel density estimation, we use this one. The syntax for the interface with the operating system i.e., where to read and write permanent data files, is the syntax of Microsoft® Windows®.

### Organization of the Text

This book is consecutively subdivided in a statistical part and a SAS–specific part. For better legibility the SAS–specific part, including the diagrams generated with SAS, always starts with a computer symbol, representing the beginning of a session at the computer, and ends with a printer symbol for the end of this session.

This SAS–specific part is again divided in a diagram created with SAS, the program, which generated the diagram, and explanations to this program. In order to achieve a further differentiation between SAS–commands and individual nomenclature, SAS–specific commands are written in CAPITAL LETTERS, whereas individual notations are written in lower-case letters.

In addition, these programs as well as the data files can be downloaded from `http://statistics-with-sas.ku-eichstaett.de/`. Further information about data records and how to read them into the SAS–system is published there as well.

## Stimulating Books

The excellent book *Mathematical Statistics and Data Analysis* by John Rice actually stimulated the writing of this book. We used it successfully in our lectures, but we found nevertheless that there is still a missing link between theory and applications, namely the incorporation of a widely used non academic computer package for the computation of the examples.

Another stimulating text was *Applied Multivariate Data Analysis*, Volume I and II, by J.D. Jobson. It lists the software packages used in the examples, but gives no further details.

*Applied Multivariate Statistics with SAS Software* by R. Khattree and D.N. Naik is an excellent text with the focus on statistical methods and their applications with SAS, but it does not provide the mathematically rigorous development of the statistical theory.

## Acknowledgments

We would like to thank Rainer Becker, who gave expert SAS advice at an early stage of this project. We especially thank Sreenivasan Ravi for the great deal of time that he spent on reading the manuscript and preparing many helpful comments. We thank numerous students at the Catholic University of Eichstaett–Ingolstadt and the University of Wuerzburg for pointing out errors in earlier drafts. Helma Höfter typed much of the very first draft and Hildegund Schulz–Merkel continued and completed the work in a most helpful manner. Peter Zimmermann provided expert LaTeX advice at every stage.

## A Warning

Though we derive seemingly precise results in this book, we may never forget that *statistics is never having to say you're certain.*

| | |
|---|---|
| Würzburg | Michael Falk |
| Eichstätt | Frank Marohn |
| | Bernward Tewes |

# Chapter 1

# Elements of Exploratory Data Analysis

The computer enables us to analyze a given data set by applying a variety of statistical procedures with different tuning parameters, thus revealing hidden structures in the data within seconds. Statistics provides us with the methods. But though we derive seemingly precise results, we may never forget that *statistics is never having to say you're certain.*

## 1.1  Histograms and Kernel Density Estimators

Starting with histograms, or more generally, kernel density estimators, in this section we compile several procedures which can serve as first steps in the analysis of a given data set. The procedures typically provide basic information about the data. For instance, the data can be measurements made to answer questions similar to those in the examples given below.

**1.1.1 Example.** Three different ways of cultivating two types of wheat are to be compared. The comparison is based on the yearly total production of a specified area. We may be interested in analyzing the effects of the two variables 'cultivation' and 'wheat' on the crop yield.

**1.1.2 Example.** A music hall is to be built. In order to plan its size and seating arrangements, the prospective audience has to be studied and the population has to be analyzed. The analysis aims at modeling the individual inclination of an inhabitant to attend the music hall events as a variable, which depends on explanatory variables such as gender, income or educational level together with an interfering disturbance.

**1.1.3 Example** (CNS Data; Läuter and Pincus (1989), page 21). Mental deterioration is caused by degenerations of the central nervous system (cns). Certain diseases such as Alzheimer's (group 1), Pick's (group 2) and senile mental deterioration (group 3) are be to identified by morphological deviations of the cns when compared to mentally healthy persons (age 50 – 60 years (group 4) and age 61 – 103 years (group 5)). The comparison is based on cells in the ammonite formation of the cns, which is concerned with the ability to learn and memorize complex facts. The number of cells of type neuron $n$, astrocyte $a$, oligodendrocyte $o$, microglia $m$ and glia $g$ in this formation of deceased patients was counted and the ratios ($an = a/n, on = o/n$ etc.) were computed. The following table lists the results. Those patients with an identification number between 1 and 20 belong to group 1, 21 to 39 belong to group 2, 40 to 58 to group 3, 59 to 78 to group 4 and 79 to 98 to group 5.

| NO | GROUP | AN | ON | MN | GN | AO |
|----|-------|------|------|------|------|------|
| 1 | 1 | 2.04 | 0.29 | 0.15 | 2.84 | 7.21 |
| 2 | 1 | 1.70 | 0.23 | 0.14 | 2.07 | 7.74 |
| 3 | 1 | 1.95 | 0.31 | 0.13 | 2.38 | 6.63 |
| 4 | 1 | 2.24 | 0.32 | 0.15 | 2.71 | 7.32 |
| 5 | 1 | 2.35 | 0.32 | 0.13 | 2.81 | 7.53 |
| 6 | 1 | 2.57 | 0.36 | 0.14 | 3.03 | 7.33 |
| 7 | 1 | 2.28 | 0.31 | 0.13 | 2.71 | 7.45 |
| 8 | 1 | 2.56 | 0.41 | 0.13 | 3.11 | 6.41 |
| 9 | 1 | 2.27 | 0.32 | 0.13 | 2.72 | 7.79 |
| 10 | 1 | 2.77 | 0.46 | 0.16 | 3.38 | 6.37 |
| 11 | 1 | 4.69 | 0.70 | 0.28 | 5.68 | 7.12 |
| 12 | 1 | 3.93 | 0.46 | 0.26 | 4.70 | 9.03 |
| 13 | 1 | 4.72 | 0.65 | 0.27 | 5.66 | 7.43 |
| 14 | 1 | 4.74 | 0.77 | 0.28 | 5.69 | 6.29 |
| 15 | 1 | 4.21 | 0.74 | 0.24 | 4.79 | 5.85 |
| 16 | 1 | 4.95 | 0.85 | 0.26 | 6.06 | 5.91 |

.
.
.

**Figure 1.1.1.** Printout of cns data.

```
***    Program 1_1_1   ***;
GOPTIONS RESET=GLOBAL;
TITLE1 'Printout';
TITLE2 'CNS Data';
LIBNAME datalib 'c:\data';

PROC PRINT DATA=datalib.cns NOOBS;
   VAR no group an on mn gn ao;
RUN; QUIT;
```

Here and in the following programs we assume that the data are already stored as a SAS data set. This is a data set that contains not only the data but also instructions to SAS about what they mean and how to display them. The rows of a SAS data set are referred to as *observations* and the columns as *variables*. Raw data are not directly usable by SAS procedures. One can use DATA steps to convert raw data to SAS data sets, cf Exercise 10 or Program 4_1_2 or one can use the add–on product SAS/FSP. This is a collection of procedures such as FSEDIT that facilitates SAS data set handling.

The statement 'GOPTIONS RESET=GLOBAL' resets current values of the graphics options to the default values.

The LIBNAME statement assigns a name (here 'datalib') to the link to the location (here 'c:\data') of permanent SAS data sets for the rest of the program. The syntax of this location depends on the operating system of the computer. Here we use the syntax of Microsoft Windows.

The statement 'PROC *procedure* DATA=*data set*' invokes a procedure, which analyzes the data set defined by the option 'DATA='. Without this option, SAS uses the most recently created data set.

The procedure PRINT lists the data set. Various options and statements can be used to dress up the display and to provide a first visual check that the data appear correctly, cf. SAS Procedures Guide. From SAS version 8 on the Procedures Guide is accessible through SAS OnlineDoc, delivered on CD in html format by SAS Institute. The observation numbers (OBS) are printed by default as a separate column in the output. This column can be suppressed by the option NOOBS. The number of each patient in this example is already a variable in the data set. We suppress, therefore, the automatic observation number by NOOBS. The VAR statement determines the order (from left to right) in which variables are displayed. If not specified, all variables are printed in the order of their definition.

The RUN statement tells SAS to stop reading statements and procedures and to execute the unit of work. It is recommended to enter RUN whenever a unit of work ends.

SAS statements start in general with a keyword and end with a semicolon ';'. Comment statements start with an asterisk '*'. Embedded comments have the structure '/* text of comment */'. Both are ignored by SAS.

Suppose that the data consist of $k$ measurements from each of $n$ objects

$$y_{11}, \cdots, y_{1k}$$
$$\vdots \qquad \vdots$$
$$y_{n1}, \cdots, y_{nk},$$

where the $i$–th row corresponds to the measurements from the $i$–th object. As $n$ and/or $k$ get large, in order to get an idea about the structure of the data, we need to summarize the data. The structures may be

– the location
– the spread
– the distribution

of the data. We will first discuss these items for each of the $k$ measurements individually. To this end we choose a particular measurement with column suffix $m \in \{1, \ldots, k\}$ and put

$$\begin{pmatrix} x_1 \\ \vdots \\ x_n \end{pmatrix} := \begin{pmatrix} y_{1m} \\ \vdots \\ y_{nm} \end{pmatrix}.$$

We now analyze the *one-dimensional data* $x_1, \ldots, x_n$. First we order the data $x_1, \ldots, x_n$ according to their size, thus obtaining the ordered values $x_{1:n} \leq x_{2:n} \leq \cdots \leq x_{n:n}$. In particular $x_{1:n} = \min\{x_1, \ldots, x_n\}$ is the minimum of the data and $x_{n:n} = \max\{x_1, \ldots, x_n\}$ is their maximum. The interval $[x_{1:n}, x_{n:n}]$ is the shortest one containing all data; its length $x_{n:n} - x_{1:n}$ is called the *range*.

These quantities give a first impression of the location and spread of the data. To obtain an impression of their distribution as well, we cover the interval $[x_{1:n}, x_{n:n}]$ by $d$ disjoint intervals or *cells*

$$(a_0, a_1], (a_1, a_2], \ldots, (a_{d-1}, a_d],$$

which we denote by $I_1, \ldots, I_d$. We require $a_0 < a_1 < \cdots < a_d$, $a_0 < x_{1:n} \leq x_{n:n} \leq a_d$. By putting $n_s :=$ number of those $x_1, \ldots, x_n$, which are elements of $I_s = |\{x_i, \ i = 1, \ldots, n : x_i \in I_s\}|$ and plotting $n_s$ against $I_s, 1 \leq s \leq d$, we obtain a *histogram*.

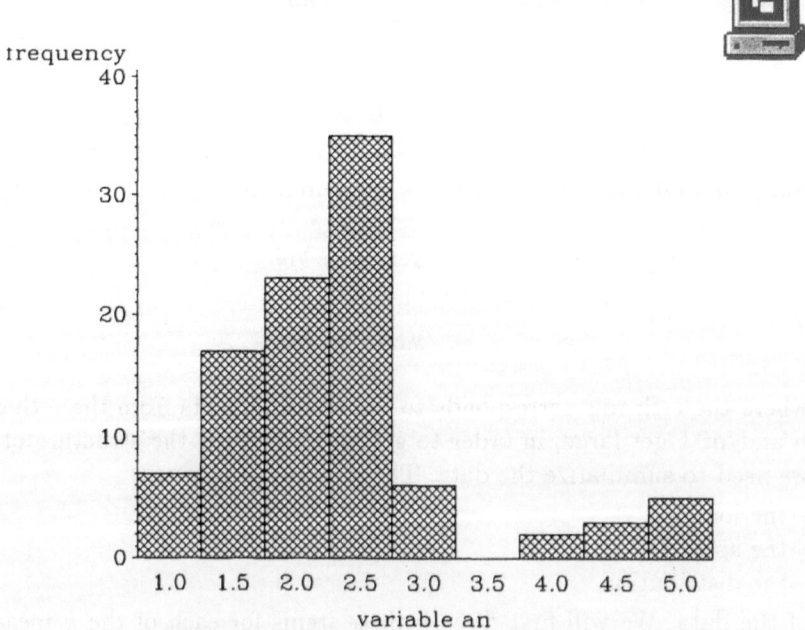

**Figure 1.1.2.** Histogram of the values *an* of the cns data.

```
***    Program 1_1_2    ***;
TITLE1 'Histogram';
TITLE2 'CNS Data';
LIBNAME datalib 'c:\data';

AXIS1 LABEL=('frequency');
AXIS2 LABEL=('variable an');
PROC GCHART DATA=datalib.cns;
    VBAR an / MIDPOINTS=1 TO 5 BY 0.5    SPACE=0
            RAXIS=AXIS1    MAXIS=AXIS2;
RUN; QUIT;
```

The procedure GCHART produces vertical and horizontal bar charts, block charts, pie charts and star charts. The VBAR statement creates a vertical bar chart of the chart–variable 'an' (=a/n). The option 'MID-POINTS=*value list*' specifies values for midpoints on the bar chart and, thus, determines the cells of the histogram. Since by default the bars are separated by a small space, the option 'SPACE=0' is needed to get a histogram display. The options 'RAXIS=AXIS1' and 'MAXIS=AXIS2' as-sign axis characteristics to the response (vertical) and to the midpoint (horizontal) axis as defined in the pertaining AXIS statements.

The global AXIS statement is followed by a number and those options you want to use such as LABEL in the example. This option controls the labels for the axis as well as such appearance items as the color ('C=*color*'), size ('H=*height*'), type ('F=*font*') An example is LABEL=(H=3 F=CENTX C=RED J=R 'frequency').

The histogram can give a misleading impression of the distribution of the data, in particular if the cells have different widths. It is, therefore, reasonable to weight the number $n_s$ of observations in each $I_s$ by its width $a_s - a_{s-1}$ and to plot

$$f_n(t) := \frac{n_s}{n} \frac{1}{a_s - a_{s-1}}, \qquad t \in I_s, \quad s = 1, \ldots, d,$$

against $t$. Note that $\int f_n(x)\,dx = 1$, i.e., $f_n$ is a *probability density*.

The choice of a cell width is a typical trade–off situation: If it is chosen too small with the extremal situation of one observation per cell, the histogram does not summarize the data, but it is rather a plot of the observations. If the cell width is chosen too large with the extremal situation that all data are elements of just one single cell, then the histogram is a constant function. In this case it is *overfitted* and does not reflect local characteristics of the data set. The information derived from a histogram about the distribution of the data is, therefore, quite sensitive to the choice of the cell width. This is a typical trade–off situation in graphical data analysis, which has a strong experimental character.

## The Empirical Distribution Function

Histograms or *bar charts* are often used to summarize data for which there are
no *model assumptions*. A typical example is the foreign–trade surplus between
nations. Note, however, that charts summarizing data can often be manipulated
quite easily so as to serve one's purpose, an example is given in Figure 5.1.4;
for a review of artful or unfair means we refer to Huff (1992). If the data
are assumed to be a *random sample* of observations generated by a density
function, then the histogram turns out to be a density estimator as we show in
the following.

**1.1.4 Definition.** Let $x_1, \ldots, x_n \in \mathbb{R}$. The function

$$F_n(t) := \frac{|\{x_i, i = 1, \ldots, n : x_i \leq t\}|}{n} = \frac{1}{n} \sum_{i=1}^{n} 1_{(-\infty, t]}(x_i), \qquad t \in \mathbb{R},$$

is the *empirical distribution function* of $x_1, \ldots, x_n$. By $1_A(\cdot)$ we denote the
*indicator function* of a set $A$, i.e., $1_A(x) = 1$ if $x \in A$ and $1_A(x) = 0$ else.

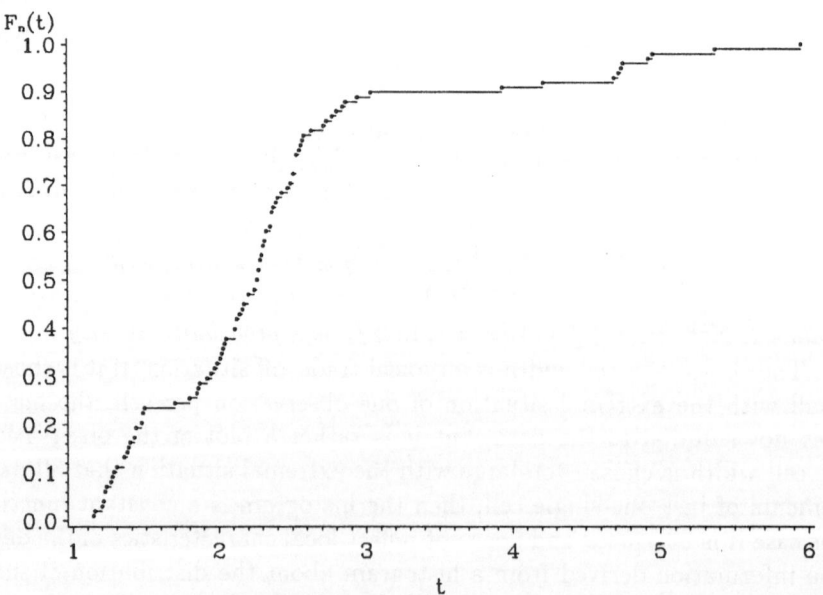

**Figure 1.1.3.** Empirical distribution function of the values *an;
cns* data.

```
***    Program 1_1_3    ***;
TITLE1 'Empirical Distribution Function';
TITLE2 'CNS Data';
LIBNAME datalib 'c:\data';

PROC FREQ DATA=datalib.cns NOPRINT;
   TABLES an/OUT=freqdata;

DATA data1;
   SET freqdata;
   Fn+PERCENT/100;

SYMBOL1 C=RED V=DOT H=0.4 I=STEPL;
AXIS1 LABEL=(H=2 'F' H=1 'n' H=2 '(t)');
AXIS2 LABEL=('t');
PROC GPLOT;
   PLOT Fn*an / VAXIS=AXIS1 HAXIS=AXIS2;
RUN; QUIT;
```

An empirical distribution function can be displayed quite easily using the procedure GPLOT. It requires, however, some preliminaries. The procedure FREQ produces a frequency table for the variable 'an' specified in the TABLES statement. Two options influence the output: 'OUT=freqdata' generates a data set containing the different values of an, as well as the absolute and relative (in percent) frequencies, which are stored in the automatic variables COUNT and PERCENT. 'NOPRINT' suppresses the default printed output.

The subsequent DATA step creates a temporary data set data1 and copies the data in freqdata to data1 by the SET statement. The unusual syntax of the following statement 'Fn+PERCENT/100' creates the variable Fn defined by the cumulative sum of PERCENT divided by 100. These values correspond to the empirical distribution function $F_n$ because the data in freqdata are sorted in ascending order of an.

The global SYMBOL statement defines the appearance of the plotting symbol and plot lines as well as specific interpolation methods. 'C=BLACK V=DOT H=0.3' specifies black dots of 0.3 unit height for the data points. 'I=STEPL' specifies that the dots are stepwise interpolated with the data points being displayed on the left side of the step. This visualizes the continuity from above of a distribution function. The PLOT statement in the GPLOT procedure is of the form 'PLOT *vertical variable * horizontal variable / options*'. Here the options assign axis characteristics as defined in the AXIS statements.

Suppose now that $x_1, \ldots, x_n$ are *realizations* of a random sample $X_1, \ldots, X_n$ of *independent random variables* which were generated according to some *distribution function* $F$, i.e.,

$$P\{X_i \le t_i, \ i = 1, \ldots, n\} = \prod_{i=1}^{n} P\{X_i \le t_i\} = \prod_{i=1}^{n} F(t_i), \qquad t_1, \ldots, t_n \in \mathbb{R}.$$

Then we know by the *Glivenko–Cantelli Theorem* cf Section 12.3 in Fridstedt and Gray (1997)

$$P\Big\{ \sup_{t\in \mathbb{R}} |F_n(t) - F(t)| \longrightarrow_{n\to\infty} 0 \Big\} = 1,$$

or

$$\sup_{t\in \mathbb{R}} |F_n(t) - F(t)| \longrightarrow_{n\to\infty} 0 \quad \text{almost surely (a.s.)},$$

where $F_n(t) = n^{-1} \sum_{i=1}^n 1_{(-\infty,t]}(X_i)$ is the empirical distribution function of $X_1, \ldots, X_n$. If $F$ is differentiable with $F' = f$, then $f$ is the density function of $F$, i.e.,

$$\int_a^b f(x)\, dx = F(b) - F(a), \qquad a, b \in \mathbb{R}.$$

Hence, we have $P\{X_i \in (a, b]\} = F(b) - F(a) = \int_a^b f(y)\, dy$ for $a < b$, and we obtain a.s.

$$\frac{F_n(t+h) - F_n(t)}{h} \longrightarrow_{n\to\infty} \frac{F(t+h) - F(t)}{h} \longrightarrow_{h\to 0} F'(t) = f(t)$$

for $t \in \mathbb{R}$. For large $n$ and small $h$ we will, therefore, expect the approximation

$$\frac{F_n(t+h) - F_n(t)}{h} \approx f(t).$$

With the particular choice $t = a_{s-1}$, $h = a_s - a_{s-1}$ we consequently obtain for $x \in I_s$

$$\frac{F_n(t+h) - F_n(t)}{h} = \frac{F_n(a_s) - F_n(a_{s-1})}{a_s - a_{s-1}}$$

$$= \frac{n_s}{n} \frac{1}{a_s - a_{s-1}} = f_n(x) \approx f(t).$$

Therefore, the weighted histogram $f_n(\cdot)$ with small cell widths and large sample sizes turns out to be an estimator of the underlying density $f(\cdot)$.

## Kernel Estimator

If we knew that the density $f$ underlying a random sample is a smooth function, especially that it is continuous or differentiable, then we would waste this prior information when using $f_n$ as an estimator of $f$. We will, therefore, derive in the sequel a smooth density estimator as follows: First, we smooth the empirical distribution function $F_n$ such that the resulting estimator $\hat{F}_n$ of the underlying distribution function $F$ is differentiable. Its derivative $\hat{F}'_n(t) =: \hat{f}_n(t)$ is then an estimator of $f(t)$. If $\hat{F}_n$ is in addition continuously differentiable or even has derivatives of higher order, then $\hat{f}_n$ is a corresponding smooth function.

**1.1.5 Definition.** Let $K : \mathbb{R} \to [0,1]$ be a distribution function. Put for $h > 0$ and $t \in \mathbb{R}$

$$\hat{F}_n(t) := \int K\left(\frac{t-x}{h}\right) F_n(dx) = \frac{1}{n} \sum_{i=1}^{n} K\left(\frac{t-X_i}{h}\right).$$

The estimator $\hat{F}_n(t)$ is the *convolution* of $K(\cdot/h)$ and the empirical distribution function $F_n$. It is called the *kernel estimator* of $F$ with *bandwidth* $h$ and *kernel* $K$.

Since $K$ is a distribution function, it satisfies in particular $\lim_{x \to \infty} K(x) = 1$ and $\lim_{x \to -\infty} K(x) = 0$. The kernel estimator $\hat{F}_n$ is consequently a distribution function, that is for a small bandwidth $h$ close to the empirical distribution function:

$$\hat{F}_n(t) = \frac{1}{n} \sum_{i=1}^{n} K\left(\frac{t-X_i}{h}\right) \longrightarrow_{h \to 0} F_n(t)$$

for any $t \notin \{X_1, \ldots, X_n\}$. Thus, $\hat{F}_n(t)$ is an estimator of $F(t)$. If the kernel $K$ is differentiable, then we obtain with $K' = k$

$$\hat{F}'_n(t) = \frac{1}{nh} \sum_{i=1}^{n} k\left(\frac{t-X_i}{h}\right) = \int \frac{1}{h} k\left(\frac{t-x}{h}\right) F_n(dx) =: \hat{f}_n(t).$$

Unlike $F_n$, the function $\hat{F}_n$ is differentiable everywhere. One might expect that it estimates the underlying distribution function $F$ and that, in addition, its derivative $\hat{f}_n = \hat{F}'_n$ approximates the density $f$.

We generalize in the following the above concept of kernel density estimators by considering functions $k : \mathbb{R} \to \mathbb{R}$ that only have to satisfy the condition $\int k(x)\, dx = 1$. We do not require $k$ to be the derivative of a distribution function $K$. This provides us with a huge set of auxiliary functions $k$, which we nevertheless call kernels again.

**1.1.6 Definition.** Let $k : \mathbb{R} \to \mathbb{R}$ satisfy $\int k(x)\, dx = 1$. The function

$$\hat{f}_n(t) := \frac{1}{nh} \sum_{i=1}^{n} k\left(\frac{t-X_i}{h}\right) = \int \frac{1}{h} k\left(\frac{t-x}{h}\right) F_n(dx), \qquad t \in \mathbb{R},$$

is called *univariate kernel density estimator* with *kernel* $k$ and *bandwidth* $h > 0$.

A kernel estimator is readily affected by the choice of $h$. If $h$ is too small, the shape of $\hat{f}_n(\cdot)$ is serrated with peaks at the data. If $h$ is too large, $\hat{f}_n(\cdot)$ is oversmoothed. This is in complete accordance with the problem of choosing a suitable interval length for constructing a histogram. This analogy can easily be explained as follows: For the particular kernel $k(x) = (1/2)1_{[-1,1)}(x)$ we obtain

$$\hat{f}_n(t) = (nh)^{-1} \sum_{i=1}^{n} k((t - X_i)/h)$$

$$= (F_n(t + h) - F_n(t - h))/(2h)$$

$$= \frac{\text{number of observations in the interval } (t - h, t + h]}{n \times \text{ length of } (t - h, t + h]},$$

i.e., with this particular kernel, $\hat{f}_n(t)$ turns out to be of the type of a standardized histogram. Kernel density estimators can, therefore, be viewed as generalized histograms. They are easy to handle and have become a standard tool in the data analyst's toolbox. The following two auxiliary results show how we can reduce the *bias* $E(\hat{f}_n(t)) - f(t)$ and the variance $E(\{\hat{f}_n(t) - E(\hat{f}_n(t))\}^2)$ of $\hat{f}_n(t)$ by imposing further conditions on the kernel $k$.

**1.1.7 Lemma.** *Suppose that the underlying density $f$ is twice differentiable in a neighborhood of $t$ and that its second derivative is continuous at $t$. If the kernel $k$ vanishes outside a bounded interval, i.e., $k(x) = 0$ for large $|x|$, and if it satisfies $\int k(x)\, dx = 1$, $\int xk(x)\, dx = 0$, then we have for $h \to 0$ the expansion*

$$E\left(\hat{f}_n(t)\right) = f(t) + f''(t) \frac{h^2}{2} \int x^2 k(x)\, dx + o(h^2).$$

**Proof:** By the linearity of the expectation and Taylor's formula we obtain

$$E\left(\hat{f}_n(t)\right) = \frac{1}{h} E\left(k\left(\frac{t - X_1}{h}\right)\right) = \frac{1}{h} \int k\left(\frac{t - x}{h}\right) f(x)\, dx$$

$$= \int k(x) f(t - hx)\, dx$$

$$= \int k(x)\left(f(t) - f'(t)hx + f''(\xi)\frac{(hx)^2}{2}\right) dx,$$

where $\xi$ is between $t$ and $t - hx$. The integrability conditions of the kernel $k$ then imply

$$E(\hat{f}_n(t)) = f(t) \underbrace{\int k(x)\, dx}_{=1} - hf'(t) \underbrace{\int k(x)x\, dx}_{=0}$$

$$+ \frac{h^2}{2} f''(t) \int k(x)x^2\, dx + \underbrace{\frac{h^2}{2} \int k(x)x^2 \underbrace{\left(f''(\xi) - f''(t)\right)}_{=o(1)} dx}_{=o(h^2)}. \quad \square$$

The condition $\int x k(x)\, dx = 0$ is, for instance, automatically satisfied if $k$ is symmetric about the origin, i.e., $k(x) = k(-x)$ for $x \in \mathbb{R}$, and if it vanishes outside a bounded interval.

**1.1.8 Lemma.** *Suppose that $f$ is continuous at $t$. If $k$ vanishes outside a bounded interval and satisfies the condition $\int k^2(x)\, dx < \infty$, then we have for $h \to 0$*

$$E\Big((\hat{f}_n(t) - E(\hat{f}_n(t)))^2\Big) = \frac{1}{nh} f(t) \int k^2(x)\, dx + o\Big(\frac{1}{nh}\Big).$$

**Proof:** Exercise 9. □

The following expansion of the *mean squared error* of $\hat{f}_n(t)$ is now an immediate consequence.

**1.1.9 Corollary.** *Under the joint conditions of Lemma 1.1.7 and 1.1.8 we have for the mean squared error $MSE(\hat{f}_n(t))$ of $\hat{f}_n(t)$ the expansion*

$$
\begin{aligned}
MSE(\hat{f}_n(t)) &:= E\Big((\hat{f}_n(t) - f(t))^2\Big) \\
&= E\Big((\hat{f}_n(t) - E(\hat{f}_n(t)))^2\Big) + \Big(E(\hat{f}_n(t)) - f(t)\Big)^2 \\
&= \frac{1}{nh} f(t) \int k^2(x)\, dx + h^4 \Big(\frac{f''(t)}{2} \int k(x) x^2\, dx\Big)^2 + o\Big(\frac{1}{nh} + h^4\Big).
\end{aligned}
$$

## The Optimal Bandwidth

The two leading terms in the above expansion of $MSE(\hat{f}_n(t))$ are commonly used for the definition of an optimal bandwidth.

**1.1.10 Remark.** Conceive the sum of the two leading terms in the expansion of $MSE(\hat{f}_n(t))$ as a function of the bandwidth and put

$$g(h) := \frac{1}{nh} f(t) \int k^2(x)\, dx + h^4 \Big(\frac{f''(t)}{2} \int k(x) x^2\, dx\Big)^2, \qquad h > 0.$$

The function $g$ is minimized with the *optimal bandwidth* (Exercise 9)

$$h_n^* := \frac{1}{n^{1/5}} \frac{(f(t) \int k^2(x)\, dx)^{1/5}}{(f''(t) \int k(x) x^2\, dx)^{2/5}}.$$

The *rate* of this optimal bandwidth depending on the sample size is $n^{-1/5}$, the constant term in $h_n^*$ is, however, unknown. It is reasonable to plot the density estimator utilizing various bandwidths and to compare the results visually. If the graph pertaining to a bandwidth $h$ shows several maxima, then these may be due to a small value of $h$ or may indicate a *stratification* of the observations.

In the latter case, the sample was drawn from different populations with different location parameters, which would be an important insight into the structure of the sample, cf Exercise 10. An example is the histogram of the $an$–values in Figure 1.1.2 of the *cns* data in Example 1.1.3. The two peaks of this histogram might indicate a stratification of the observations, possibly with the two strata or *classes* of healthy and ill patients. In this case, the variable $a/n$ would show a certain capacity to separate these two classes, i.e., to *discriminate* between them (cf Exercise 11).

## Cross Validation

While in the case of a twice continuously differentiable underlying density the choice of a kernel $k$ is only a minor problem, the choice of a suitable bandwidth is a major one. The optimal bandwidth $h_n^*$ involves the unknown target value $f(t)$. Various suggestions such as the following one called *cross validation* were made to break this circular argument.

Define for $j = 1, \ldots, n$ the modified kernel density estimator $\hat{f}_{nj}$ based on the observations $X_i$, $i \neq j$, by

$$\hat{f}_{nj}(t) := ((n-1)h)^{-1} \sum_{i=1, i\neq j}^{n} k((t - X_i)/h), \qquad t \in I\!R.$$

We expect the approximation

$$\hat{f}_{nj}(X_j) \approx f(X_j).$$

This cross approach now suggests to use that bandwidth $h_{n,cv}^*$ for the original kernel density estimator, which maximizes the estimated *likelihood function*

$$\prod_{j=1}^{n} \hat{f}_{nj}(X_j) \approx \prod_{j=1}^{n} f(X_j).$$

This variation of the *maximum–likelihood method* requires the above modification of $\hat{f}_n(X_j)$ by $\hat{f}_{nj}(X_j)$: If $k$ vanishes outside a bounded interval and if $k(0)$ is positive, then we have

$$\lim_{h \to 0} \hat{f}_n(X_j) = \lim_{h \to 0} (nh)^{-1}k(0) = \infty.$$

The approximate likelihood function $\prod_{j=1}^{n} \hat{f}_n(X_j)$ consequently has no global maximum, whereas $\prod_{j=1}^{n} \hat{f}_{nj}(X_j)$ has one.

The mathematical results evaluating this or other procedures are necessarily of an asymptotic nature with the sample size increasing to infinity. But the data analyst's toolbox benefits from this research, cf the monograph by Simonoff

(1996). For a description of several different bandwidth selection methods which are also available in SAS, version 8, we refer to Jones et al. (1996).

In particular we encounter here the idea of *resampling*: The initial sample $X_1, \ldots, X_n$ is multiplied to $n$ subsamples by leaving the $j$–th observation out, $j = 1, \ldots, n$. These $n$ subsamples are then evaluated to deduce information about the initial sample. This is also the idea behind *jackknifing* or, more generally, behind *bootstrapping*, cf the survey article by Manteiga et al. (1994) or the monograph by Chernick (1999).

## The Epanechnikov Kernel, the Normal Kernel

A popular kernel is the *Epanechnikov kernel*

$$k_E(x) = \frac{3}{4\sqrt{5}}\left(1 - \frac{x^2}{5}\right) \qquad \text{if } |x| \leq \sqrt{5}, \quad k_E(x) = 0 \text{ elsewhere.}$$

Among all kernels $k$, which vanish outside the interval $[-\sqrt{5}, \sqrt{5}]$ and which satisfy

$$\int k(x)\,dx = 1, \quad \int x^2 k(x)\,dx = 1,$$

it minimizes $\int k^2(x)\,dx$ (Exercise 12). It minimizes, therefore, the value $g(h)$ in Remark 1.1.10 among all kernels in this class.

Another popular kernel is the density of the standard normal distribution

$$k(x) = \frac{1}{\sqrt{2\pi}} \exp\left(-\frac{x^2}{2}\right), \quad x \in \mathbb{R}.$$

This is a very smooth function, it is arbitrarily often differentiable and satisfies $\int k(x)dx = 1, \int x^2 k(x)dx = 1$.

**1.1.11 Example** (Beeswax Data; White et al. (1960)). Chemical properties of beeswax such as the melting point were investigated to detect the presence of synthetic waxes that had been added to beeswax. If every pure beeswax had the same melting point, its determination would be a reasonable way to detect dilutions. The melting point and other chemical properties of beeswax, however, vary from one beehive to another. Samples of pure beeswax from 59 sources were taken and several chemical properties such as the melting points (in °C) were measured. A kernel density estimator can now be used to obtain a first insight into the variability of the measurements. The data are taken from Rice (1995), page 346.

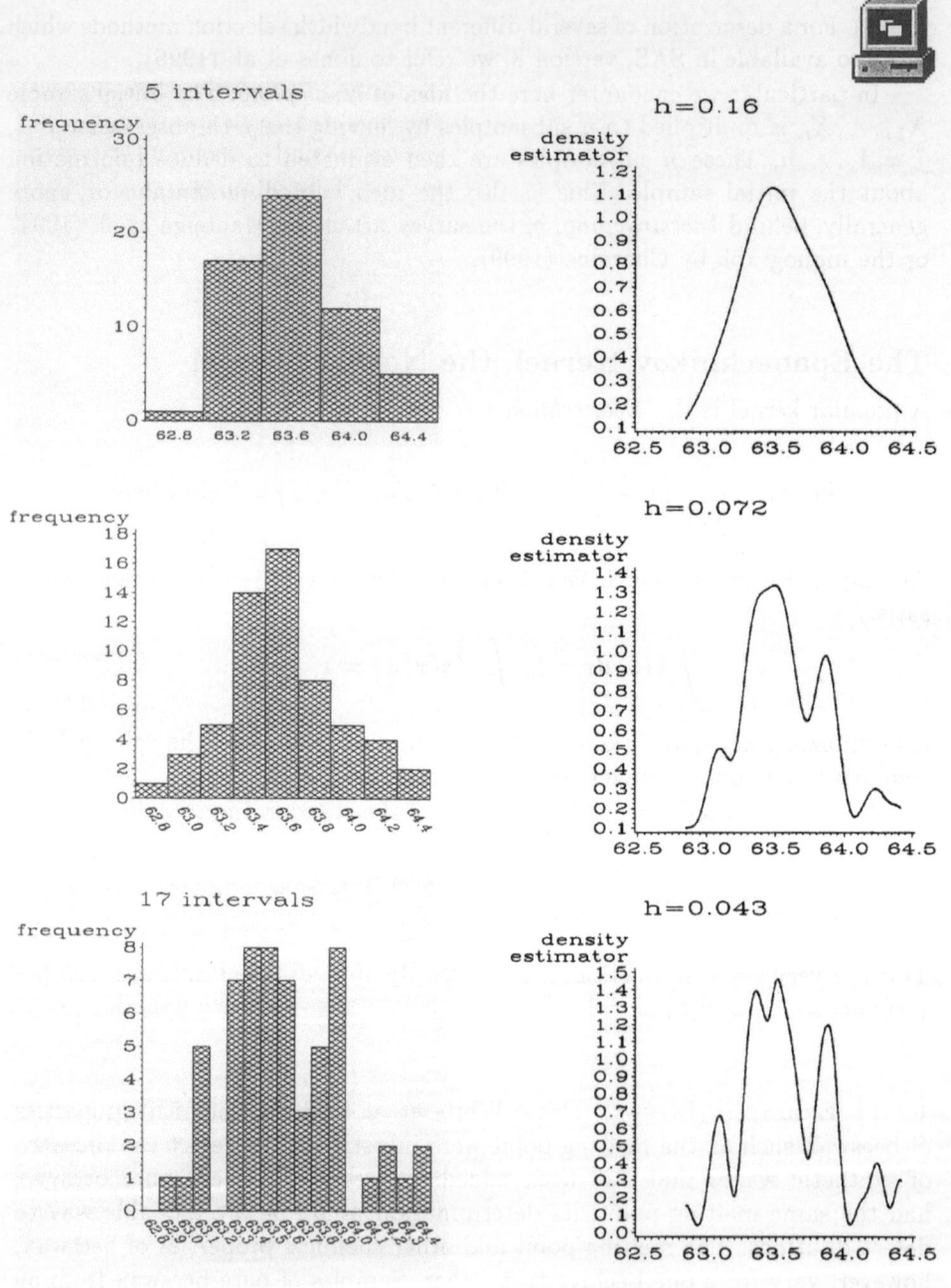

**Figure 1.1.4.** Histograms and kernel density estimators of the melting points of the beeswax data.

```
***    Program 1_1_4   ***;
TITLE1 'Kernel Density Estimators';
TITLE2 'Beeswax Data';
LIBNAME datalib 'c:\data';

PROC KDE DATA=datalib.beeswax BWM=.3 OUT=data1;
   VAR degree;

AXIS1 LABEL=(J=R 'density' J=R 'estimator');
SYMBOL1 C=GREEN I=JOIN V=NONE L=1 WIDTH=2;
PROC GPLOT DATA=data1;
   PLOT DENSITY*degree / VAXIS=AXIS1;
RUN; QUIT;
```

This program computes kernel density estimators using the normal kernel. It requires version 8 of SAS. PROC KDE invokes this procedure. Several different bandwidth selection methods are available, the default method follows the Sheather–Jones plug in formula. 'BWM=*number*' specifies the bandwidth multiplier for the kernel density estimate which actually uses the bandwidth *number* × default.

The SYMBOL statement defines the appearance of the plot, using the following options:

I=JOIN     The dots are joined by a straight line.

V=NONE    The dots themselves are not displayed by a special symbol.

L=1      The line type of the plot is solid.

WIDTH=2   The thickness of the interpolated lines is twice the default value.

The procedure GPLOT plots the data in the form *vertical variable* ∗ *horizontal variable* with the characteristics of its appearance as defined in the SYMBOL statement.

The histograms are plotted analogous to Program 1_1_2. The number of cells or the width of the bars can be specified in the VBAR statement using the options LEVELS=n or WIDTH=r. Note that the display of Figure 1.1.4 with all plots and charts in one graphic is a bit more complex, see, for instance, Program 3_1_1.

## Stem-and-Leaf Plots

One disadvantage of a histogram or of a kernel density estimator is the intrinsic loss of information; the original data cannot be reconstructed. *Stem-and-leaf plots* (Tukey (1977)) overcome this disadvantage by providing information about the shape of the density of the observations and by keeping simultaneously their numerical values. The following printout of the beeswax data in the preceding Example 1.1.11 explains the architecture of a stem–and–leaf plot. The first three digits of the melting points form the stem. The leaves on each stem are the fourth digit of all numbers with that stem. The boxplot, which is included in this printout, will be discussed in Section 1.5.

Variable=DEGREE

```
         Stem Leaf                        #      Boxplot
          644 02                          2         |
          643                                       |
          642 147                         3         |
          641 2                           1         |
          640                                       |
          639 22223                       5         |
          638 334668                      6      +-----+
          637 88                          2      |     |
          636 0013689                     7      |     |
          635 0000113668                 10      *--+--*
          634 01335                       5      |     |
          633 001446669                   9      +-----+
          632 77                          2         |
          631 033                         3         |
          630 358                         3         |
          629                                       |
          628 5                           1         |
              ----+----+----+----+
         Multiply Stem.Leaf by 10**-1
```

Figure 1.1.5. Stem–and–leaf plot of the beeswax data.

```
***    Program 1_1_5    ***;
TITLE1 'Stem-and-Leaf Plot';
TITLE2 'Beeswax Data';
LIBNAME datalib 'c:\data';

OPTIONS PAGESIZE=60;
PROC UNIVARIATE PLOT DATA=datalib.beeswax;
   VAR degree;
RUN; QUIT;
```

A stem–and–leaf plot can be displayed using the procedure UNIVARIATE together with the option PLOT. PROC UNIVARIATE readily computes means, standard deviations and extreme values, and other single-variable measures of location and spread for the variables defined by the VAR statement. If this statement is not specified, all numerical variables are analyzed.

## 1.2 Measures of Location

A *measure of location* is a measurement of the *center* of a set of numbers $x_1, \ldots, x_n$.

**1.2.1 Definition.** The most popular measure of location is the *arithmetic mean*

$$\bar{x}_n := \frac{1}{n} \sum_{i=1}^{n} x_i.$$

**1.2.2 Example** (Platinum Data; Hampson and Walker (1961)). To determine the heat of sublimation of platinum in kcal/mole, 26 measurements were taken. Each one was an attempt to measure the "true" heat, but there is a certain variability in the measurements. We would expect in this situation that a measure of location such as the arithmetic mean is more reliable than the 26 individual measurements, since the individual errors should have the tendency to cancel each other. A discussion of these data is given in Rice (1995), page 361.

```
Analysis Variable : KCAL


 N        Mean        Std Dev      Minimum       Maximum
-----------------------------------------------------------
26   137.0346154    4.4542961   133.7000000   148.8000000
-----------------------------------------------------------
```

Figure 1.2.1. First descriptive statistics of the platinum data.

```
***   Program 1_2_1   ***;
TITLE1 'First Descriptive Statistics';
TITLE2 'Platinum Data';
LIBNAME datalib 'c:\data';

PROC MEANS DATA=datalib.platinum;
   VAR kcal;
RUN; QUIT;
```

Just like UNIVARIATE, the procedure MEANS computes means, standard deviations, extrema and other single–variable statistics for the variables defined by the VAR statement. The output is, however, much more compact as only relatively few statistics are computed by default.

Additional statistics such as the variance are computed by adding the pertaining option to the PROC MEANS statement. When a nondefault statistic is required, all the statistics MEANS should compute must be specified. For further processing, the results can be written to a SAS data set us-

ing the OUTPUT statement together with    1_6_1.
the 'OUT=*data set*' option, see Program

A common *statistical model* for the variability of a sequence of $n$ measurements
is

$$X_i = \mu + \varepsilon_i, \qquad i = 1, \ldots, n.$$

By $X_i$ we denote the random outcome of the $i$–th measurement, which we model
to be the sum of the "true" target value $\mu$ and a measurement error $\varepsilon_i$. The
numbers $x_i$ are in this case assumed to be realizations of the random variables
$X_i$, $i = 1, \ldots, n$. The errors $\varepsilon_i$ are commonly assumed to be independent and
identically distributed random variables with mean 0, i.e., $E(\varepsilon_i) = 0$, $i =$
$1, \ldots, n$. If a systematic error is inherent in the measurements, i.e., $E(\varepsilon_i) =$
$\beta \neq 0$, then our measurement procedure is biased:

$$E(X_i) = \mu + \beta, \qquad i = 1, \ldots, n.$$

If the parameter $\beta$ is unknown, then we can deduce no information about $\mu$
from this biased procedure.

It is often quite informative to plot the measurements in the order in which
the experiments were done. The following figure gives an example.

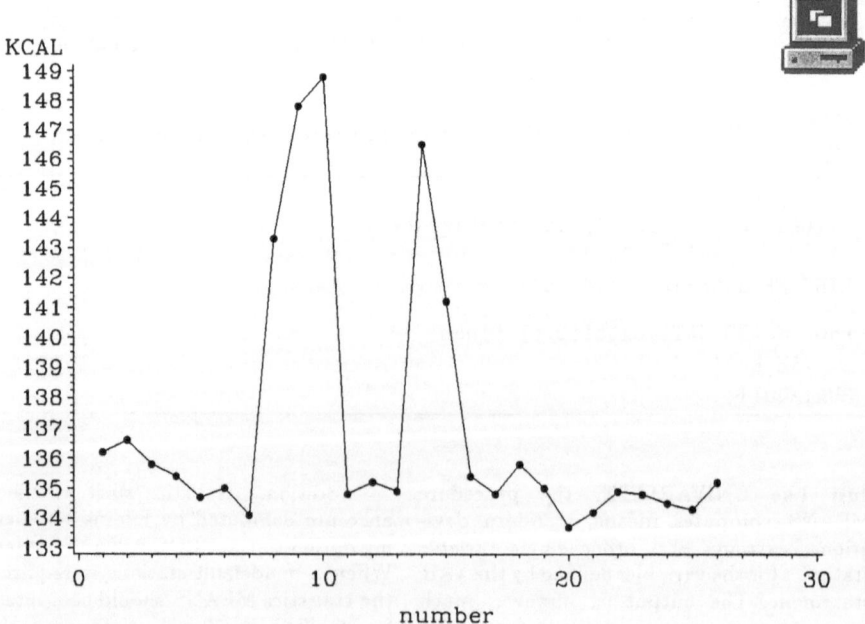

**Figure 1.2.2.** Plot of the platinum data in the order of their outcome.

```
***     Program 1_2_2    ***;
TITLE1 'Order of Outcome';
TITLE2 'Platinum Data';
LIBNAME datalib 'c:\data';

DATA data1;
    SET datalib.platinum;
    number=_N_;

AXIS1 LABEL=('number');
SYMBOL1 C=RED V=DOT I=JOIN;
PROC GPLOT DATA=data1;
    PLOT kcal*number / HAXIS=AXIS1;
RUN; QUIT;
```

Figure 1.2.2 is a plot of the platinum data in the order of their outcome. For this plot, the order of outcome of the data is assigned to the new variable 'number'. This is done in the first DATA step by the statement 'number=_N_'. The automatic SAS variable _N_ contains the observation number of each observation in the dataset datalib.platinum, which is the desired information. The procedure GPLOT then generates the desired plot.

Most striking is the fact that five observations are extremely large. Such observations, which are far away from the bulk of the data, are called *outliers*. Their presence can actually be caused by an error in the transmission of the original measurements such as a misprint, in which case these observations do not belong to the sample. But they might be due to some high variability of the measurement procedure itself as well, in which case they are genuine members of the sample. The identification of outliers is, therefore, a major problem in the analysis of data, in particular as outliers may heavily influence the numerical output. The above plot makes us further doubt that our model

$$X_i = \mu + \varepsilon_i, \qquad i = 1, \dots, n,$$

with independent and identically distributed measurement errors $\varepsilon_i$ is correct, as the outliers occur in groups of three and two subsequent observations. For a detailed discussion of the platinum data we refer to Section 10.4 in Rice (1995).

## Robust Measurements of Location

The arithmetic mean is obviously extremely sensitive to outliers: One out of $n$ observations can serve as a *leverage point* making $\bar{x}_n$ arbitrarily large or small. The *breakdown point* $\varepsilon(\bar{x}_n)$ of the arithmetic mean is, therefore, $1/n$.

The automatic or blind use of $\bar{x}_n$ can consequently produce serious errors. This fact led to an increasing interest in *robust* procedures with higher breakdown points, cf Davies and Gather (1993) and Jurečková and Sen (1996) and the literature cited therein.

**1.2.3 Definition.** If the sample size is odd, i.e., $n = 2m + 1$, then the central observation $x_{m+1:n}$ is called the *sample median* of $x_1, \ldots, x_n$; if the sample size is even, i.e., $n = 2m$, the sample median is commonly the average $(x_{m:n} + x_{m+1:n})/2$ of the *two* central observations $x_{m:n}$ and $x_{m+1:n}$. By $x_{1:n} \leq \cdots \leq x_{n:n}$ we denote as in Section 1.1 the ordered values of $x_1, \ldots, x_n$.

The sample median $med(x_1, \ldots, x_n)$ has the characteristic property that the number of observations which are smaller coincides with the number of those that are larger. In this sense it is a center of the set of observations. The platinum data have, for example, the median 135.1, see Figure 1.3.1. In order to move the median to $\infty$ or $-\infty$, one obviously has to move all larger or smaller observations to $\infty$ or $-\infty$, respectively. The breakdown point of the median is, therefore, given by

$$
\varepsilon(med(x_1, \ldots, x_n)) = \begin{cases} \dfrac{m+1}{n} = \dfrac{1}{2} + \dfrac{1}{2n}, & n = 2m+1 \\[2ex] \dfrac{m}{n} = \dfrac{1}{2}, & n = 2m \end{cases} \longrightarrow_{n \to \infty} 1/2.
$$

The sample median is, therefore, quite a robust measure of location. Since its computation involves, however, only the location of most of the data but not their numerical values, a certain loss of information will be inherent in its use. The trimmed mean is a compromise.

**1.2.4 Definition.** Put for $\alpha \in [0, 1/2)$

$$
\bar{x}_{n,\alpha} := \frac{x_{[n\alpha]+1:n} + x_{[n\alpha]+2:n} + \cdots + x_{n-[n\alpha]:n}}{n - 2[n\alpha]},
$$

where $[n\alpha]$ denotes the greatest integer, which is less than or equal to $n\alpha$, i.e., $[n\alpha] = \max\{k \in \mathbb{N} \cup \{0\} : k \leq n\alpha\}$. The number $\bar{x}_{n,\alpha}$ is called $\alpha$ *trimmed mean.*

The $\alpha$ trimmed mean $\bar{x}_{n,\alpha}$ is easily computed by removing the $[n\alpha]$ smallest and the $[n\alpha]$ largest observations from the sample and averaging the ones left. Its robustness increases with $\alpha$: The breakdown point is $\varepsilon(\bar{x}_{n,\alpha}) = ([n\alpha] + 1)/n \longrightarrow_{n \to \infty} \alpha$. A value of $\alpha$ between 0.1 and 0.2 is generally recommended. The 20 % trimmed mean of the platinum data is 135.28. The median can be viewed as a 50 % trimmed mean. The median of the platinum data is 135.1, which is quite close to the 20% trimmed mean. With the particular choice of $\alpha = 0$, the arithmetic mean is a special case of the $\alpha$ trimmed mean. Its value for the platinum data is 137.03, which shows the heavy influence of the outliers on the arithmetic mean.

```
Analysis Variable : KCAL


 N        Mean       Std Dev       Minimum        Maximum
------------------------------------------------------------
16    135.2812500    0.5564396    134.7000000    136.6000000
------------------------------------------------------------
```

**Figure 1.2.3.** $\alpha$ trimmed mean of the platinum data, $\alpha = 0.2$.

```
***     Program 1_2_3    ***;
TITLE1 'Alpha Trimmed Mean';
TITLE2 'Platinum Data';
LIBNAME datalib 'c:\data';

PROC SORT DATA=datalib.platinum OUT=sorted;
   BY kcal;
PROC MEANS DATA=sorted(FIRSTOBS=6 OBS=21);
   VAR kcal;
RUN; QUIT;
```

The $\alpha$ trimmed mean is computed as the ordinary mean of a dataset, where the $[n\alpha]$ smallest and $[n\alpha]$ largest observations were dropped. This is achieved by first sorting the data using PROC SORT. The sorted data are written to the temporary file 'sorted'. With $\alpha = 0.2$ and n=26 we have $[n\alpha]$=5. The data set options FIRSTOBS and OBS in the MEANS procedure specify the first observation to be read from the data file sorted and the last one. By FIRSTOBS=6 and OBS=21 the first 5 and the last 5 observations are excluded from further processing, since sorted contains 26 observations.

The most frequent numerical value among $x_1, \ldots, x_n$ is called the *mode*. It is another measure of location. If there are two or more values with the same maximum frequency, then each of them is called mode. The distribution of the data is correspondingly called *uni-, bi-* or *multimodal*. The breakdown point of the mode obviously depends on the actual sample. It can range between $2/n$ in the case of only two identical observations $x_i$ and $([n/2]+1)/n$ in the case of $n$ identical observations.

If $x_1, \ldots, x_n$ are assumed to be realizations of $n$ independent and identically distributed random variables $X_i$, $i = 1, \ldots, n$, whose distribution has a density $f$, then those values, for which $f$ has local maxima, are also called *modes*. The density $f$ is correspondingly called uni-, bi- or multimodal as well. The observations should accumulate near these modes of $f$ and a kernel density estimator should have local maxima close to these modes.

As we mentioned in Remark 1.1.10, two- or multimodal distributions or densities might indicate a stratification of the population, i.e., its division into two or more subpopulations.

The question which measure of location is best cannot be answered uniquely. If the observations $x_1, \ldots, x_n$ are all generated by the same density $f$, and this is symmetric about some point $\vartheta \in \mathbb{R}$, i.e., $f(\vartheta + x) = f(\vartheta - x)$, $x \in \mathbb{R}$, then each of $\bar{x}_n$, $\mathrm{med}(x_1, \ldots, x_n)$ and $\bar{x}_{n,\alpha}$ estimates the center of symmetry $\vartheta$. In any case it is reasonable to compute different measures of location and to compare the results.

# 1.3    Measures of Spread

In the following we will define various *measures of spread* of a sample $x_1, \ldots, x_n$.

**1.3.1 Definition.** The most popular measure of spread is the *sample standard deviation*

$$s_n := s_n(x_1, \ldots, x_n) := \left( \frac{1}{n-1} \sum_{i=1}^{n} (x_i - \bar{x}_n)^2 \right)^{1/2}.$$

Let $X_1, \ldots, X_n$ be independent and identically distributed random variables with $E(X_1^2) < \infty$. Then $s_n^2(X_1, \ldots, X_n)$ is an unbiased estimator of the variance $\sigma^2 = E((X_1 - E(X_1))^2)$, i.e., $E(s_n^2(X_1, \ldots, X_n)) = \sigma^2$ (Exercise 16). The *law of large numbers* implies in addition that $s_n^2(X_1, \ldots, X_n) \to \sigma^2$ as $n \to \infty$ with probability one. Like $\bar{x}_n$, $s_n$ is highly sensitive to outliers; we have $\varepsilon(s_n) = 1/n$.

The ratio $s_n/\bar{x}_n$ of the standard deviation and the mean is called *coefficient of variation* (*CV*). It measures the spread of the data in multiples of their mean and is commonly given in percent. A small *CV* indicates, therefore, that the observations are close to their mean. The theoretical justification of the *CV* is given by the 2- or 3-$\sigma$-rule, see also Section 1.5: Let $X_1, \ldots, X_n$ be independent replicates of a normal distributed random variable $X$ with mean $\mu \neq 0$ and variance $\sigma^2 > 0$. The law of large numbers implies with $\bar{X}_n = n^{-1} \sum_{i=1}^{n} X_i$

$$CV = \frac{s_n(X_1, \ldots, X_n)}{\bar{X}_n} \xrightarrow{n \to \infty} \frac{\sigma}{\mu} \quad \text{with probability one.}$$

The probability that $X$ attains a value in the interval $[\bar{X}_n(1 - rCV), \bar{X}_n(1 + rCV)]$ is by the approximation

$$P\left\{X \in [\bar{X}_n(1 - rCV), \bar{X}_n(1 + rCV)]\right\}$$

$$\approx P\left\{X \in \left[\mu\left(1 - r\frac{\sigma}{\mu}\right), \mu\left(1 + r\frac{\sigma}{\mu}\right)\right]\right\} = P\left\{-r \leq \frac{X - \mu}{\sigma} \leq r\right\}$$

about 0.97 for $r = 2$ and 0.99 for $r = 3$, independent of $\mu$ and $\sigma$. The break-down point $\varepsilon(CV)$ of the coefficient of variation is obviously $1/n$, since the denominator $\bar{x}_n$ of the $CV$ converges to zero by a proper move of only one observation. Note that the numerator $s_n$ is bounded away from zero if there are at least two different observations in the sample.

## Robust Measures of Spread

The following two measures of spread are much more robust than $s_n$ or $CV$.

**1.3.2 Definition.** The measure of spread

$$IQR := IQR(x_1, \ldots, x_n) := x_{[\frac{3}{4}n]:n} - x_{[\frac{1}{4}n]:n}$$

is called *interquartile range*; $x_{[\frac{1}{4}n]:n}$ and $x_{[\frac{3}{4}n]:n}$ are the *lower* and *upper quartile*. The breakdown point of the $IQR$ obviously satisfies $\varepsilon(IQR) \to 1/4$ as $n \to \infty$.

**1.3.3 Definition.** Put $y_i := |x_i - med(x_1, \ldots, x_n)|$ for $i = 1, \ldots, n$. The number

$$MAD := MAD(x_1, \ldots, x_n) := med(y_1, \ldots, y_n)$$

is the *median absolute deviation from the median* $(MAD)$.

The breakdown point of the $MAD$ equals that of the median of $x_1, \ldots, x_n$. In the case of independent and identically normal distributed random variables $X_1, \ldots, X_n$ with

$$P\{X_i \leq t\} = \frac{1}{\sqrt{2\pi}\sigma} \int_{-\infty}^{t} \exp\left(-\frac{(x - \mu)^2}{2\sigma^2}\right) dx, \qquad t \in \mathbb{R},$$

for $i = 1, \ldots, n$ and arbitrary numbers $\mu \in \mathbb{R}$, $\sigma > 0$, the ratios

$$\frac{IQR(X_1, \ldots, X_n)}{1.35}, \quad \frac{MAD(X_1, \ldots, X_n)}{0.675}$$

are reasonable estimates of $\sigma$ (see Exercise 34). For the platinum data we obtain, for instance,

$$s_n = 4.45, \quad \frac{IQR}{1.35} = \frac{1.4}{1.35} = 1.04, \quad \frac{MAD}{0.675} = \frac{0.65}{0.675} = 0.96.$$

The two robust estimates, which are based on the spread of the data in their central portion, are quite close to each other, whereas $s_n$ is obviously heavily influenced by the outliers. This makes us doubt that the platinum data were generated independently by the same normal distribution. Measures of the goodness of fit of a normal distribution to a sample will be introduced in the next section.

The UNIVARIATE Procedure
Variable: KCAL

### Moments

| | | | |
|---|---|---|---|
| N | 26 | Sum Weights | 26 |
| Mean | 137.034615 | Sum Observations | 3562.9 |
| Std Deviation | 4.45429611 | Variance | 19.8407538 |
| Skewness | 1.86943435 | Kurtosis | 2.16550277 |
| Uncorrected SS | 488736.65 | Corrected SS | 496.018846 |
| Coeff Variation | 3.25048974 | Std Error Mean | 0.87355934 |

### Basic Statistical Measures

| Location | | Variability | |
|---|---|---|---|
| Mean | 137.0346 | Std Deviation | 4.45430 |
| Median | 135.1000 | Variance | 19.84075 |
| Mode | 134.8000 | Range | 15.10000 |
| | | Interquartile Range | 1.40000 |

### Tests for Location: Mu0=0

| Test | -Statistic- | | -----p Value------ | |
|---|---|---|---|---|
| Student's t | t | 156.8693 | Pr > \|t\| | <.0001 |
| Sign | M | 13 | Pr >= \|M\| | <.0001 |
| Signed Rank | S | 175.5 | Pr >= \|S\| | <.0001 |

| Quantile | Estimate |
|----------|----------|
| 100% Max | 148.8 |
| 99% | 148.8 |
| 95% | 147.8 |
| 90% | 146.5 |
| 75% Q3 | 136.2 |
| 50% Median | 135.1 |
| 25% Q1 | 134.8 |
| 10% | 134.2 |
| 5% | 134.1 |
| 1% | 133.7 |
| 0% Min | 133.7 |

Extreme Observations

| -----Lowest---- | | ----Highest---- | |
|-------|------|-------|------|
| Value | Obs | Value | Obs |
| 133.7 | 20 | 141.2 | 15 |
| 134.1 | 7 | 143.3 | 8 |
| 134.2 | 21 | 146.5 | 14 |
| 134.3 | 25 | 147.8 | 9 |
| 134.5 | 24 | 148.8 | 10 |

**Figure 1.3.1.** Univariate descriptive statistics of the platinum data.

```
***    Program 1_3_1    ***;
TITLE1 'Univariate Descriptive Statistics';
TITLE2 'Platinum Data';
LIBNAME datalib 'c:\data';

PROC UNIVARIATE DATA=datalib.platinum;
   VAR kcal;
RUN; QUIT;
```

If no additional option is specified, PROC UNIVARIATE produces the above output. Available options are, for instance, NORMAL, which performs tests of the hypothesis that the data were generated from a normal distribution or ROBUSTSCALE, which provides robust estimates such as the MAD.

## 1.4   Skewness and Kurtosis

The visual evaluation of the spread of a sample by means of a density estimator can be quite informative. As mentioned earlier, several local maxima might for instance indicate a stratification of the sample. Also first conclusions about the symmetry or the peakedness or flatness of the distribution can be made. The most popular measures for these are the *skewness* and the *kurtosis*, which are computed from the first four *central moments*. Both measure deviations from the normal distribution.

**1.4.1 Definition.** Let $X$ be a random variable with $E(|X|^3) < \infty$. Denote by $\mu := E(X)$ the expectation of $X$ and by $\sigma^2 := E((X - \mu)^2)$ its variance. We require $\sigma^2 > 0$. The number

$$b_1 := E\left(\left(\frac{X - \mu}{\sigma}\right)^3\right)$$

is the *skewness* and

$$b_2 := E\left(\left(\frac{X - \mu}{\sigma}\right)^4\right) - 3$$

is the *kurtosis* of $X$.

The skewness can be used for evaluating the asymmetry of a distribution: Suppose that $X$ has the density $f$. Then we have

$$b_1 = \frac{1}{\sigma^3} \int_{-\infty}^{\infty} (x - \mu)^3 f(x)\, dx = \frac{1}{\sigma^3} \int_0^{\infty} x^3 \Big(f(\mu + x) - f(\mu - x)\Big)\, dx.$$

A positive or negative value of $b_1$ indicates that $f(\mu + x) - f(\mu - x)$ has for $x \to \infty$ the tendency to be positive or negative, respectively. This means that $f(\mu + x)$ decreases for $x \to \infty$ at a slower rate than $f(\mu - x)$ and at a faster rate in the second case. The density $f$ is correspondingly called *skewed to the right* or *skewed to the left*. For a discussion of the reliability of the skewness measure we refer to Li and Morris (1991).

If $X$ has a normal distribution, then its symmetry implies $b_1 = 0$ and $b_2 = 0$ (see Exercise 20). The kurtosis can be used as a measure of the deviation of the underlying density $f$ from the density $\varphi$ of the standard normal distibution: Suppose without loss of generality that $\mu = 0$ and $\sigma^2 = 1$, which is an immediate consequence of the definition of $b_2$ (see Exercise 18). We have

$$b_2 = \int_{-\infty}^{\infty} x^4 f(x)\, dx - 3 = \int_{-\infty}^{\infty} x^4 (f(x) - \varphi(x))\, dx,$$

where

$$\varphi(x) := \frac{1}{\sqrt{2\pi}} \exp(-x^2/2), \qquad x \in \mathbb{R},$$

is the density of the standard normal distribution. A positive value $b_2 > 0$ now indicates that $f(x)$ decreases for $|x| \to \infty$ at a slower rate than $\varphi$, whereas $b_2 < 0$ indicates that $f(x)$ decreases at a faster rate than $\varphi(x)$. In the first case, the density $f$ has *heavy tails*, in the second case it has *light tails*. For a critical review of the development of the concept of kurtosis we refer to Balanda and MacGillivray (1988).

The empirical counterparts of $b_1$ and $b_2$ are

$$\hat{b}_1 := \frac{\frac{1}{n}\sum_{i=1}^{n}(x_i - \bar{x}_n)^3}{s_n^3}$$

and

$$\hat{b}_2 := \frac{\frac{1}{n}\sum_{i=1}^{n}(x_i - \bar{x}_n)^4}{s_n^4} - 3,$$

where $\bar{x}_n$ is the arithmetic mean and $s_n$ is the sample standard deviation of $x_1, \ldots, x_n$. We can use $\hat{b}_1$, $\hat{b}_2$ for testing, whether $x_1, \ldots, x_n$ are realizations of normal distributed random variables. If this is true, then $\hat{b}_1$ and $\hat{b}_2$ ought to be close to 0. If, however, $|\hat{b}_1|$ or $|\hat{b}_2|$ are *significantly* large, then we would have to reject the hypothesis of normal data, cf the comments on probability plots in Section 1.6. Note that SAS uses slightly different normalizing constants in the definition of $\hat{b}_1$ and $\hat{b}_2$.

## 1.5  Boxplots

A *boxplot* (Tukey (1977)) is a graphical device from the data analyst's toolbox, which simultaneously displays the median as a measure of location, the interquartile range $IQR$ as a measure of spread, outliers above $x_{[\frac{3}{4}n]:n} + 1.5$ $IQR$ and below $x_{[\frac{1}{4}n]:n} - 1.5$ $IQR$, and gives also indications of symmetry or skewness. The following Figure 1.5.1 is a boxplot of the values of the variable *an* of the cns data in Example 1.1.3.

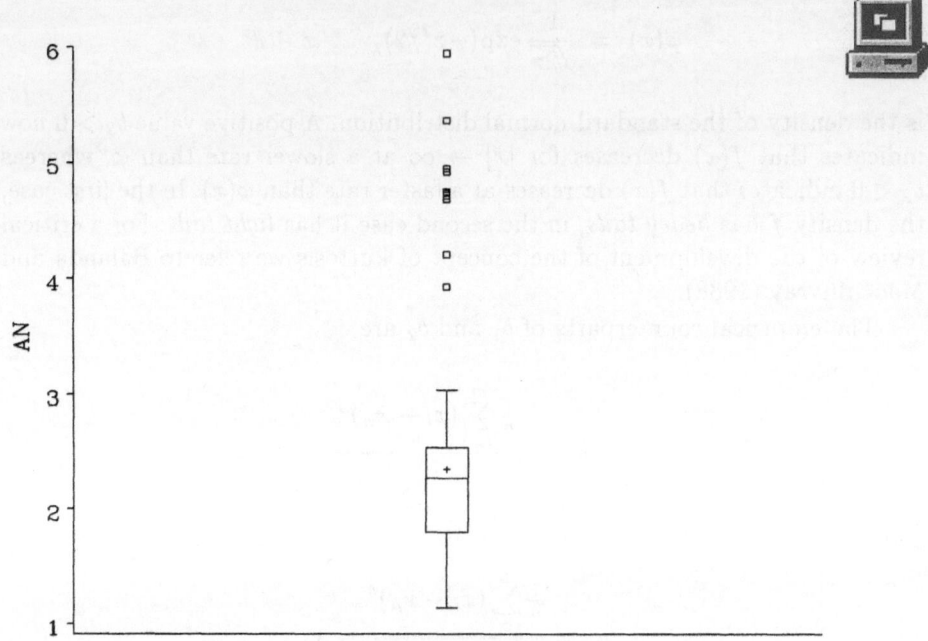

**Figure 1.5.1.** Boxplot of the values of *an* of the cns data.

```
***    Program 1_5_1   ***;
TITLE1 'Boxplot of an';
TITLE2 'CNS Data';
LIBNAME datalib 'c:\data';

DATA data1;
   SET datalib.cns;
   t=1;

AXIS1 LABEL=NONE MAJOR=NONE VALUE=NONE;
PROC BOXPLOT DATA=data1;
   PLOT an*t / BOXSTYLE=SCHEMATIC HAXIS=AXIS1;
RUN; QUIT;
```

A simple boxplot can be plotted using the UNIVARIATE procedure together with the option PLOT, see Figure 1.1.5. A high resolution boxplot can be drawn using PROC BOXPLOT (new in version 8 of SAS). This is done by defining an additional horizontal variable in the first data set, which is named 't' and which is assigned the constant value 1 for all observations. The above AXIS1 statement suppresses any labeling of the axis.

PROC BOXPLOT with the statement PLOT an∗t plots the values of an against the constant value 1, while the option BOXSTYLE=SCHEMATIC ensures an output, which corresponds to the schematic box-and-whisker plot described in Chapter 2 of Tukey (1977).

A boxplot can be motivated as follows, using normal distributed data $x_1, \ldots, x_n$. Denote by

$$\Phi(t) := \frac{1}{\sqrt{2\pi}} \int_{-\infty}^{t} \exp(-x^2/2) \, dx, \qquad t \in \mathbb{R},$$

the distribution function of the standard normal distribution. The normal distribution with mean $\mu$ and variance $\sigma^2 > 0$ then has the distribution function $\Phi_{\mu,\sigma}(t) := \Phi((t-\mu)/\sigma)$, $t \in \mathbb{R}$. Suppose now that $x_1, \ldots, x_n$ are realizations of $n$ independent and identically normal distributed random variables with mean $\mu$ and variance $\sigma^2$. Then we have for $q \in (0,1)$ approximately

$$x_{[qn]:n} \approx \Phi_{\mu,\sigma}^{-1}(q) = \sigma \Phi^{-1}(q) + \mu,$$

see Lemma 1.6.5 and Corollary 1.6.8 below. From the remarks after Definition 1.3.3 we obtain, thus, the approximations

$$x_{[\frac{3}{4}n]:n} + 1.5 \, IQR \approx \sigma \Phi^{-1}(3/4) + \mu + 1.5 \, IQR$$
$$\approx \sigma \underbrace{\Phi^{-1}(3/4)}_{\approx 0.675} + \mu + 1.5 \times 1.35\sigma \approx \mu + 2.7\sigma$$

and

$$x_{[\frac{1}{4}n]:n} - 1.5 \, IQR \approx \mu - 2.7\sigma,$$

where $P\{X \leq \mu + 2.7\sigma\} = P\{X \geq \mu - 2.7\sigma\} \approx 0.9965$ if $X$ is standard normal distributed with mean $\mu$ and variance $\sigma^2$. In the case of approximately normal distributed data one expects, therefore, that almost all of them will be elements of the interval

$$I := \left[ x_{[\frac{1}{4}n]:n} - 1.5 \, IQR, \ x_{[\frac{3}{4}n]:n} + 1.5 \, IQR \right].$$

The data in this interval are displayed by the box together with the two *whiskers*, which end at the largest and smallest observation in $I$. The box itself runs from the lower quartile $x_{[n/4]:n}$ to the upper quartile $x_{[3n/4]:n}$. The horizontal bar in the box displays the median. The location of the median visualizes a skewness of the data: If it is located near the lower or upper end of the box, then the sample is skewed to the right or to the left, respectively. Those data that do not belong to the interval $I$ are marked as outliers by a dot. Note that a boxplot displays only robust measures with high breakdown points. This prevents outliers from hiding themselves behind nonrobust measures, which might have been already influenced by these outliers. This *masking* of outliers would, for example, be eased if we replaced the above interval $I$ by

$$I_{mask} := [\bar{x}_n - 2.7s_n, \ \bar{x}_n + 2.7s_n],$$

where $\bar{x}_n$ is the arithmetic mean and $s_n$ is the standard deviation of $x_1, \ldots, x_n$, cf Definitions 1.2.1 and 1.3.1. These measures are highly sensitive to outliers and can, therefore, heavily influence the interval $I_{mask}$.

**1.5.1 Example** (Sunspot Data; Andrews and Herzberg (1985), Chapter 11). The monthly averages of the daily counts of sunspots were recorded from January 1931 until December 1983. The boxplots in the following Figure 1.5.2 are all based on the 12 data for each of the years from 1931 to 1983. Certain periodicities in the data are obvious. The assumption that all monthly averages were independent and identically distributed random variables would surely be wrong, as shown by this graph. The analysis of the sunspot data requires instead elements from *time series analysis*, cf Exercise 30 in Chapter 3.

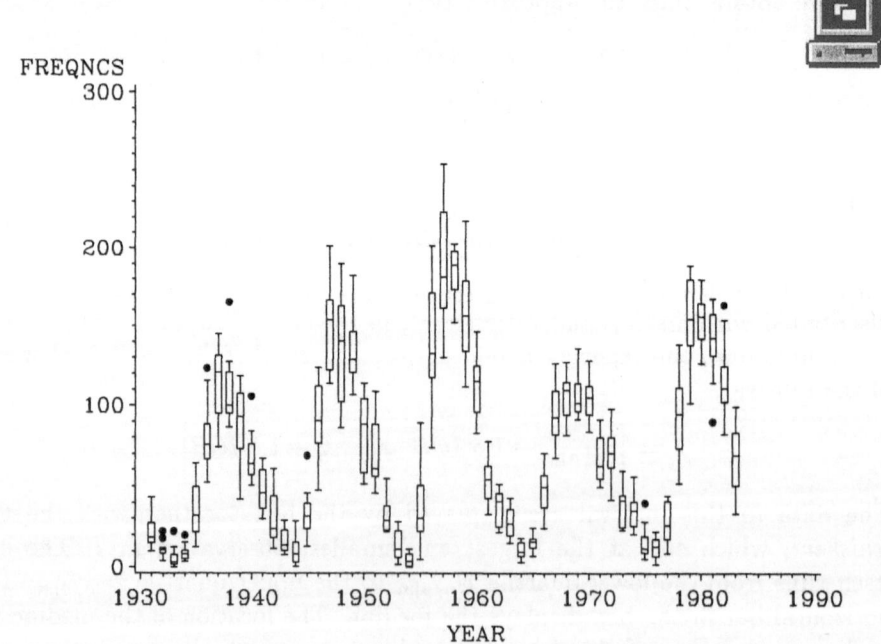

**Figure 1.5.2.** Boxplots of the sunspot data.

```
*** Program 1_5_2 ***;
TITLE1 'Boxplots';
TITLE2 'Sunspot Data';
LIBNAME datalib 'c:\data';

SYMBOL1 C=GREEN V=DOT I=BOXT;
PROC GPLOT DATA=datalib.sunspot;
   PLOT freqncy*year;
RUN; QUIT;
```

An alternative to PROC BOXPLOT is PROC GPLOT with the I=BOXT option in a SYMBOL statement (already in Version 6 of SAS). The syntax of the PLOT statement is almost the same as in PROC BOXPLOT (see Program 1_5_1), so that 'PLOT freqncy*year;' generates a boxplot of the twelve observations of the variable 'freqncy' for each year.

# 1.6  Quantile Plots

Quantile plots are graphical devices from the data analyst's toolbox. They can be used for evaluating the goodness of fit of a theoretical distribution to a given sample. In this case they are called probability plots such as the normal probability plot. But they can also be used for the evaluation of a linear relationship between random variables, in which case they are called quantile–quantile plots. In each case quantiles are displayed, either quantiles of a theoretical distribution function or sample quantiles. As the latter ones are given by *order statistics*, we will compile in the following some theoretical results first. For a thorough study of order statistics we refer to the monograph by Reiss (1989). The special case of extreme order statistics and their practical significance for the modeling and investigation of *rare events* is extensively treated in Reiss and Thomas (2001). This approach is in particular important if extreme observations are not erroneous outliers but genuinely belong to the sample. This decision is often hard to make, see the discussion in Section 1.6 of Reiss and Thomas (2001).

### Order Statistics and the Quantile Transformation

In the following we denote by $U_{1:n} \leq \cdots \leq U_{n:n}$ the ordered values or order statistics pertaining to the $n$ random variables $U_1, \ldots, U_n$.

**1.6.1 Lemma.** *Let $U_1, \ldots, U_n$ be independent and uniformly distributed random variables on (0,1), i.e., $P\{U_i \leq s\} = s$, $s \in (0,1)$. Then we have*

$$E(U_{i:n}) = \frac{i}{n+1}, \qquad i = 1, \ldots, n.$$

**Proof:** See Exercise 26.                                                  □

**1.6.2 Definition.** Let $F$ be a distribution function on $\mathbb{R}$. By

$$F^{-1}(q) := \inf\{t \in \mathbb{R} : F(t) \geq q\}, \qquad q \in (0,1),$$

we denote the *generalized inverse* or *quantile function* of $F$.

The fact that a distribution function is continuous from the right implies the following basic result.

**1.6.3 Lemma.** *We have for an arbitrary distribution function $F$ on $\mathbb{R}$ the equivalence*

$$F^{-1}(q) \leq t \Leftrightarrow q \leq F(t), \qquad q \in (0,1), \quad t \in \mathbb{R}.$$

**Proof:** See Exercise 27.                                                  □

**1.6.4 Corollary.** *Let $X$ be a random variable with distribution function $F$ and let $U$ be a uniformly distributed random variable on (0,1).*

*(i) $F(X)$ is uniformly distributed on (0,1) if $F$ is continuous:*

$$P\{F(X) \leq s\} = s, \qquad s \in (0,1).$$

*(ii) The random variable $F^{-1}(U)$ has for an arbitrary $F$ the distribution function $F$:*

$$P\{F^{-1}(U) \leq t\} = F(t), \qquad t \in \mathbb{R}.$$

*This is the quantile transformation.*

**Proof:** Exercise 31.                                                       □

Suppose we want to investigate the stochastic behavior of an arbitrary random variable $Z$ with distribution function $F$. By the quantile transformation we can now assume without loss of generality that it is given as

$$Z = F^{-1}(U),$$

where $U$ is a uniformly distributed random variable on $(0,1)$. We thus divide $Z$ into a *random* part, which is uniformly distributed on (0,1), and into the *nonrandom* transformation $F^{-1}$. Results on the stochastic behavior of arbitrary random variables can thus be deduced from that on uniformly distributed random variables. Corollary 1.6.8 on the convergence of order statistics will, for instance, be deduced this way.

**1.6.5 Lemma.** *Let $Y$ be a random variable with distribution function $G$. Denote by $F$ the distribution function of $X := \sigma Y + \mu$, where $\sigma > 0$, $\mu \in \mathbb{R}$ are arbitrary numbers, i.e., $F(t) = G((t - \mu)/\sigma)$, $t \in \mathbb{R}$. Then we have for $q \in (0,1)$*

$$F^{-1}(q) = \sigma G^{-1}(q) + \mu.$$

**Proof:** Exercise 32. □

For the motivation of quantile plots we need the following results on the convergence of $U_{k:n}$. Lemma 1.6.1 suggests the approximation $U_{k:n} \approx k/(n+1)$. Actually, we have the following upper bound.

**1.6.6 Lemma.** *Let $U_1, \ldots, U_n$ be independent and uniformly distributed random variables on (0,1). Then we have with probability one the bound*

$$\max_{1 \le k \le n} |U_{k:n} - k/n| \le \sup_{t \in [0,1]} |F_n(t) - t|,$$

*where $F_n$ is the empirical distribution function of $U_1, \ldots, U_n$.*

**Proof:** With probability one the random variables $U_1, \ldots, U_n$ attain different values and, thus, $U_{1:n} < \cdots < U_{n:n}$, see Exercise 33. In this case we have for $t = U_{k:n}$

$$F_n(t) = k/n,$$

which immediately implies the assertion. □

**1.6.7 Corollary.** *We have with probability one*

$$\max_{1 \le k \le n} \left| U_{k:n} - \frac{k}{n+1} \right| \longrightarrow_{n \to \infty} 0.$$

**Proof:** By the Glivenko–Cantelli Theorem we have with probability one

$$\sup_{t \in [0,1]} |F_n(t) - t| \longrightarrow_{n \to \infty} 0$$

and, thus, the assertion is immediate from Lemma 1.6.6. □

**1.6.8 Corollary.** *Let $X_1, \ldots, X_n$ be independent and identically distributed random variables with distribution function $F$. Suppose that $F^{-1}$ is continuous on the interval $(a, b) \subset (0, 1)$. Then we have with probability one*

$$\max_{k_1 \le k \le k_2} \left| X_{k:n} - F^{-1}\left(\frac{k}{n+1}\right) \right| \longrightarrow_{n \to \infty} 0$$

*if $k_1 = k_1(n) \le k_2 = k_2(n)$ are sequences of integers which satisfy*

$$a < \liminf_{n \in I\!N} \frac{k_1}{n} \le \limsup_{n \in I\!N} \frac{k_2}{n} < b.$$

**Proof:** The function $F^{-1}$ is uniformly continuous on each compact interval $[a + \varepsilon, b - \varepsilon]$ contained in $(a, b)$. Put $X_{k:n} = F^{-1}(U_{k:n})$. The assertion is now a consequence of Corollary 1.6.7. □

## Quantile Plots

We assume in the sequel that the random variables $X_1, \ldots, X_n$ are independent and identically distributed with distribution function $F$, which is of the form

$$F(t) = G\left(\frac{t - \mu}{\sigma}\right), \qquad t \in I\!R,$$

where $G$ is known, the numbers $\mu$ and $\sigma$ are, however, unknown. The number $\mu \in I\!R$ is a *location parameter* and $\sigma > 0$ is a *scale parameter*. By the quantile transformation in Corollary 1.6.4 (ii) we can assume that the random variables $X_i$ are of the form $X_i = F^{-1}(U_i)$, $i = 1, \ldots, n$, where $U_1, \ldots, U_n$ are independent and uniformly distributed random variables on $(0,1)$. Since $F^{-1} : (0, 1) \to I\!R$ is a monotone increasing function, the order of the $U_{i:n}$ is kept under the transformation $F^{-1}$. We obtain, therefore, from Lemma 1.6.5 for the order statistics $X_{i:n}$ the representation

$$X_{i:n} = F^{-1}(U_{i:n}) = \sigma G^{-1}(U_{i:n}) + \mu, \qquad i = 1, \ldots, n.$$

If we plot $X_{k:n}$ against $G^{-1}(k/(n + 1))$, i.e., if we plot the points

$$(G^{-1}(k/(n + 1)), X_{k:n}), \qquad k = 1, \ldots, n,$$

then we obtain a *quantile plot*. Another quite common name is *probability plot*. In the case of a continuous quantile function $G^{-1} : (0, 1) \to I\!R$ we expect from Corollary 1.6.8 the approximation

$$X_{k:n} \approx F^{-1}\left(\frac{k}{n + 1}\right) = \sigma G^{-1}\left(\frac{k}{n + 1}\right) + \mu.$$

The points

$$\left(G^{-1}\left(\frac{k}{n + 1}\right), X_{k:n}\right) \approx \left(G^{-1}\left(\frac{k}{n + 1}\right), \sigma G^{-1}\left(\frac{k}{n + 1}\right) + \mu\right)$$

should then be close to the line $s = \sigma t + \mu \approx S_n t + \bar{X}_n$, $t \in I\!R$. By $S_n = s_n(X_1, \ldots, X_n)$ we denote the empirical standard deviation of $X_1, \ldots, X_n$ and by $\bar{X}_n$ their arithmetic mean, cf Sections 1.2 and 1.3. If this approximately linear relationship is obviously not true, then the quantile plot makes us doubt that the distribution function $G$ actually underlies the sample. With the particular choice of the standard normal distribution function $G = \Phi$, we would, for example, doubt the assumption of normal data. Location and scale parameters are here only of minor importance. The evaluation of a quantile plot evidently requires a certain experience.

A *normal probability plot*, which is defined by the choice $G = \Phi$, can reveal, in addition, some information about the skewness and the kurtosis of the sample. In the case of a convex probability plot, the differences $x_{k+1:n} - x_{k:n}$ of subsequently ordered data have the tendency to grow at an increasing rate. This indicates that the distribution of the data is skewed to the right. If the probability plot is concave, then the ratios $(x_{k+1:n} - x_{k:n})/(x_{k:n} - x_{k-1:n})$ have the tendency to be less than one. This indicates that the distribution of the data is skewed to the left. A probability plot, which is first concave and then convex, indicates a distribution with heavy tails. If it is first convex and then concave, this is an indication of light tails, cf Section 1.4.

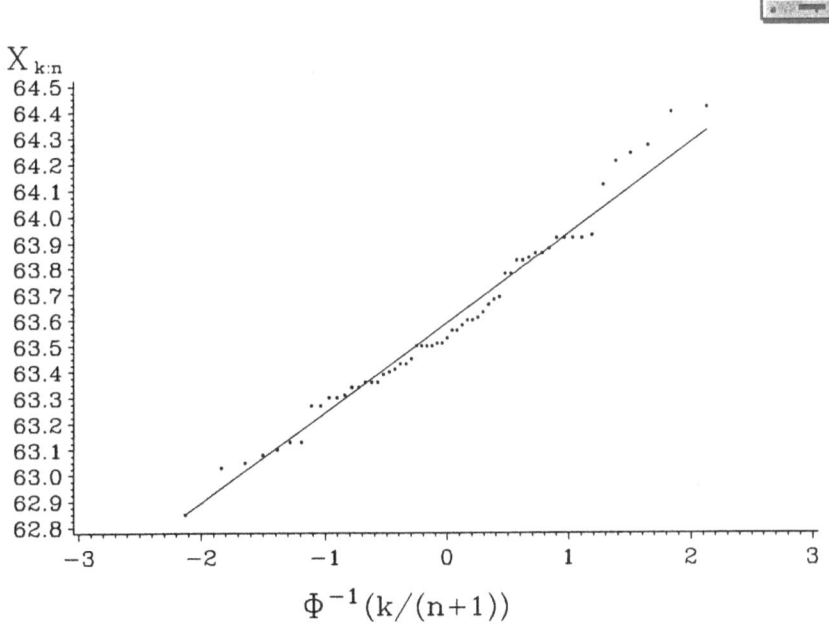

**Figure 1.6.1.** Normal probability plot of the beeswax data in Example 1.1.11.

```
***    Program 1_6_1    ***;
TITLE1 'Normal Probability Plot';
TITLE2 'Beeswax Data';
LIBNAME datalib 'c:\data';

PROC SORT DATA=datalib.beeswax OUT=sortdata;
    BY degree;
PROC MEANS DATA=datalib.beeswax NOPRINT MEAN STD N;
    VAR degree;
    OUTPUT OUT=meandata MEAN=mu STD=sigma N=n;
DATA data1;
    SET meandata;
    DO k=1 TO n;
        SET sortdata POINT=k;
        quantile=PROBIT(k/(n+1));
        model=sigma * quantile+mu;
        OUTPUT;
    END; STOP;

SYMBOL1    V=DOT I=NONE C=GREEN H=0.3;
SYMBOL2    V=NONE C=RED    I=JOIN L=1 W=0.5;
AXIS1 LABEL=(H=3 'X' H=1.5 ' k:n');
AXIS2 LABEL=NONE;
FOOTNOTE1 H=3    F=CGREEK       'F'
            H=2    M=(+0,+1.2) F=COMPLEX '-1'
            H=2.5 M=(+0,-1)      '(k/(n+1))';

PROC GPLOT DATA=data1;
    PLOT degree*quantile=1 model*quantile=2
        / OVERLAY VAXIS=AXIS1 HAXIS=AXIS2;
RUN; QUIT;
```

A simple normal probability plot can be displayed using PROC UNIVARIATE with the option PLOT, cf Program 1_1_5. The above high resolution normal probability plot is the combination of two plots. First, the variable 'degree' containing the original data, is plotted against the variable 'quantile', which contains the values $\Phi^{-1}(k/(n+1))$. These are computed by the PROBIT function. The second plot displays the straight line sample mean + sample standard deviation × quantile against quantile.

The original data are sorted first and the result is written to 'sortdata'. Using PROC MEANS with the options MEAN, STD and N, we compute the sample mean 'mu', the sample standard deviation 'sigma' and the number of available observations 'n' of the variable 'degree'. The statement 'OUTPUT OUT = ...' creates the temporary file 'meandata' which contains mu, sigma and n. Then the two files 'meandata' and 'sortdata' have to be merged. To this end, a temporary file 'data1' is created and 'meandata' is moved to this file. The DO loop loads sequentially each of the n observations of 'sortdata': To get an individual observation of the original data the SET statement is used with the 'POINT=k' option selecting the k-th observation. The pertaining value of quantile and of the straight line is computed, and the OUTPUT statement writes these values sequentially to the temporary file 'data1'. Since the POINT option is used, the DATA step must be finished after the END statement by a STOP statement to close the DO loop.

The options '=1' and '=2' in the PLOT statement invoke the pertaining SYMBOL statement for each plot. The two plots are overlaid by the option OVERLAY.

## Quantile–Quantile Plots

Quantile–Quantile plots are used for comparing the distributions of two samples. Let $X_1, \ldots, X_n$ be independent and identically distributed random variables with common distribution function $F$, and let $Y_1, \ldots, Y_n$ be independent and identically distributed random variables with common distribution function $G$. A typical example in biostatistics is the case, where the $X_i$ are observations of a group that has received some treatment and the $Y_i$ are observations of a control group that did not receive the treatment. If this treatment has an additive effect $\mu$, such as an average increase of weight by $\mu = 2$ lb, then we have

$$F(t) = G(t - \mu), \qquad t \in \mathbb{R}.$$

Lemma 1.6.5 implies

$$F^{-1}(q) = G^{-1}(q) + \mu, \qquad q \in (0, 1).$$

Then the pertaining *quantile–quantile plot*, where we plot $X_{k:n}$ against $Y_{k:n}$, $k = 1, \ldots, n$, will by Corollary 1.6.8 now approximately give the line $s = t + \mu$, $t \in \mathbb{R}$:

$$(Y_{k:n}, X_{k:n}) \approx \left( G^{-1}\left(\frac{k}{n+1}\right), \; F^{-1}\left(\frac{k}{n+1}\right) \right)$$

$$= \left( G^{-1}\left(\frac{k}{n+1}\right), \; G^{-1}\left(\frac{k}{n+1}\right) + \mu \right).$$

If the treatment has a multiplicative effect, such as an average increase of weight by 5%, then we have

$$F(t) = G(t/\sigma), \qquad t \in \mathbb{R},$$

for some $\sigma > 0$. In the 5% example we have $\sigma = 105/100$. Lemma 1.6.5 now yields

$$F^{-1}(q) = \sigma G^{-1}(q), \qquad q \in (0, 1).$$

We obtain, thus, for the quantile–quantile plot the approximation

$$(Y_{k:n}, X_{k:n}) \approx \left( G^{-1}\left(\frac{k}{n+1}\right), \; \sigma G^{-1}\left(\frac{k}{n+1}\right) \right),$$

i.e., the points $(Y_{k:n}, X_{k:n})$ approximate the line $s = \sigma t$, $t \in \mathbb{R}$.

If we have in the above situation $\mu > 0$ or $\sigma > 1$, then $F$ is called *stochastically larger* than $G$, since the random variables $X_i$ have in this case the tendency to attain larger values than the $Y_i$.

## 1.7   Variance–Stabilizing Transformations

Let $X$ be a random variable with mean $\mu$ and variance $\sigma^2 = \sigma^2(\mu)$, which we assume to be a function of $\mu$. Suppose, for example, that $X$ has a *binomial distribution* with the parameters $n \in I\!N$ and $p \in [0,1]$, denoted by $B(n,p)$:

$$P\{X = j\} = \binom{n}{j} p^j (1-p)^{n-j} =: B(n,p)(\{j\}), \qquad j = 0,1,\dots,n.$$

We have in this case

$$\mu = E(X) = np, \;\; Var(X) = np(1-p) = E(X)\Big(1 - \frac{E(X)}{n}\Big) = \sigma^2(\mu).$$

We try now to find a transformation $T$ such that the variance of $T(X)$ is approximately constant and thus, it is in particular independent of $\mu$. Put $Y = T(X)$, where $T$ is a twice differentiable function. Suppose that $E(Y^2) < \infty$. Taylor's formula suggests the approximation

$$
\begin{aligned}
Var(Y) &= E\left(\left\{T(X) - E(T(X))\right\}^2\right) \\[2mm]
&= E\left(\left\{T(X) - T(\mu) - E(T(X) - T(\mu))\right\}^2\right) \\[2mm]
&= E\left(\Big\{T'(\mu)(X-\mu) + T''(\xi)\frac{(X-\mu)^2}{2}\right. \\[2mm]
&\qquad\quad \left. -E\Big(T'(\mu)(X-\mu) + T''(\xi)\frac{(X-\mu)^2}{2}\Big)\Big\}^2\right) \\[2mm]
&= E\left(\Big\{T'(\mu)(X-\mu) + T''(\xi)\frac{(X-\mu)^2}{2} - E\Big(T''(\xi)\frac{(X-\mu)^2}{2}\Big)\Big\}^2\right) \\[2mm]
&\approx E\left(\left\{T'(\mu)(X-\mu)\right\}^2\right) = (T'(\mu))^2 \sigma^2(\mu),
\end{aligned}
$$

where $\xi$ is some number between $X$ and $\mu$. If the transformation $T$ has the property that $(T'(\mu))^2\sigma^2(\mu)$ is approximately a constant as a function of $\mu$, then the variance of $Y = T(X)$ should be nearly independent of $\mu$ . Such a function $T$ is called a *variance–stabilizing transformation*.

For the binomial distribution $B(n,p)$ we obtain, for instance, the condition

$$(T'(np))^2 np(1-p) \approx \text{const} \quad \text{or} \quad (T'(np))^2 np \approx \text{const}$$

if $1 - p$ is approximately 1, i.e., if $p$ is close to zero. The second approximation is obviously satisfied by

$$T(x) = \sqrt{x} \quad \text{with} \quad \text{const} = 1/4.$$

The function $T(x) = \sqrt{x}$ is in the case of the $B(n, p)$ distribution with a small $p$ consequently a variance–stabilizing transformation with

$$E(\sqrt{X}) \approx \sqrt{np}, \; Var(\sqrt{X}) \approx 1/4.$$

Applying this variance–stabilizing transformation to histograms, we obtain hanging rootograms that we describe in the following.

## Hanging Histograms and Rootograms

A hanging rootogram is a graphical tool to evaluate the goodness–of–fit of a model distribution to a given sample. Let $X_1, \ldots, X_n$ be independent and identically distributed random variables which we assume to have the common distribution function $F$. We want to investigate, whether this particular $F$ actually underlies the sample. To this end we consider a histogram of the $X_i$. The expected number of observations, which fall into the $j$–th cell $I_j$ of a histogram with a total of $m$ cells, is

$$E\Big(\sum_{i=1}^{n} 1_{I_j}(X_i)\Big) = n\, P\Big\{X_1 \in I_j\Big\} =: np_j, \qquad j = 1, \ldots, m.$$

The frequencies

$$N_j := \sum_{i=1}^{n} 1_{I_j}(X_i), \qquad j = 1, \ldots, m$$

of the $m$ cells are binomial distributed random variables with the parameters $n$ and $p_j$, $j = 1, \ldots, m$. They are *not* independent, as their total sum is $n$. If we compare $N_j$ with the expected value $np_j$, by plotting $N_j - np_j$, then we obtain a *hanging histogram*. Note that the probabilities $p_j$ are derived from the distribution function $F$. If this distribution function actually underlies the data, then each $N_j - np_j$ has the expectation zero. The variances $np_j(1 - p_j)$ of the $N_j$ will, however, be different.

If we apply now the variance–stabilizing transformation $T(x) = \sqrt{x}$ to each $N_j$ and plot $\sqrt{N_j} - \sqrt{np_j}$ against $I_j$, then the variances of the transformed frequencies are all approximately 1/4 and the expectations are nearly 0, provided that each $p_j$ is small and our model for $F$ is correct. This is the concept of a *hanging rootogram*. By the 2– or 3–$\sigma$ rule we will expect that the distance $\sqrt{N_j} - \sqrt{np_j}$ of the random variable from its approximate expectation zero is not greater than two or three times its standard deviation. Since this standard deviation is approximately $\sqrt{1/4} = 1/2$, distances larger than 1 up to 1.5 indicate outliers. The occurrence of a relatively large number of outliers would make us doubt that our model distribution function $F$ is the correct one.

Suppose now that we want to investigate the problem, whether the underlying distribution belongs to a certain parametric *family* of distributions, such as the normal ones rather than that it coincides with a particular one. A tool for this investigation can be a hanging rootogram based on studentized observations, which we describe in the following.

Let $X_1, \ldots, X_n$ be arbitrary random variables. Then $\bar{X}_n := n^{-1} \sum_{i=1}^{n} X_i$ denotes the sample mean and $S_{X,n}^2 := s_n^2(X_1, \ldots, X_n) = (n-1)^{-1} \sum_{i=1}^{n} (X_i - \bar{X}_n)^2$ the sample variance. The *studentized* random variables

$$Y_i := \frac{X_i - \bar{X}_n}{S_{X,n}}, \qquad i = 1, \ldots, n,$$

have sample mean zero and sample variance one:

$$\bar{Y}_n = 0, \ S_{Y,n}^2 = 1,$$

provided $S_{X,n} > 0$, i.e., that not all $X_i$ coincide. This *studentization* eliminates a possible location parameter $\mu$ of the $X_i$ and sets a scaling parameter to one. The multivariate version of this studentization, called *sphering the data* will be introduced in Section 6.

Suppose, for example, that we want to test the hypothesis that the $X_i$ are independent and identically distributed normal random variables. The $Y_i$ will in this case approximately behave like independent standard normal random variables:

$$Y_i = \frac{X_i - \bar{X}_n}{S_{X,n}} \approx \frac{X_i - \mu}{\sigma}, \qquad i = 1, \ldots, n,$$

where $\mu$ and $\sigma^2$ denote the expectation and the variance of $X_i$. The hanging rootogram pertaining to the $Y_i$ has for the cell

$$I_j = (a_{j-1}, a_j]$$

the value $\sqrt{N_j} - \sqrt{np_j}$, where $N_j$ is the number of $Y_i$ in the cell $I_j$ and

$$p_j = \Phi(a_j) - \Phi(a_{j-1}) = \frac{1}{\sqrt{2\pi}} \int_{a_{j-1}}^{a_j} \exp(-x^2/2)\, dx$$

is approximately the probability $P\{Y_i \in I_j\}$ of this occurrence. If the $X_i$ actually follow a normal distribution, then the observations $\sqrt{N_j} - \sqrt{np_j}$ will usually have values between $-1.5$ and $1.5$ by the 2- or 3-$\sigma$ rule. Only in the case of quite a large $n$ we have to expect with high probability values outside the interval $[-1.5, 1.5]$ as well.

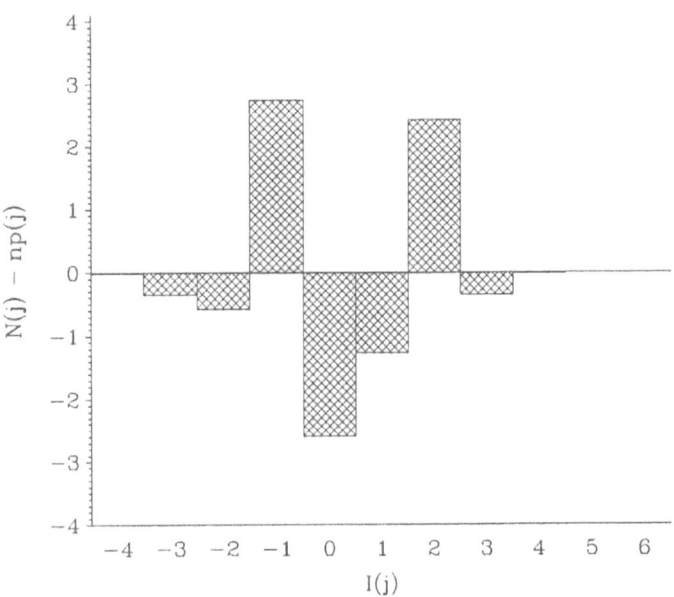

**Figure 1.7.1.** Hanging histogram of the beeswax data in Example 1.1.11.

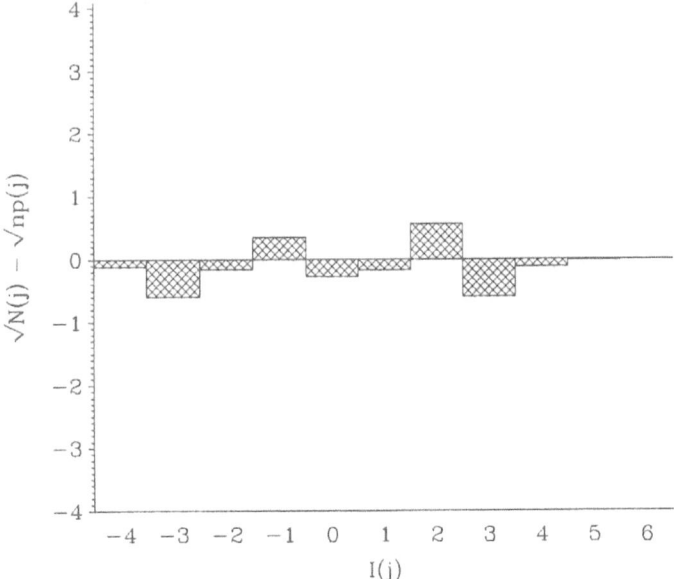

**Figure 1.7.2.** Hanging rootogram of the beeswax data.

```
***    Program 1_7_1   ***;
TITLE1 'Hanging Histogram and Hanging Rootogram';
TITLE2 'Beeswax Data';
LIBNAME datalib 'c:\data';

PROC MEANS DATA=datalib.beeswax NOPRINT MEAN STD N;
  VAR degree;
  OUTPUT OUT=meandata MEAN=mu STD=sigma N=n;

DATA data1; SET meandata;
  DO i=-4.5 TO 5.5 BY 1;
    model=n*(PROBNORM(i+1)-PROBNORM(i)); * theoretical frequencies;
    freqncy=0;
    DO k=1 TO n;
      SET datalib.beeswax POINT=k;
      degree=(degree-mu)/sigma;           * studentization;
      IF i<degree<=i+1 THEN freqncy+1;    * indicator function
    END;
    diff =freqncy-model;                  * value for 'histogram';
    rdiff=SQRT(freqncy)-SQRT(model);      * value for 'rootogram';
    center=i+0.5;                         * interval center;
    OUTPUT;
  END; STOP;

AXIS1 LABEL=('I(j)');
AXIS2 LABEL=(A=90 'N(j) - np(j)')
    ORDER=(-4 TO 4 BY 1);
AXIS3 LABEL=(A=90 F=MATH 'a' F=COMPLEX 'N(j) - '
        F=MATH 'a' F=COMPLEX 'np(j)')
    ORDER=(-4 TO 4 BY 1);

PROC GCHART DATA=data1;
  TITLE1 'Hanging Histogram';
  TITLE2 'Beeswax Data';
  VBAR center / DISCRETE SUMVAR=diff SPACE=0
      RAXIS=AXIS2   MAXIS=AXIS1;
RUN;
  TITLE1 'Hanging Rootogram';
  TITLE2 'Beeswax Data';
  VBAR center / DISCRETE SUMVAR=rdiff SPACE=0
      RAXIS=AXIS3   MAXIS=AXIS1;
RUN; QUIT;
```

The drawing of a hanging histogram or rooto-gram requires several steps, which are conveniently done in the DATA step.

The cell boundaries i= -4.5, -3.5,..., 4.5, 5.5 are determined in the first DO loop. The expected cell frequencies for standard normal distribution are computed by the standard normal distribution function PROBNORM multiplied by the number of observations n.

The sample frequency of each cell is computed in the next DO loop. The beeswax data are sequentially read, studentized with mean and standard deviation delivered by PROC MEANS, and the counter of the cell (i, i+1] is increased by 1 if the studentized observation is an element of this interval.

The difference between the sample frequency and the theoretical frequency as well as the difference of the square roots of these values together with the interval center i+0.5 is computed for each cell and written to the temporary file 'data1' by the OUTPUT statement.

The options SUM and SUMVAR=diff must be specified in order to let PROC GCHART print a bar chart, where the heights of the bars are given by the values of the variable 'diff'. The option DISCRETE suppresses the automatic definition of cells by SAS.

## Exercises

1. (pH Data; Schierl and Göttlein (1987), taken from Pruscha (1989), p. 143). A study by the faculty of forestry of the University of Munich investigated among others the effect of watering and liming on the hydrogen–ion concentration in forest floor. The measurements were taken in pH, used in expressing both acidity and alkalinity on a scale with values from 0 to 14, where 7 represents neutrality, numbers less than 7 increasing acidity, and numbers greater than 7 increasing alkalinity. The factor *watering* has three different levels, *no* (no additional watering), *ac* (additional acid watering), and *w* (additional neutral watering) and the factor *liming* has the two levels *n* (no additional liming) and *y* (additional liming). To compare the effects, 6 relatively homogeneous plots of land were selected and 16 observations were taken from each of the 6 possible cross effects.

| Watering *no* | | Watering *ac* | | Watering *w* | |
|---|---|---|---|---|---|
| n | y | n | y | n | y |
| 4.31 | 7.17 | 3.80 | 7.16 | 4.42 | 7.84 |
| 4.59 | 7.17 | 4.27 | 7.19 | 4.25 | 7.25 |
| 4.13 | 6.89 | 4.19 | 7.45 | 4.32 | 7.18 |
| 4.25 | 6.49 | 4.31 | 7.49 | 4.19 | 7.31 |
| 4.15 | 6.89 | 3.95 | 7.39 | 4.17 | 7.65 |
| 4.28 | 7.05 | 4.24 | 6.93 | 4.46 | 7.46 |
| 4.20 | 7.32 | 3.82 | 7.08 | 4.22 | 7.43 |
| 4.66 | 5.84 | 4.07 | 6.96 | 4.90 | 6.96 |
| 3.45 | 6.40 | 4.06 | 7.13 | 3.73 | 7.05 |
| 3.71 | 6.14 | 4.21 | 7.19 | 3.86 | 6.49 |
| 3.87 | 6.36 | 3.58 | 6.27 | 4.14 | 7.57 |
| 4.11 | 5.36 | 3.70 | 6.60 | 3.87 | 6.60 |
| 3.68 | 6.39 | 3.79 | 7.03 | 3.75 | 7.77 |
| 3.73 | 5.19 | 3.74 | 6.88 | 3.70 | 7.05 |
| 3.54 | 6.53 | 3.67 | 6.18 | 4.01 | 7.32 |
| 3.65 | 6.18 | 3.85 | 6.28 | 3.85 | 7.16 |

Import in that order those raw data, which belong to the combination $n|no$, $n|ac$, $n|w$ of *liming*|*watering*. The data can be read line by line if the INPUT statement ends with @@, cf Exercise 10. Plot histograms with different cell width.

**2.** (Properties of distribution functions) Consider a random variable $X$ with distribution function $F$, i.e.,

$$P\{X \leq x\} = F(x), \qquad x \in \mathbb{R}.$$

Show that

(i) $\lim_{x \to \infty} F(x) = 1$, $\quad \lim_{x \to -\infty} F(x) = 0$.

(ii) $F$ is increasing.

(iii) $F$ is continuous from the right and has limits from the left.

**3.** Consider a random variable $X$ with distribution function $F$. Prove the following facts.

(i) $F$ is continuous at $x \in \mathbb{R} \Leftrightarrow P\{X = x\} = 0$.

(ii) The set of those points, where $F$ is not continuous, is countable. Hint: Consider the sets $A_n = \{x \in \mathbb{R} : F(x) - F(x - 0) > 1/n\}, n \in \mathbb{N}$.

**4.** Let $F_n(t) = n^{-1} \sum_{i=1}^{n} 1_{(-\infty, t]}(x_i)$, $t \in \mathbb{R}$, be the empirical distribution function of the points $x_1, \ldots, x_n$. Show that the equation

$$\int f(x) \, dF_n(x) = \frac{1}{n} \sum_{i=1}^{n} f(x_i)$$

holds for any function $f : \mathbb{R} \to \mathbb{R}$.

**5.** The convolution $F$ of two distribution functions $F_1$ and $F_2$ is defined by

$$F(x) = \int_{-\infty}^{\infty} F_1(x - y) \, dF_2(y), \quad x \in \mathbb{R}.$$

Prove the following assertions.

(i) $F$ is a distribution function, i.e., $F$ satisfies (i)–(iii) in Exercise 2.

(ii) Consider independent random variables $X_1$, $X_2$ with distribution functions $F_1$, $F_2$. The random variable $X_1 + X_2$ then has the distribution function $F$. Hint: Fubini's Theorem.

(iii) If $X_1$ has a continuous distribution function, so has $X_1 + X_2$.

**6.** The convolution $f$ of two probability densities $f_1$ and $f_2$ is defined by

$$f(x) = \int_{-\infty}^{\infty} f_1(x - y) f_2(y) \, dy, \quad x \in \mathbb{R}.$$

Show that

(i) $f$ is a probability density.

(ii) Let $X_1$ and $X_2$ be independent random variables with density $f_1$ and $f_2$. $X_1+X_2$ then has the density $f$. Hint: Fubini's Theorem.

**7.** Show that the following kernels satisfy the conditions

$$\int k(x)\,dx = 1, \quad \int k^2(x)\,dx < \infty, \quad \int xk(x)\,dx = 0, \quad \int x^2 k(x)\,dx < \infty :$$

(i) $\qquad k_u(x) := \begin{cases} 1/2 & \text{for } |x| \leq 1 \\ 0 & \text{for } |x| > 1 \end{cases}$ $\qquad$ "uniform kernel"

(ii) $\qquad k_\Delta(x) := \begin{cases} 1 - |x| & \text{for } |x| \leq 1 \\ 0 & \text{for } |x| > 1 \end{cases}$ $\qquad$ "triangular kernel"

(iii) $\qquad k_\varphi(x) := \dfrac{1}{\sqrt{2\pi}} \exp(-x^2/2), \; x \in \mathbb{R}$ $\qquad$ "normal kernel".

**8.** Let $X$ and $Y$ be independent random variables with $E(X^2) < \infty$, $E(Y^2) < \infty$. Prove the equation

$$E((X-Y)^2) = Var(X) + E((E(X) - Y)^2).$$

By putting $X = \hat{f}_n(t)$ and $Y = f(t)$ we obtain the decomposition $MSE = Var + Bias^2$.

**9.** Let $X_1, \ldots, X_n$ be random variables with the common density $f$ and let

$$\hat{f}_n(t) = \frac{1}{nh} \sum_{i=1}^{n} k\left(\frac{t - X_i}{h}\right)$$

be a kernel density estimator with kernel $k$ and bandwidth $h > 0$.

(i) Show that we obtain with the kernel $k(x) = (1/2)\, 1_{[-1,1)}(x)$

$$\hat{f}_n(t) = \frac{1}{2h}(F_n(t + h) - F_n(t - h)),$$

where $F_n$ is the empirical distribution function.

(ii) Consider a density $f$ which is $r$ times differentiable near $t$ and whose $r$-th derivative $f^{(r)}$ is continuous at $t$, $r \geq 2$. Let the kernel $k$ vanish outside a bounded interval and satisfy $\int k(x)\,dx = 1$, $\int x^j k(x)\,dx = 0$, $j = 1, \ldots, r - 1$. Prove that

$$E(\hat{f}_n(t)) = f(t) + (-1)^r h^r \frac{f^{(r)}(t)}{r!} \int x^r k(x)\,dx + o(h^r), \qquad h \to 0.$$

(iii) Prove Lemma 1.1.8.

(iv) Show that the mean squared error $MSE(\hat{f}_n(t))$ has under the conditions of (ii) and Lemma 1.1.8 the expansion

$$MSE(\hat{f}_n(t)) = \frac{1}{nh}f(t)\int k^2(x)\,dx + h^{2r}\left(\frac{f^{(r)}(t)}{r!}\int k(x)x^r\,dx\right)^2 + o\left(\frac{1}{nh}+h^{2r}\right).$$

(v) Show that under the condition of (iv)

$$h^* = n^{-1/(2r+1)}(2r)^{-1/(2r+1)}\left(\frac{f(t)\int k^2(x)\,dx}{((f^{(r)}(t)/r!)\int x^r k(x)\,dx)^2}\right)^{1/(2r+1)}$$

is the optimal bandwidth, cf Remark 1.1.10.

(vi) Compute the *MSE* pertaining to $h^*$.

**10.** (pH data; continuation of Exercise 1). Import the raw data belonging to the factor combination $y|no$, $y|ac$, $y|w$ (in this order). Plot histograms and density estimates with different bandwidths, now based on all data. Do stratifications become visible? (Use the option SUBGROUP in the VBAR statement.)

Further results can be obtained by applying PROC MEANS to each of the six possible factor combinations *liming/watering*. This can be done by nested DO loops as follows:

```
DATA pH_vals;
   DO liming = 'n', 'y';
     DO watering = 'no', 'ac', 'w';
       DO I = 1 TO 16;
         INPUT pH @@;
         OUTPUT;
       END;
     END;
   END;
     CARDS;
Now come the data ...
;
PROC MEANS; BY liming watering NOTSORTED;
PROC GCHART;
   VBAR pH/MIDPOINTS = 3 TO 8 BY 0.5 SUBGROUP = liming;
RUN;
```

The option NOTSORTED is required, since SAS expects observations in lexicographical order, and in this order $n$ comes after $a$.

**11.** Is the bimodality in Figure 1.1.2 due to the two strata healthy and ill patients? Use again the option SUBGROUP, cf Exercise 10.

**12.** Prove that the Epanechnikov kernel $k_E$ minimizes $\int k^2(x)dx$ among all kernels $k : [-\sqrt{5}, \sqrt{5}] \to I\!R$ with $\int k(x)dx = 1 = \int x^2 k(x)\,dx$.
Hint: $\int_{-\sqrt{5}}^{\sqrt{5}} k^2(x)dx \to \min \Leftrightarrow \int_{-\sqrt{5}}^{\sqrt{5}}(k(x) - k_E(x))^2 dx \to \min$.

**13.** (Integral representation of the expectation by means of distribution function) Let $X$ be a random variable with distribution function $F$ and $E(X) < \infty$. Establish the equations

$$E(X 1_{(0,\infty)}(X)) = \int_{(0,\infty)} (1 - F(y))\, dy, \quad E(X 1_{(-\infty,0)}(X)) = -\int_{(-\infty,0)} F(y)\, dy.$$

Hint: Fubini's Theorem.

**14.** (i) (Minimum property of the expectation) Let $X$ be a square integrable random variable. Show that the function $f(c) := E((X - c)^2)$ attains its minimum in $c_0 := E(X)$.

(ii) (Minimum property of the median.) A median of a random variable $X$ is any real number $m = m(X)$ with the property

$$P\{X \leq m\} \geq \frac{1}{2} \leq P\{X \geq m\}.$$

A median always exists (take, for instance, $m := \inf\{x \in \mathbb{R} : P\{X \leq x\} \geq 1/2\}$). It is, however, in general not uniquely determined (give an example). Show that

$$E(|X - m|) \leq E(|X - c|), \quad c \in \mathbb{R}.$$

**15.** Show that $s_n^2 = \frac{1}{2n(n-1)} \sum_{1 \leq i,j \leq n} (x_i - x_j)^2$.

**16.** Let $X_1, \ldots, X_n$ be independent and identically distributed random variables with $E(X_1^2) < \infty$. Prove that the empirical variance

$$S_n^2 := s_n^2(X_1, \ldots, X_n) = \frac{1}{n-1} \sum_{i=1}^n (X_i - \bar{X}_n)^2, \quad \bar{X}_n := \frac{1}{n} \sum_{i=1}^n X_i$$

is an unbiased estimator of the variance of $X_1$, i.e., $E(S_n^2) = Var(X_1)$.

**17.** Let $g : (0,\infty) \to (0,\infty)$ be a monotone increasing function. Prove that for arbitrary $\varepsilon > 0$

$$P\{|X| \geq \varepsilon\} \leq \frac{1}{g(\varepsilon)} E(g(|X|)).$$

With $g(x) = x$ this is the *Markov inequality* and with $g(x) = x^2$ we obtain the *Chebyshev inequality*

$$P\{|X - E(X)| \geq \varepsilon\} \leq \frac{1}{\varepsilon^2} Var(X).$$

Hint: Consider first the case $g(x) = x$.

**18.** Show that

(i) skewness $b_1$ and kurtosis $b_2$ of a random variable $X$ are invariant under the standardization $(X - E(X))/Var(X)^{1/2}$.

(ii) $b_2 \geq b_1^2 - 2$. Hint: Integrate the variable $(X^2 - b_1 X - 1)^2$.

**19.** (Platinum data) Write a SAS program computing the $MAD$ of the platinum data in Example 1.2.2.

**20.** Show that the normal distribution $N(\mu, \sigma^2)$ with mean $\mu \in \mathbb{R}$ and variance $\sigma^2 > 0$ has the density $\sigma^{-1} \varphi((x - \mu)/\sigma)$, $x \in \mathbb{R}$. Prove that the central moments of $N(\mu, \sigma^2)$ satisfy:

(i) $\int x^{2k+1} \, dN(0, \sigma^2)(x) = 0, \qquad k \in \mathbb{N} \cup \{0\}$.

(ii) $\int x^{2k} \, dN(0, \sigma^2)(x) = 1 \cdot 3 \cdot 5 \cdot \ldots \cdot (2k - 1)\sigma^{2k}, \qquad k \in \mathbb{N}$.

(iii) $\int |x|^{2k+1} \, dN(0, \sigma^2)(x) = \dfrac{2^{k+1}}{\sqrt{2\pi}} k! \sigma^{2k+1}, \qquad k \in \mathbb{N} \cup \{0\}$.

**21.** (CNS data) Draw histograms of the variables *an, on, mn, gn, ao*, separately for
– ill
– healthy patients.

Which variable indicates illness best? Draw boxplots using PROC GPLOT, separately for each variable and class.

**22.** (Platinum data) Check by a stem–and–leaf plot and a boxplot, whether the measurements of the heat of sublimation of platinum were generated by a normal distribution. What do you conclude from skewness and kurtosis?

**23.** Let $X_1, \ldots, X_n$ be independent random variables with $P\{X_i = 1\} = p = 1 - P\{X_i = 0\}$, $p \in [0, 1]$, $i = 1, \ldots, n$. Prove that the sum $\sum_{i=1}^{n} X_i$ is binomial distributed with the parameters $n$ and $p$, cf Section 1.7.

**24.** Let $X_1, \ldots, X_n$ be arbitrary random variables and denote by $X_{1:n} \leq \cdots \leq X_{n:n}$ the pertaining order statistics. Prove the equation $P\{X_{r:n} \leq t\} = P\{F_n(t) \geq r/n\}$, where $F_n(t) = n^{-1} \sum_{i=1}^{n} 1_{(-\infty, t]}(X_i)$, $t \in \mathbb{R}$, is the empirical distribution function of $X_1, \ldots, X_n$.

**25.** (Distribution function and density of order statistics) Denote by $X_{i:n}$ the $i$–th order statistic pertaining to $n$ independent random variables $X_1, \ldots, X_n$ with common distribution function $F$, $i = 1, \ldots, n$. Prove that

(i) $P\{X_{i:n} \leq t\} = \sum_{j=i}^{n} \binom{n}{j} F(t)^j (1 - F(t))^{n-j}$. In particular we obtain for the minimum $P\{X_{1:n} \leq t\} = 1 - (1 - F(t))^n$ and for the maximum $P\{X_{n:n} \leq t\} \leq F^n(t)$. Hint: Exercises 23 and 24.

(ii) If $F$ is differentiable with the derivative $f$, then $X_{i:n}$ has the density

$$f_{i:n}(t) := i \binom{n}{i} F(t)^{i-1} (1 - F(t))^{n-i} f(t), \quad t \in \mathbb{R}.$$

**26.** (Expectation and variance of order statistics)

(i) Let $U_1, \ldots, U_n$ be independent and uniformly distributed random variables on (0,1). Show that

$$E(U_{i:n}) = \frac{i}{n+1}, \quad Var(U_{i:n}) = \frac{1}{n+2}\frac{i}{n+1}\left(1 - \frac{i}{n+1}\right).$$

(ii) Let $X_1, \ldots, X_n$ be independent random variables with common distribution function $F$ and density $f$. If $F^{-1}$ is differentiable, motivate that

$$E(X_{i:n}) \approx F^{-1}(i/(n+1))$$

and

$$Var(X_{i:n}) \approx \left(f(F^{-1}(i/(n+1)))\right)^{-2} Var(U_{i:n}).$$

(iii) Use (i) and (ii) to motivate the approximation

$$Var(F_n^{-1}(q)) \approx \frac{q(1-q)}{nf^2(F^{-1}(q))}, \quad q \in (0,1).$$

(iv) Let $X_1, \ldots, X_n$ be independent and $N(\mu, \sigma^2)$ distributed random variables. Use (iii) to derive an approximation of the sample median and compare this approximation with the variance of the sample mean.

Hint to (i): The integrals lead to the *beta function*

$$B(r, s) := \int_0^1 t^{r-1}(1-t)^{s-1}\, dt, \quad r, s > 0.$$

Apply then a well-known relationship between the beta function and the *gamma function*, which interpolates the factorial function, cf Exercise 5 in Chapter 2. Hint to (ii): Use Taylor's formula.

**27.** Let $F : \mathbb{R} \to [0, 1]$ be an arbitrary distribution function. Show that

(i) $F^{-1}(q) \leq x \Leftrightarrow q \leq F(x), \quad q \in (0,1), \ x \in \mathbb{R}$,

(ii) $F^{-1}$ is increasing and continuous from the left.

(iii) $F$ is continuous if, and only if, $F(F^{-1}(q)) = q, \ q \in (0,1)$.

Hint to (i): Check first for the implication "$\Rightarrow$" that the infimum is a minimum by the continuity from the right of $F$.

**28.** Let $F$ be a distribution function and $F^{-1}$ its quantile function. Prove that

(i) $F$ is continuous if $F^{-1}$ is strictly increasing.

(ii) $F^{-1}$ is continuous if $F$ is strictly increasing on the interval $(\alpha(F), \omega(F))$, where
$\alpha(F) := \inf\{x \in \mathbb{R} : F(x) > 0\}$, $\omega(F) := \sup\{x : F(x) < 1\}$.

**29.** Let $X$ be a discrete random variable with support $\{x_1, \ldots, x_n\}$, i.e., $p_i := P\{X = x_i\} > 0$, $i = 1, \ldots, n$, $\sum_{i=1}^n p_i = 1$. Compute the quantile function of $X$.

**30.** (CNS data) Check whether the measurements of the variable *mn* come from a normal distribution. Compute skewness and kurtosis and draw a
  - boxplot
  - normal probability plot.

**31.** Prove Corollary 1.6.4.

**32.** Prove Lemma 1.6.5.

**33.** Show that $P\{X = Y\} = 0$ if $X$ and $Y$ are independent random variables with continuous distribution functions. Hint: Exercise 3 (i) and Fubini's Theorem.

**34.** Let $X_1, \ldots, X_n$ be independent and normal distributed random variables with expectation $\mu$ and variance $\sigma^2$. Motivate that

$$\frac{IQR}{1.35} \quad \text{and} \quad \frac{MAD}{0.675}$$

are (robust) estimators of $\sigma$.

Hint: Lemma 1.6.5 and Corollary 1.6.8. The distribution function $\Phi$ of the standard normal distribution is symmetric, i.e., $\Phi(x) = 1 - \Phi(-x)$, $x \in \mathbb{R}$.

**35.** Show that the following transformations are variance stabilizing:

  (i) Let $X$ be *Poisson* distributed with parameter $\lambda > 0$, i.e., $P\{X = k\} = e^{-\lambda} \lambda^k / k!$, $k = 0, 1, 2, \ldots$ In this case $T(x) = \sqrt{x}$ is a variance stabilizing transformation.

  (ii) Let $X$ be a $B(n, p)$ distributed random variable. Then $T(x) := \arcsin \sqrt{x}$, $x \in (0, 1)$, is a variance stabilizing transformation of the random variable $X/n$.

**36.** Let $X$ be a random variable with $E(X) = \mu > 0$ and $Var(X) = c\mu^2$, $c > 0$. This is, for example, satisfied with $c = 1$ by an exponential distributed random variable $X$. A random variable $X$ is *exponential* distributed with parameter $\lambda > 0$ if $P\{X \leq t\} = 1 - \exp(-\lambda t)$, $t > 0$. Find a variance stabilizing transformation.

**37.** (CNS data) Check by means of hanging histograms and rootograms whether the measurements of the variable *ao* are normal.

**38.** (CNS data) Does age have an influence on the variable *an*? Use quantile–quantile plots.

# Chapter 2

# Some Mathematical Statistics for the Normal Distribution

We will derive in this chapter several elementary facts for the normal distribution and for basic statistics based on normal observations. The use of a normal model for a set of data is traditionally suggested by the central limit theorem and is, therefore, the most extensively studied one. This is, for example, the reason, why we marked in a boxplot in Section 1.5 those observations as outliers that were extremely large or small in a normal model. Equally, the definition of skewness or kurtosis in Section 1.4 is based on a comparison with the shape of a standard normal density.

## 2.1 The Normal Distribution and Derived Distributions

We will compile in the sequel several basic properties of the normal distribution and of some distributions derived from the normal one. These are the chi-square, the $F$ and the $t$ distribution, which will all be crucial in statistical applications. By $\boldsymbol{A}^T = (a_{ji})$ we denote the usual transpose of an $m \times n$–matrix $\boldsymbol{A} = (a_{ij})$ with $m$ rows and $n$ columns, so that $\boldsymbol{A}^T$ is an $n \times m$–matrix. Our default representation of a vector $\boldsymbol{x}$ is the column style, and consequently $\boldsymbol{x}^T$ is a row vector.

**2.1.1 Definition.** Let $X_1, \ldots, X_n$ be independent, standard normal distributed random variables. We, thus, have in particular

$$
P\{X_i \leq t\} = \Phi(t) = \int_{-\infty}^{t} \varphi(y)\, dy
$$

$$
= \int_{-\infty}^{t} (2\pi)^{-1/2} \exp(-y^2/2)\, dy, \qquad t \in \mathbb{R}, \quad i = 1, \ldots, n.
$$

We denote this standard normal distribution of each $X_i$ by $N(0,1)$; see, for example, Section 3.4 in Fristedt and Gray (1997) for mathematical details. The random vector $\boldsymbol{X} := (X_1, \ldots, X_n)^T$, which takes values in $\mathbb{R}^n$, is called $n$–*dimensional standard normal distributed*.

It is well–known from probability theory (cf Section 9.5 in Fristedt and Gray (1997)) that the random vector $\boldsymbol{X}$ has the density

$$\varphi(y_1, \ldots, y_n) := \prod_{i=1}^{n} \varphi(y_i) = (2\pi)^{-n/2} \exp\left(-\frac{1}{2} \sum_{i=1}^{n} y_i^2\right)$$

$$= (2\pi)^{-n/2} \exp\left(-\frac{1}{2} y^T I_n y\right), \qquad y = (y_1, \ldots, y_n)^T \in I\!R^n,$$

where

$$I_n := \begin{pmatrix} 1 & & 0 \\ & \ddots & \\ 0 & & 1 \end{pmatrix}$$

is the $n \times n$–identity matrix. By $\boldsymbol{AB}$, $\boldsymbol{Ay}$, $\boldsymbol{y^T A}$, etc. we denote the usual product of two matrices $\boldsymbol{A}$ and $\boldsymbol{B}$, the product of matrix $\boldsymbol{A}$ and $\boldsymbol{y}$ and that of matrix $\boldsymbol{y^T}$ and $\boldsymbol{A}$, etc. We have

$$P\{\boldsymbol{X} \leq t\} = \int_{-\infty}^{t_1} \cdots \int_{-\infty}^{t_n} \varphi(y_1, \ldots, y_n)\, dy_1 \cdots dy_n$$

$$= \prod_{i=1}^{n} \int_{-\infty}^{t_i} \varphi(y)\, dy = \prod_{i=1}^{n} \Phi(t_i),$$

where the inequality $\boldsymbol{X} \leq t$ or $(X_1, \ldots, X_n)^T \leq (t_1, \ldots, t_n)^T$ is meant componentwise, i.e., $X_i \leq t_i$ for $i = 1, \ldots, n$, $t = (t_1, \ldots, t_n)^T$.

**2.1.2 Definition.** Consider an $n$–dimensional standard normal random vector $\boldsymbol{X} = (X_1, \ldots, X_n)^T$. Let $\boldsymbol{\mu} = (\mu_1, \ldots, \mu_m)^T$ be an arbitrary vector in $I\!R^m$ and let $\boldsymbol{A} = (a_{ij})$ be an $m \times n$–matrix with $\text{rank}(\boldsymbol{A}) = m \leq n$. The random vector $\boldsymbol{Y} = (Y_1, \ldots, Y_m)^T$, defined by

$$\boldsymbol{Y} = \boldsymbol{AX} + \boldsymbol{\mu} = \left(\sum_{j=1}^{n} a_{ij} X_j + \mu_i\right)_{1 \leq i \leq m},$$

is called *$m$–dimensional normal distributed with mean vector $\boldsymbol{\mu}$ and covariance matrix $\boldsymbol{\Sigma} := \boldsymbol{AA^T}$*. We denote its distribution on $I\!R^m$ by $\boldsymbol{N(\mu, \Sigma)}$.

It is easy to verify that the vector of the expectations satisfies

$$E(\boldsymbol{Y}) := (E(Y_1), \ldots, E(Y_m))^T = \boldsymbol{\mu}$$

and that the $m \times m$–*covariance matrix* of $\boldsymbol{Y}$ is given by

$$Cov(\boldsymbol{Y}) := (E((Y_i - \mu_i)(Y_j - \mu_j)))_{1 \leq i,j \leq m} = \boldsymbol{AA^T} = \boldsymbol{\Sigma}$$

(Exercise 1). Check that $\mathbf{\Sigma}$ is symmetric and positive definite. Since the rank of $\mathbf{A}$ coincides with the rank of $\mathbf{A}\mathbf{A}^T$, cf Lemma 3.3.3, the $m \times m$–matrix $\mathbf{\Sigma} = \mathbf{A}\mathbf{A}^T$ has full rank $m$ and is, therefore, invertible with the inverse matrix $\mathbf{\Sigma}^{-1}$. Additionally, each symmetric and positive definite $m \times m$–matrix $\mathbf{\Sigma}$ is the covariance matrix of an $m$–dimensional normal distributed vector. This follows from the existence of a (symmetric and invertible) $m \times m$–matrix $\mathbf{A}$ with the property $\mathbf{\Sigma} = \mathbf{A}\mathbf{A}^T = \mathbf{A}^T\mathbf{A}$, cf Exercise 18 in Chapter 6.

In the particular case $m = n = 1$ we have $\mathbf{A} = a \in \mathbb{R} \setminus \{0\}$, $\mathbf{\Sigma} = a^2 > 0$ and $\boldsymbol{\mu} = \mu \in \mathbb{R}$. The random variable

$$Y = aX + \mu$$

is one dimensional normal distributed with mean $\mu$ and variance $a^2$, denoted by $N(\mu, a^2)$. Here $X$ is a one dimensional standard normal random variable. By the symmetry $P\{X \leq t\} = P\{X \geq -t\}$, $t \in \mathbb{R}$, of the standard normal distribution with respect to zero, the density of the normal distribution $N(\mu, a^2)$ depends only on $\sigma := |a| > 0$ and $\mu$:

$$
\begin{aligned}
\varphi_{\mu,\sigma}(t) &:= \frac{d}{dt}P\{Y \leq t\} = \frac{d}{dt}P\{X \leq \frac{t-\mu}{\sigma}\} = \frac{d}{dt}\Phi\left(\frac{t-\mu}{\sigma}\right)\\
&= \frac{1}{\sigma}\varphi\left(\frac{t-\mu}{\sigma}\right) = \frac{1}{\sqrt{2\pi}\sigma}\exp\left(-\frac{(t-\mu)^2}{2\sigma^2}\right), \qquad t \in \mathbb{R}.
\end{aligned}
$$

One replaces, therefore, $a \in \mathbb{R}$ usually by $\sigma > 0$ and writes $N(\mu, \sigma^2)$. The following theorem provides us with the density of the $m$–dimensional normal distribution $\mathbf{N}(\boldsymbol{\mu}, \mathbf{\Sigma})$.

**2.1.3 Theorem.** *The random vector* $\mathbf{Y} = (Y_1, \ldots, Y_m)^T$ *in Definition 2.1.2 has the $m$–dimensional density*

$$\varphi_{\boldsymbol{\mu},\,\mathbf{\Sigma}}(\mathbf{y}) := \frac{1}{(2\pi)^{m/2}(\det(\mathbf{\Sigma}))^{1/2}}\exp\left(-\frac{1}{2}(\mathbf{y}-\boldsymbol{\mu})^T\mathbf{\Sigma}^{-1}(\mathbf{y}-\boldsymbol{\mu})\right)$$

*for $\mathbf{y} \in \mathbb{R}^m$, i.e., we have for $\mathbf{t} = (t_1, \ldots, t_m)^T \in \mathbb{R}^m$*

$$
\begin{aligned}
P\{\mathbf{Y} \leq \mathbf{t}\} &= P\{Y_i \leq t_i,\ i = 1, \ldots, m\}\\
&= \int_{-\infty}^{t_1} \cdots \int_{-\infty}^{t_m} \varphi_{\boldsymbol{\mu},\,\mathbf{\Sigma}}((y_1, \ldots, y_m)^T)\,dy_1 \cdots dy_m.
\end{aligned}
$$

**Proof:** See, for example, Section 5.7 in Stone (1996).     $\square$

Recall that a distribution is uniquely determined by its density.

**2.1.4 Corollary.** *Consider an $n$–dimensional standard normal random vector $\mathbf{X}$ and let $\mathbf{A}$ be an arbitrary orthogonal $n \times n$–matrix, i.e., $\mathbf{A}^T\mathbf{A} = \mathbf{A}\mathbf{A}^T = \mathbf{I}_n$. The random vector $\mathbf{Y} = \mathbf{A}\mathbf{X}$ is then $n$–dimensional standard normal as well.*

**2.1.5 Corollary.** *Let $Y_1, \ldots, Y_n$ be independent random variables, each being normal $N(\mu_i, \sigma_i^2)$ distributed, $i = 1, \ldots, n$. Their sum*

$$Y := \sum_{i=1}^{n} Y_i$$

*is also normal $N(\mu_1 + \cdots + \mu_n, \sigma_1^2 + \cdots + \sigma_n^2)$ distributed. This is the convolution theorem of the normal distribution.*

**Proof:** We can suppose the representation

$$Y_i = \sigma_i X_i + \mu_i, \qquad i = 1, \ldots, n,$$

where $\boldsymbol{X} := (X_1, \ldots, X_n)^T$ is $n$–dimensional standard normal vector. Then we can write

$$Y = \boldsymbol{A} \boldsymbol{X} + \mu,$$

where $\boldsymbol{A}$ is the $1 \times n$–matrix $(\sigma_1, \ldots, \sigma_n)$ and $\mu := \mu_1 + \cdots + \mu_n$. With $\Sigma = \boldsymbol{A}\boldsymbol{A}^T = \sigma_1^2 + \cdots + \sigma_n^2 =: \sigma^2$ we obtain from Theorem 2.1.3 that $Y$ has density

$$\varphi_{\mu,\sigma}(y) = \frac{1}{(2\pi)^{1/2}\sigma} \exp\left(-\frac{(y-\mu)^2}{2\sigma^2}\right), \qquad y \in \mathbb{R},$$

i.e., it is normal $N(\mu, \sigma^2)$ distributed.                                               $\square$

While analyzing normal data, we will encounter the chi–square, the $F$ and the $t$ distributions in a natural way. We will now introduce these distributions and compute their densities.

## The Chi–Square Distribution

The chi–square distribution is the distribution of the sum of squares of independent standard normal random variables.

**2.1.6 Definition.** Let $X_1, \ldots, X_n$ be independent and $N(0, 1)$ distributed random variables. The distribution of the sum $Y := X_1^2 + \cdots + X_n^2$ is called the *chi-square distribution with $n$ degrees of freedom*, denoted by $\chi_n^2$.

**2.1.7 Theorem.** *The $\chi_n^2$ distribution has density*

$$g_n(y) := \frac{1}{2^{\frac{n}{2}}\Gamma(\frac{n}{2})} y^{\frac{n}{2}-1} \exp\left(-\frac{y}{2}\right), \qquad y > 0,$$

*and $g_n(y) = 0$ elsewhere. By $\Gamma(x) = \int_0^\infty t^{x-1} \exp(-t)\, dt$, $x > 0$, we denote the gamma function.*

**Proof:** The assertion follows by induction. The case $n = 1$ is elementary, observe that $\Gamma(1/2) = \sqrt{\pi}$. Suppose that the assertion holds for $n - 1$. The convolution theorem for densities, cf Exercise 6, Chapter 1, applied to $(X_1^2 + \cdots + X_{n-1}^2) + X_n^2$, then yields

$$g_n(y) = \int_{-\infty}^{\infty} g_{n-1}(x)\, g_1(y-x)\, dx$$

$$= \frac{y^{(n/2)-1}\exp(-y/2)}{2^{n/2}\Gamma((n-1)/2)\Gamma(1/2)} \int_0^1 x^{(n-1)/2-1}(1-x)^{(1/2)-1}\, dx$$

using the substitution $x \mapsto yx$. The second integral is the *beta function* $B(r,s) = \int_0^1 x^{r-1}(1-x)^{s-1}\, dx$ with the parameters $r = (n-1)/2$ and $s = 1/2$. The equality $B(r,s) = \Gamma(r)\Gamma(s)/\Gamma(r+s)$ with arbitrary $r,s > 0$, cf Exercise 5, now implies the assertion. $\square$

Note that $\chi_2^2$ is the exponential distribution with parameter $\lambda = 1/2$. One expects by the central limit theorem that the shape of the density of the $\chi_n^2$ distribution approaches that of the density of a normal distribution as $n$ increases.

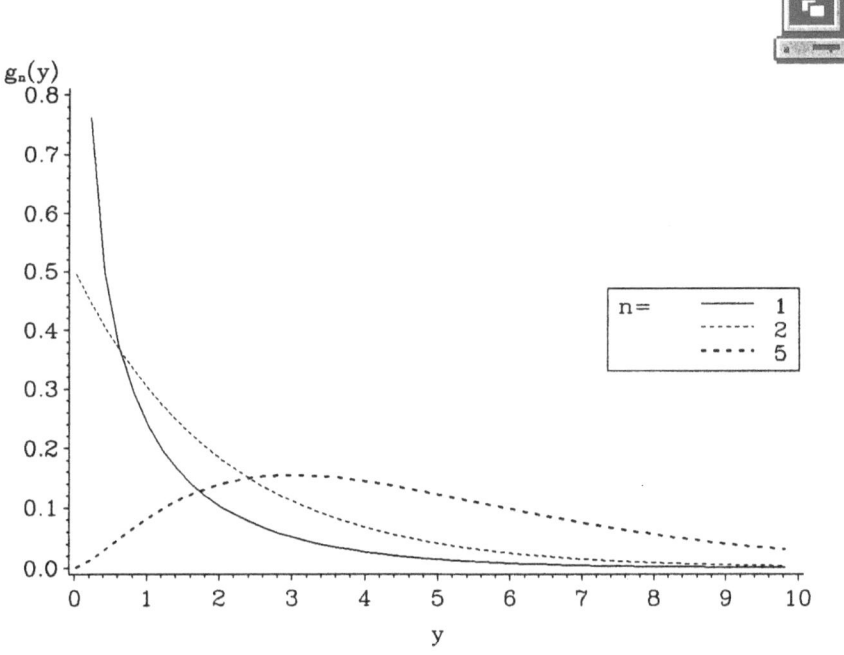

**Figure 2.1.1.** Densities of the $\chi_n^2$ distribution with $n = 1, 2$ and 5 degrees of freedom.

```
*** Program 2_1_1 ***;
TITLE1 'Densities of the 'M=(+0,+0)     F=CGREEK 'x'
              H=2.3     M=(+0.5,+1.5) F=CENTX '2' M=(-1.5,-2.5)'n'
              H=4       M=(+0.5,+1)   F=CENTX 'Distribution';
TITLE2 'with n=1, 2 and 5 degrees of freedom';

DATA data1;
    DO n=1, 2, 5;                * degrees of freedom;
        DO x=0.02 TO 10 BY 0.2;  * grid points;
            chi_n=((x**(n/2-1))*(exp(-x/2)))/((2**(n/2))*GAMMA(n/2));
            OUTPUT; END; END;
SYMBOL1 C=RED   V=NONE I=JOIN L=1;
SYMBOL2 C=BLUE  V=NONE I=JOIN L=2;
SYMBOL3 C=GREEN V=NONE I=JOIN L=33 W=2;
AXIS1 LABEL=(H=2 'g' H=1 'n' H=2 '(y)');
AXIS2 LABEL=('y');
LEGEND1 LABEL=('n= ') POSITION=(MIDDLE RIGHT INSIDE)
    FRAME DOWN=3;
PROC GPLOT DATA=data1(WHERE=(1.1>chi_n));
    PLOT chi_n*x=n / VAXIS=AXIS1 HAXIS=AXIS2 LEGEND=LEGEND1;
RUN; QUIT;
```

A function is plotted by computing its values at a grid of points and joining the results, for example, by straight lines. In the above DATA step the values of the chi–square densities with $n = 1, 2$ and 5 degrees are computed at the grid points x = 0.02 to 10 by step 0.2 (upper bound actually 9.82). The implemented SAS function GAMMA makes this computation easy.

For the plot three SYMBOL statements, two AXIS statements and a LEGEND statement controlling the form and position of the legend are specified. The additional '=n' in the PLOT statement assigns the pertaining SYMBOL definitions to each plot. To avoid too large values for the density with 1 degree of freedom near 0, cases with values greater than 1.1 are eliminated by the WHERE data set option in the PROC GPLOT statement.

**2.1.8 Remark.** For a $\chi_n^2$ distributed random variable $X$, we have

$$E(X) = n, \; Var(X) = 2n$$

(Exercise 6). Let $Y$ be another random variable that is independent of $X$ and which has the $\chi_m^2$ distribution. The sum $X + Y$ is then obviously $\chi_{n+m}^2$ distributed. This is the *convolution theorem* for the chi–square distribution.

## The $F$ Distribution

The $F$ distribution controls the stochastic behavior of the ratios of independent and chi–square distributed random variables.

**2.1.9 Definition.** Let $X, Y$ be independent random variables, with $X$ being $\chi_n^2$ and $Y$ being $\chi_m^2$ distributed. The distribution of the fraction $(Y/m)/(X/n)$ is called the $F$ *distribution with* $(m, n)$ *degrees of freedom*, denoted by $F_{m,n}$.

**2.1.10 Theorem.** *The $F_{m,n}$ distribution has density*

$$f_{m,n}(y) = \frac{\Gamma(\frac{m+n}{2})}{\Gamma(\frac{m}{2})\Gamma(\frac{n}{2})} m^{\frac{m}{2}} n^{\frac{n}{2}} \frac{y^{\frac{m}{2}-1}}{(n+my)^{\frac{m+n}{2}}}, \qquad y > 0,$$

*and $f_{m,n}(y) = 0$ elsewhere.*

**Proof:** By Fubini's Theorem we have for $t > 0$

$$F_{m,n}(t) = P\left\{ Y \leq \frac{m}{n} tX \right\}$$

$$= \int_0^\infty \int_0^{\frac{m}{n}tx} g_n(x) g_m(y) \, dy \, dx$$

$$= \int_0^\infty \int_0^t g_n(x) g_m\left(xy\frac{m}{n}\right) \frac{m}{n} x \, dy \, dx$$

$$= \int_0^t \int_0^\infty g_n(x) g_m\left(yx\frac{m}{n}\right) \frac{m}{n} x \, dx \, dy,$$

where $g_n$ and $g_m$ are defined as in Theorem 2.1.7. The function

$$(0, \infty) \ni y \longmapsto \int_0^\infty g_n(x) g_m\left(yx\frac{m}{n}\right) \frac{m}{n} x \, dx = f_{m,n}(y)$$

is, consequently, a density of the $F_{m,n}$ distribution.                □

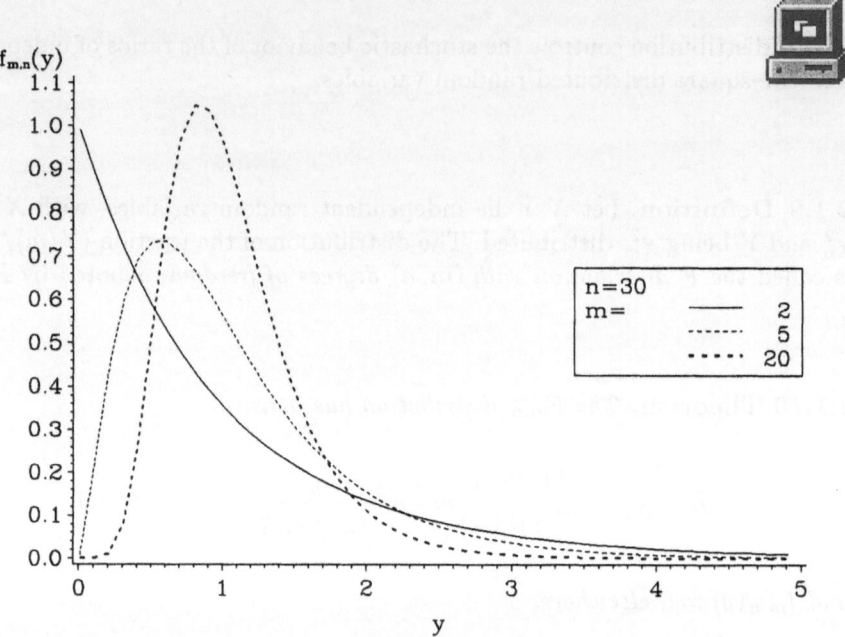

**Figure 2.1.2.** Densities of the $F_{m,n}$ distribution with $m = 2, 5, 20$ and $n = 30$ degrees of freedom.

Observe that the random variable $1/X$ is $F_{n,m}$ distributed if $X$ is $F_{m,n}$ distributed. This implies the equality

$$F_{n,m}^{-1}(q) = \frac{1}{F_{m,n}^{-1}(1-q)}, \qquad q \in (0,1),$$

for the quantile functions (Exercise 10). In view of this, tables of the quantiles of the $F$ distribution are usually given only for $q \geq 1/2$.

## The $t$ Distribution

The $t$ distribution controls the stochastic behavior of the ratio of a standard normal random variable and the root of an independent and chi–square distributed one.

**2.1.11 Definition.** Consider independent random variables $X, Y$ with $X$ being standard normal and $Y$ being $\chi_n^2$ distributed. The distribution of the ratio $X/\sqrt{Y/n}$ is the $t$ *distribution with n degrees of freedom*, denoted by $t_n$.

**2.1.12 Theorem.** *The $t_n$ distribution has density*

$$h_n(y) = \frac{\Gamma(\frac{n+1}{2})}{\Gamma(\frac{n}{2})\sqrt{\pi n}} \left(1 + \frac{y^2}{n}\right)^{-\frac{n+1}{2}}, \qquad y \in \mathbb{R}.$$

**Proof:** Consider a random variable $Z$, which follows the $t_n$ distribution. Its square $Z^2$ is $F_{1,n}$ distributed (Exercise 11). Since the distribution of $Z$ is symmetric about zero, i.e., $P\{Z \leq t\} = P\{Z \geq -t\}$ for $t \in \mathbb{R}$, we obtain for $t \geq 0$:

$$P\{0 \leq Z \leq t\}$$

$$= \frac{1}{2} P\{|Z| \leq t\}$$

$$= \frac{1}{2} P\{Z^2 \leq t^2\}$$

$$= \frac{1}{2} \int_0^{t^2} f_{1,n}(y) \, dy$$

$$= \int_0^t f_{1,n}(y^2) y \, dy.$$

The function $|y| f_{1,n}(y^2)$, $y \in \mathbb{R}$ is, therefore, the density of the distribution of $Z$. Since $\Gamma(1/2) = \sqrt{\pi}$, the proof is now complete. $\qquad \square$

Note that if $Y$ is $\chi_n^2$ distributed and $n$ is large, then $Y/n \approx 1$ by the law of large numbers. Consequently, the ratio $X/\sqrt{Y/n}$ will be close to $X$. Therefore, we may expect that the density $h_n(y)$ will converge to the density $\varphi(y)$ of the standard normal distribution as $n$ tends to infinity. This can easily be verified to be true (Exercise 15).

**2.1.13 Remark.** The $t_1$ distribution is the standard *Cauchy distribution*. This is, consequently, the distribution of the ratio $X/|Y|$, where $X$ and $Y$ are independent standard normal random variables. The Cauchy distribution has by Theorem 2.1.12 the density

$$h_1(y) = \frac{1}{\pi(1 + y^2)}, \qquad y \in \mathbb{R}.$$

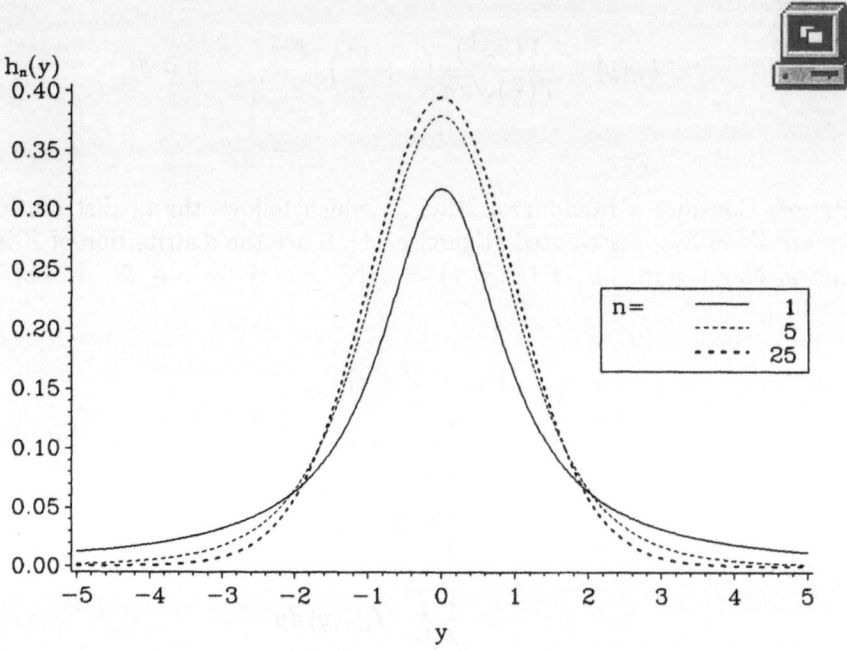

**Figure 2.1.3.** Densities of the $t_n$ distribution with $n = 1, 5$ and 25 degrees of freedom.

The shape of the density of the Cauchy distribution seems to be very much like that of a normal distribution. It has, however, very heavy tails since even the first moment $\int_{-\infty}^{\infty} y\, h_1(y)\, dy$ does not exist. If it is erroneously assumed that the data generated by a Cauchy distribution come from a normal distribution, then the data will appear to contain outliers.

## 2.2  Sample Mean and Sample Variance

We consider in the sequel independent and identically $N(\mu, \sigma^2)$ distributed random variables $X_1, \ldots, X_n$. The following Theorem 2.2.1, which is crucial for the analysis of normal data, provides the distribution of the *sample mean*

$$\bar{X}_n = \frac{1}{n} \sum_{i=1}^{n} X_i,$$

the distribution of the *sample variance*

$$S_n^2 = \frac{1}{n-1} \sum_{i=1}^{n} (X_i - \bar{X}_n)^2$$

and the joint distribution of $(\bar{X}_n, S_n)$.

**2.2.1 Theorem.** *Under the above assumptions, for $n \geq 2$, we have:*

(i) $\bar{X}_n$ and $S_n^2$ are independent,

(ii) $(n-1)S_n^2/\sigma^2$ is $\chi_{n-1}^2$ distributed,

(iii) $\bar{X}_n$ is $N(\mu, \sigma^2/n)$ distributed, $\sqrt{n}(\bar{X}_n - \mu)/S_n$ is $t_{n-1}$ distributed.

**Proof:** Let $A$ be an orthogonal $n \times n$-matrix, i.e., $A^T A = I_n$, whose first row is $(n^{-1/2}, \ldots, n^{-1/2})$, cf Exercise 16. Put $X_i^* := (X_i - \mu)/\sigma$, $i = 1, \ldots, n$, and $(Y_1, \ldots, Y_n)^T := A(X_1^*, \ldots, X_n^*)^T$. From Corollary 2.1.4 we obtain that the random variables $Y_1, \ldots, Y_n$ are independent and standard normal distributed. Hence,

$$Y_1 = \frac{1}{\sqrt{n}} \sum_{i=1}^{n} X_i^* = \sqrt{n} \Big( \frac{1}{n} \sum_{i=1}^{n} \frac{X_i - \mu}{\sigma} \Big) = \sqrt{n} \frac{\bar{X}_n - \mu}{\sigma}$$

and $Y_2^2 + \cdots + Y_n^2$ are independent as well. We have, moreover,

$$\begin{aligned}
Y_2^2 + \cdots + Y_n^2 &= Y_1^2 + \cdots + Y_n^2 - Y_1^2 \\
&= (Y_1, \ldots, Y_n)(Y_1, \ldots, Y_n)^T - Y_1^2 \\
&= (A(X_1^*, \ldots, X_n^*)^T)^T (A(X_1^*, \ldots, X_n^*)^T) - Y_1^2 \\
&= (X_1^*, \ldots, X_n^*) A^T A (X_1^*, \ldots, X_n^*)^T - Y_1^2 \\
&= (X_1^*, \ldots, X_n^*)(X_1^*, \ldots, X_n^*)^T - Y_1^2 \\
&= \sum_{i=1}^{n} X_i^{*2} - n \Big( \frac{1}{n} \sum_{j=1}^{n} X_j^* \Big)^2 \\
&= \sum_{i=1}^{n} \Big( X_i^* - \frac{1}{n} \sum_{j=1}^{n} X_j^* \Big)^2 \\
&= \frac{1}{\sigma^2} \sum_{i=1}^{n} (X_i - \bar{X}_n)^2 = \frac{n-1}{\sigma^2} S_n^2,
\end{aligned}$$

which is part (i) of the assertion. Part (ii) follows from the representation

$$\frac{n-1}{\sigma^2} S_n^2 = Y_2^2 + \cdots + Y_n^2$$

and the Definition 2.1.6 of the chi–square distribution, while (iii) is a consequence of the representation $\bar{X}_n = (\sigma/\sqrt{n})Y_1 + \mu$, the relation

$$\frac{\sqrt{n}(\bar{X}_n - \mu)/\sigma}{S_n/\sigma} = \frac{Y_1}{\left((Y_2^2 + \cdots + Y_n^2)/(n-1)\right)^{1/2}}$$

and the Definition 2.1.11 of the $t$ distribution.                                                □

## 2.3   Comparing Two Independent Samples

We consider in this section a sample $X_1, \ldots, X_n$ of independent and identically normal $N(\mu_X, \sigma^2)$ distributed random variables together with another sample $Y_1, \ldots, Y_m$ that is independent of the first one and which consists of independent and identically $N(\mu_Y, \sigma^2)$ distributed random variables. We assume that the two samples have the same variances $\sigma^2$ and that the means of the two samples may be different.

Such a model is, for instance, quite popular in biostatistics, where the members of some group get some particular treatment, whereas the members of a control group get none. The measurements in each group then yield the two samples, which are to be compared, see also the remarks on quantile–quantile plots in Section 1.6.

**2.3.1 Example** (Crystal Data; Andrews and Herzberg (1985), Chapter 44). Six different physiochemical variables such as the pH level $(pH)$ and the concentration of calcium $(Ca)$ were measured on 79 urine specimen, 34 of which contain certain crystals. Do the specimen with crystals have a significantly higher or lower concentration of calcium or a higher or lower pH level than those without crystals? See Figure 2.3.1 and Example 2.3.7.

We model the measurements of the pH level or of the calcium concentration in the two groups with and without crystals by $X_1, \ldots, X_n$ and $Y_1, \ldots, Y_m$. Then we expect that the pertaining means $\mu_X$ and $\mu_Y$ differ if the crystallization actually has an effect on the measurements. Since we have to base the possible decision

$$H_1 : \mu_X \neq \mu_Y$$

on the measurements, it is plausible to make this decision, if the two *estimators*

$$\bar{X}_n := n^{-1} \sum_{i=1}^{n} X_i \quad \text{and} \quad \bar{Y}_m := m^{-1} \sum_{j=1}^{m} Y_j$$

of $\mu_X$ and $\mu_Y$ differ significantly, i.e., if the distance $|\bar{X}_n - \bar{Y}_m|$ is *significantly large*. But now we obviously have to quantify this 'significantly large' threshold.

That means we have to determine a *critical value* $c > 0$ with the property that we will make the decision $H_1$ if $|\bar{X}_n - \bar{Y}_m| \geq c$.

| CRYSTAL | N Obs | Variable | N | Mean | Std Dev |
|---|---|---|---|---|---|
| 1 | 45 | PH | 45 | 6.0986667 | 0.7020379 |
| | | CA | 45 | 2.6248889 | 1.8629920 |
| 2 | 34 | PH | 34 | 5.9355882 | 0.7531678 |
| | | CA | 34 | 6.1429412 | 3.6372063 |

**Figure 2.3.1.** Comparison of the means of two independent samples from the crystal data with the variables *pH* level and *calcium* concentration, 1 = no crystallization, 2 = crystallization.

```
***    Program 2_3_1    ***;
TITLE1 'Comparing Two Independent Samples';
TITLE2 'Crystal Data';
LIBNAME datalib 'c:\data';

PROC MEANS DATA=datalib.crystal N MEAN STD;
    CLASS crystal;
    VAR ph ca;
RUN; QUIT;
```

The CLASS statement declares classification variables. The original data are divided into subgroups by the different levels of these CLASS variables. PROC MEANS then computes descriptive statistics separately for each subgroup. One can specify more than one variable in the CLASS statement.

# Hypothesis, Alternatives, Errors of Type I and II

It is possible that we come up with the decision $H_1$ when, in fact,

$$H_0 : \mu_X = \mu_Y$$

could be true. In this case our decision $H_1$ is wrong.

The other possible wrong decision could be $H_0$ when, in fact, $H_1$ is true. These two errors might have considerably different consequences in practice. Take, for example, a new drug, which is going to replace an older and less effective one, provided it does not have more serious unwanted side–effects. We model the possible decision $H_1$, 'the side–effects of the new drug are less serious', by $\mu_X < \mu_Y$ and the other possible decision $H_0$ 'the side–effects are more or equally serious' by $\mu_X \geq \mu_Y$. The more fatal error from the patient's point of view is, in this case, decision $H_1$ if actually $H_0$ is true.

As one usually cannot control the probabilities of the two wrong decisions simultaneously, one commonly takes that decision as the *null hypothesis* or just *hypothesis* $H_0$, whose erroneous *rejection* would have serious consequences. This is called the *error of the first kind* or *type I error*. Its probability of occurrence is controlled. The erroneous rejection of the *alternative* $H_1$ when, in fact, it is true, is called the *error of the second kind* or *type II error*. One will reasonably formulate a test in such a way that the error of the first kind, say $\alpha$, is small. A common value for $\alpha$ is .05. This would give a *test of level* $\alpha$. The following table summarizes our considerations:

| Decision | $H_0$ true | $H_0$ not true |
|:--------:|:----------:|:--------------:|
| $H_1$ | Type I error | No error |
| $H_0$ | No error | Type II error |

The next result is an immediate consequence of the convolution theorem of the normal distribution, cf Corollary 2.1.5.

**2.3.2 Lemma.** *Let* $X_1, \ldots, X_n, Y_1, \ldots, Y_m$ *be independent random variables with the* $X_i$ *being* $N(\mu_X, \sigma^2)$ *distributed and the* $Y_j$ *being* $N(\mu_Y, \sigma^2)$ *distributed. The standardized difference*

$$T_\sigma := \frac{\bar{X}_n - \bar{Y}_m - (\mu_X - \mu_Y)}{\sigma\sqrt{\frac{1}{n} + \frac{1}{m}}}$$

*is then* $N(0, 1)$ *distributed.*

If $\sigma^2$ were known, then we could test the hypothesis $H_0 : \mu_X = \mu_Y$ against the alternative $H_1 : \mu_X \neq \mu_Y$ by means of

$$\tilde{T}_\sigma := \frac{\bar{X}_n - \bar{Y}_m}{\sigma\sqrt{\frac{1}{n} + \frac{1}{m}}}.$$

If $H_0$ is true, then $\tilde{T}_\sigma$ coincides with $T_\sigma$ and is, therefore, $N(0, 1)$ distributed. If, on the other hand, $H_0$ is not true, then we have

$$\tilde{T}_\sigma = T_\sigma + \frac{\mu_X - \mu_Y}{\sigma\sqrt{\frac{1}{n} + \frac{1}{m}}} \neq T_\sigma.$$

The additional term $(\mu_X - \mu_Y)/(\sigma\sqrt{n^{-1} + m^{-1}})$ gives in this case a shift to the standard normal random variable $T_\sigma$, which has expectation zero.

## Critical Regions

We will, therefore, reject the hypothesis $H_0 : \mu_X = \mu_Y$ if $|\tilde{T}_\sigma|$ is too large, i.e., if it exceeds some high threshold. This threshold is called a *critical value* and the set of real numbers that are larger than the threshold is a *critical region* for $H_0$.

**2.3.3 Theorem.** *Consider for $\alpha \in (0,1)$ the interval $I := [\Phi^{-1}(\alpha/2), \Phi^{-1}(1 - \alpha/2)] = [-\Phi^{-1}(1 - \alpha/2), \Phi^{-1}(1 - \alpha/2)]$. The complement of $I$*

$$I^c = \left(-\infty, \Phi^{-1}(\alpha/2)\right) \cup \left(\Phi^{-1}(1 - \alpha/2), \infty\right)$$

*is a critical region of level $\alpha$ of the test statistic*

$$\tilde{T}_\sigma = \frac{\bar{X}_n - \bar{Y}_m}{\sigma\sqrt{\frac{1}{n} + \frac{1}{m}}}$$

*for the hypothesis $H_0 : \mu_X = \mu_Y$. Suppose that $H_0$ is actually true. The error probability of the first kind $P\{\tilde{T}_\sigma \in I^c\}$, i.e., the probability of the event that $\tilde{T}_\sigma$ takes values in $I^c$, which leads to the erroneous rejection of $H_0$, is equal to $\alpha$.*

**Proof:** The random variable $\tilde{T}_\sigma$ is standard normal distributed if $H_0$ is actually true. The continuity of the distribution function $\Phi$ of the standard normal distribution then implies

$$P\{\tilde{T}_\sigma \in I^c\} = P\left\{\tilde{T}_\sigma < \Phi^{-1}(\alpha/2)\right\} + P\left\{\tilde{T}_\sigma > \Phi^{-1}(1 - \alpha/2)\right\}$$

$$= \Phi\left(\Phi^{-1}(\alpha/2)\right) + 1 - \Phi\left(\Phi^{-1}(1 - \alpha/2)\right) = \alpha.$$

$\square$

The level of a critical region is just the error probability of the first kind. The error probability of the second kind depends on the difference $d := \mu_X - \mu_Y$:

$$P\{\tilde{T}_\sigma \in I\} = P_d\{\tilde{T}_\sigma \in I\} = P_d\left\{T_\sigma + \frac{d}{\sigma\sqrt{\frac{1}{n} + \frac{1}{m}}} \in I\right\}$$

$$= P_d\left\{T_\sigma \in \left[\Phi^{-1}\left(\frac{\alpha}{2}\right) - \frac{d}{\sigma\sqrt{\frac{1}{n} + \frac{1}{m}}}, \Phi^{-1}\left(1 - \frac{\alpha}{2}\right) - \frac{d}{\sigma\sqrt{\frac{1}{n} + \frac{1}{m}}}\right]\right\}$$

$$= \Phi\left(\Phi^{-1}\left(1-\frac{\alpha}{2}\right) - \frac{d}{\sigma\sqrt{\frac{1}{n}+\frac{1}{m}}}\right) - \Phi\left(\Phi^{-1}\left(\frac{\alpha}{2}\right) - \frac{d}{\sigma\sqrt{\frac{1}{n}+\frac{1}{m}}}\right)$$

$$\longrightarrow \begin{cases} 0 & \text{as } |d| \to \infty \\ 1-\alpha & \text{as } |d| \to 0 \end{cases},$$

where $P_d$ denotes the probability under $\mu_X - \mu_Y = d$. The function $\beta(d) := P_d\{\tilde{T}_\sigma \in I^c\} = 1 - P_d\{\tilde{T}_\sigma \in I\}$ is called the *power function* or just *power* of our test. For $d \neq 0$, the power should be large. The error probability of the second kind $1 - \beta(d) = P_d\{\tilde{T}_\sigma \in I\}$, $d \neq 0$, is sometimes called *operating characteristic*.

## Confidence Intervals

If the test statistic $\tilde{T}_\sigma$ now attains a value outside the interval $I = [\Phi^{-1}(\alpha/2), \Phi^{-1}(1 - \alpha/2)]$, then we do not have confidence in the assumption that the hypothesis $H_0 : \mu_X = \mu_Y$ is actually true. This leads us to the definition of *confidence intervals*. The next theorem is immediate from Lemma 2.3.2.

**2.3.4 Theorem.** *The interval*

$$\tilde{I} := \left[\bar{X}_n - \bar{Y}_m + \sigma\,\Phi^{-1}\left(\frac{\alpha}{2}\right)\sqrt{\frac{1}{n}+\frac{1}{m}},\ \bar{X}_n - \bar{Y}_m - \sigma\,\Phi^{-1}\left(\frac{\alpha}{2}\right)\sqrt{\frac{1}{n}+\frac{1}{m}}\right]$$

*is a confidence interval of level* $1-\alpha \in (0,1)$ *for* $d = \mu_X - \mu_Y$, *i.e., the random interval* $\tilde{I}$ *contains the unknown underlying parameter $d$ with probability $1-\alpha$:*

$$P_d\{d \in \tilde{I}\} = P_d\Big\{\Phi^{-1}(\alpha/2) \leq T_\sigma \leq \Phi^{-1}(1-\alpha/2)\Big\} = P_d\{T_\sigma \in I\} = 1 - \alpha.$$

## The $t$ Test

The variance $\sigma^2$ is commonly unknown and has to be estimated from the data. To this end we define the *pooled sample variance*

$$S_p^2 := \frac{(n-1)S_{X,n}^2 + (m-1)S_{Y,m}^2}{m+n-2},$$

where $S_{X,n}^2 = (n-1)^{-1}\sum_{i=1}^n (X_i - \bar{X}_n)^2$ and $S_{Y,m}^2 = (m-1)^{-1}\sum_{j=1}^m (Y_j - \bar{Y}_m)^2$ are the sample variances of $X_1, \ldots, X_n$ and $Y_1, \ldots, Y_m$.

**2.3.5 Theorem.** *Let $X_1, \ldots, X_n, Y_1, \ldots, Y_m$ be independent random variables, where the $X_i$ are $N(\mu_X, \sigma^2)$ distributed and the $Y_j$ are $N(\mu_Y, \sigma^2)$ distributed. The random variable*

$$T := \frac{\bar{X}_n - \bar{Y}_m - (\mu_X - \mu_Y)}{S_p\sqrt{\frac{1}{n}+\frac{1}{m}}}$$

*is then $t$ distributed with $m + n - 2$ degrees of freedom.*

**Proof:** Theorem 2.2.1 (i) implies that $\bar{X}_n, S^2_{X,n}, \bar{Y}_m, S^2_{Y,m}$ are independent random variables. Hence,

$$V := \frac{\bar{X}_n - \bar{Y}_m - (\mu_X - \mu_Y)}{\sigma\sqrt{\frac{1}{n} + \frac{1}{m}}} \quad \text{and} \quad W := \frac{(n-1)S^2_{X,n}}{\sigma^2} + \frac{(m-1)S^2_{Y,m}}{\sigma^2}$$

are independent as well. $V$ is by Corollary 2.1.5 $N(0,1)$ distributed and $W$ is by Theorem 2.2.1 (ii) and Remark 2.1.8 $\chi^2_{n+m-2}$ distributed. By the representation

$$T = \frac{V}{\sqrt{W/(n+m-2)}},$$

the assertion is now a consequence of Definition 2.1.11 of the $t$ distribution. $\qquad\square$

The $t$ test now uses the test statistic

$$\tilde{T} := \frac{\bar{X}_n - \bar{Y}_m}{S_p\sqrt{\frac{1}{n} + \frac{1}{m}}}$$

for testing the hypothesis $H_0 : \mu_X = \mu_Y$ that the two samples $X_1, \ldots, X_n$ and $Y_1, \ldots, Y_m$ have identical means. Recall the assumption of identical variances $\sigma^2$. The hypothesis $H_0$ is rejected if $|\tilde{T}|$ is too large. The critical values, which determine the level $\alpha$ of this test, can be deduced from the quantile function of the $t$ distribution. This is due to the fact that the test statistic $\tilde{T}$ coincides in the case of identical means with $T$ and is, thus, $t_{m+n-2}$ distributed by Theorem 2.3.5. A multiple means comparison, i.e., the simultaneous comparison of the means of more than two samples will be introduced in the framework of the analysis of variance in Chapter 5.

**2.3.6 Corollary.** *Under the assumptions of Theorem 2.3.5, the interval*

$$\tilde{I}_p := \left[\bar{X}_n - \bar{Y}_m + t^{-1}_{m+n-2}\left(\frac{\alpha}{2}\right) S_p\sqrt{\frac{1}{n} + \frac{1}{m}}, \bar{X}_n - \bar{Y}_m - t^{-1}_{m+n-2}\left(\frac{\alpha}{2}\right) S_p\sqrt{\frac{1}{n} + \frac{1}{m}}\right]$$

*is a confidence interval of level $1 - \alpha \in (0,1)$ for $d = \mu_X - \mu_Y$:*

$$P_d\{d \in \tilde{I}_p\} = 1 - \alpha, \qquad d \in \mathbb{R}.$$

Observe that the density of the $t$ distribution is symmetric about zero. Since the variance $\sigma^2$ is now unknown, the confidence interval $\tilde{I}_p$ will tend to be larger than the interval $\tilde{I}$ in the case of a known variance, thus reflecting loss of information.

## The Behrens–Fisher Problem and the Welch Test

Suppose now that the variances in the two normal distributed samples do not coincide. The definition of a test statistic for the hypothesis $H_0 : \mu_X = \mu_Y$ together with the derivation of its exact distribution in this case is called the *Behrens–Fisher problem*. This problem is not solved, but there exist approximate solutions such as the following one. An obvious estimator of the variance

$$Var(\bar{X}_n - \bar{Y}_m) = \frac{\sigma_X^2}{n} + \frac{\sigma_Y^2}{m}$$

is

$$\frac{S_{X,n}^2}{n} + \frac{S_{Y,m}^2}{m}.$$

If we replace the denominator in the test statistic $\tilde{T}$ by

$$\left(\frac{S_{X,n}^2}{n} + \frac{S_{Y,m}^2}{m}\right)^{1/2},$$

then it can be shown that the resulting variable $\tilde{T}_W$ is approximately $t$ distributed, with the degrees of freedom now being that integer $k$, which is nearest to the random number

$$\frac{(\frac{S_{X,n}^2}{n} + \frac{S_{Y,m}^2}{m})^2}{\frac{(S_{X,n}^2/n)^2}{n-1} + \frac{(S_{Y,m}^2/m)^2}{m-1}}.$$

This yields the *Welch test* (Welch (1947), Best and Rayner (1987)) as an approximate solution of the Behrens–Fisher problem. An earlier reference is Satterthwaite (1946); we refer to Steel and Torrie (1980) or Freund et al. (1986) for more details. Since the degrees of freedom $k$ is in this case a random variable, the critical region $(-\infty, t_k^{-1}(\alpha/2)) \cup (t_k^{-1}(1 - \alpha/2), \infty)$ pertaining to the test statistic $\tilde{T}_W$ is a *random* set as well. The Welch test is, therefore, an example of a *conditional* test.

## The $F$ Test

The hypothesis $H_0$ of identical variances in the independent samples $X_1, \ldots, X_n$ and $Y_1, \ldots, Y_m$ can obviously be tested by means of the ratio

$$Q := \frac{S_{Y,m}^2}{S_{X,n}^2}.$$

Numerator and denominator are each unbiased estimators of the variances in the two samples. If these coincide, then $Q$ should be close to 1. Since $S_{Y,m}^2$ and $S_{X,n}^2$ are independent, Theorem 2.2.1 (ii) together with Definition 2.1.9 of the $F$ distribution implies that the test statistic $Q$, under the hypothesis $H_0$ of equal variances, is $F_{(m-1,n-1)}$ distributed. This hypothesis will, therefore, be rejected if the deviation of $Q$ from 1 is too large. This is the $F$ *test*.

In practice, one uses the maximum of $S_{Y,m}^2$ and $S_{X,n}^2$ as the numerator in $Q$ and their minimum as the denominator, i.e., the actual test statistic is

$$Q' := \frac{\max\{S_{Y,m}^2, S_{X,n}^2\}}{\min\{S_{Y,m}^2, S_{X,n}^2\}}.$$

This statistic is not $F$ distributed, but, in the case of identical variances, for $t \geq 1$, it has the distribution function

$$F'(t) := P\{Q' \leq t\} = P\Big\{\frac{S_{X,n}^2}{S_{Y,m}^2} \leq t, \ \frac{S_{X,n}^2}{S_{Y,m}^2} \geq 1\Big\} + P\Big\{\frac{1}{t} \leq \frac{S_{X,n}^2}{S_{Y,m}^2}, \ \frac{S_{X,n}^2}{S_{Y,m}^2} < 1\Big\}$$

$$= F_{n-1,m-1}(t) - F_{n-1,m-1}(1/t) = F_{m-1,n-1}(t) - F_{m-1,n-1}(1/t).$$

The hypothesis $H_0$ is rejected if this folded $F$ statistic $Q'$ is too large.

**2.3.7 Example.** We apply the $t$ test, the Welch and the $F$ tests to the crystal data in Example 2.3.1.

The TTEST Procedure

Statistics

| Variable | CRYSTAL | | N | Lower CL Mean | Mean | Upper CL Mean | Lower CL Std Dev |
|---|---|---|---|---|---|---|---|
| PH | | 1 | 45 | 5.8878 | 6.0987 | 6.3096 | 0.5812 |
| PH | | 2 | 34 | 5.6728 | 5.9356 | 6.1984 | 0.6075 |
| PH | Diff (1-2) | | | -0.165 | 0.1631 | 0.4908 | 0.6258 |
| CA | | 1 | 45 | 2.0652 | 2.6249 | 3.1846 | 1.5423 |
| CA | | 2 | 34 | 4.8739 | 6.1429 | 7.412 | 2.9337 |
| CA | Diff (1-2) | | | -4.77 | -3.518 | -2.266 | 2.3901 |

Statistics

| Variable | CRYSTAL | | Std Dev | Upper CL Std Dev | Std Err | Minimum | Maximum |
|---|---|---|---|---|---|---|---|
| PH | | 1 | 0.702 | 0.8868 | 0.1047 | 4.9 | 7.92 |
| PH | | 2 | 0.7532 | 0.9914 | 0.1292 | 4.76 | 7.94 |
| PH | Diff (1-2) | | 0.7244 | 0.8601 | 0.1646 | | |
| CA | | 1 | 1.863 | 2.3533 | 0.2777 | 0.17 | 8.48 |
| CA | | 2 | 3.6372 | 4.7876 | 0.6238 | 0.27 | 14.34 |
| CA | Diff (1-2) | | 2.7664 | 3.2845 | 0.6286 | | |

T-Tests

| Variable | Method | Variances | DF | t Value | Pr > |t| |
|----------|--------------|-----------|------|---------|----------|
| PH | Pooled | Equal | 77 | 0.99 | 0.3249 |
| PH | Satterthwaite | Unequal | 68.4 | 0.98 | 0.3301 |
| CA | Pooled | Equal | 77 | -5.60 | <.0001 |
| CA | Satterthwaite | Unequal | 46 | -5.15 | <.0001 |

Equality of Variances

| Variable | Method | Num DF | Den DF | F Value | Pr > F |
|----------|----------|--------|--------|---------|--------|
| PH | Folded F | 33 | 44 | 1.15 | 0.6558 |
| CA | Folded F | 33 | 44 | 3.81 | <.0001 |

**Figure 2.3.2.** $t$ test, Welch and $F$ test of the crystal data using $pH$ level and calcium $Ca$, confidence intervals for means and standard deviations (of level .95 by default).

```
***    Program 2_3_2   ***;
TITLE1 't Test, Welch Test';
TITLE2 'Crystal Data';
LIBNAME datalib 'c:\data';

PROC TTEST DATA=datalib.crystal;
    CLASS crystal;
    VAR ph ca;
RUN; QUIT;
```

A CLASS statement declaring the grouping variable must accompany PROC TTEST. The grouping variable must have exactly two levels. PROC TTEST divides the observations into the two groups for the t test using the levels of this variable.

The VAR statement declares the variables whose means are to be compared. If it is omitted, all numeric variables in the input data set except for the CLASS variable are included in the analysis.

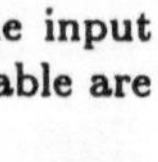

The results for the calcium sample means in the two samples (1) without crystallization and (2) with crystallization are completely different from those for the $pH$ means. We obtain for the $pH$ means

$$\bar{X}_{45} = 6.0987, \quad \bar{Y}_{34} = 5.9356,$$
$$S_{X,45} = 0.7020, \quad S_{Y,34} = 0.7532.$$

The $F'$ statistic attains the value 1.15, where the degrees of freedom are $(33, 44)$. It is, therefore, so close to 1, that one will not reject the hypothesis of identical variances. The uncertainty in this decision is quantified by the $p$–value 0.6558, which is explained below. The $t$ test with $\tilde{T} = 0.99$ and 77 degrees of freedom does not reject the hypothesis $\mu_X = \mu_Y$ of identical $pH$ means underlying the two samples (1) and (2).

The results for the comparison of the calcium means are completely different. The $F'$ statistic attains the significantly high value 3.81, which makes us doubt that the variances of the calcium values in the two samples actually coincide. The test statistic $\tilde{T}_W$ of the Welch test attains the value $-5.15$ with 46 degrees of freedom. This large deviation from zero makes us doubt that the calcium means of the two samples actually coincide. These results suggest in addition that a high calcium concentration in a specimen might be used as an indicator of crystallization. The $pH$ level, however, does not seem to have this capacity of discrimination.

## The $p$–Value

To test the hypothesis $H_0 : \mu_X = \mu_Y$ of identical $pH$ means in the two samples (1) and (2) against the *two–sided* alternative $H_1 : \mu_X \neq \mu_Y$, we use the statistic

$$\tilde{T} = \frac{\bar{X}_{45} - \bar{Y}_{34}}{S_p \sqrt{\frac{1}{45} + \frac{1}{34}}}.$$

It attains the value 0.99. Under the assumptions of Theorem 2.3.5, in the case of equal means, the probability of the event $\{|\tilde{T}| > 0.99\}$ is 0.3249. This is the $p$–value pertaining to the outcome 0.99 of the test statistic:

$$p = P\{|\tilde{T}| > 0.99\} = 0.3249.$$

The significance of the $p$–value can easily be explained by means of Corollary 1.6.4 (i). Let $T$ be an arbitrary test statistic with continuous distribution function $F_0$ under hypothesis $H_0$. Corollary 1.6.4 then implies that the random variable

$$p := 1 - F_0(T)$$

is under $H_0$ uniformly distributed on $(0,1)$. The transformation $p = 1 - F_0(T)$ of a test statistic onto its $p$–value therefore makes extremely large or small observations of the test statistic $T$ *immediately visible*. A small $p$ close to zero reveals in particular a significantly large value of $T$.

In practice one usually does not fix a level $\alpha$ when testing a hypothesis, but one rather computes the $p$–value of the pertaining test statistic. The decision, whether the hypothesis is rejected or not, is then based on this $p$–value. A $p$–value less than 0.05 is commonly considered to be significantly small. This coincides with a fixed level 0.05 test.

For a discussion of the conflict "$p$–value against fixed level", we refer to Section 4 in Lehmann (1993). It reflects the arguments between the founders of modern testing theory R.A. Fisher (1890-1962) on the one hand and J. Neyman (1894-1981) and E.S. Pearson (1895-1980) on the other, who introduced the comparison of different level $\alpha$ tests by means of their power functions. Further discussions are in the articles by Schervish (1996) and Sackrowitz and Samuel–Cahn (1999).

In our Example 2.3.7 we have $p = 0.3249$. This $p$–value is not close enough to zero to make us distrust the hypothesis $\mu_X = \mu_Y$ of identical $pH$ means in the two samples. The hypothesis of identical calcium means is, however, quite doubtful by the $p$–value $< 0.0001$ pertaining to the value $-5.15$ of the Welch statistic.

If the sample sizes $n$ and $m$ are large, then $\tilde{T}$ is approximately $t$ or normal distributed by the central limit theorem. If, however, $n$ and $m$ are small and the data are not normal, then the assumption of a $t$ distribution will be erroneous. A check for normality of the data preceding the $t$ test is, therefore, recommended, cf Section 1.6 on quantile plots.

## Logarithmic Transformation of the Data

If a sample is skewed, then it is often useful to reduce the skewness by taking the logarithms or the square roots of the data, before a test is applied which requires normal data such as the $t$ test. The log and the square root transformation both have the property that small values are spread out and large ones are condensed, cf Section 1.7.

A nonnegative random variable $X$ with the property that $\log(X)$ is normal distributed with mean $\mu$ and variance $\sigma^2$ is called *lognormal distributed*. It has the density

$$f(x) = \frac{1}{\sqrt{2\pi}\,\sigma x} \exp\left(-\frac{(\log(x) - \mu)^2}{2\sigma^2}\right), \qquad x > 0,$$

and $f(x) = 0$ elsewhere (Exercise 21). This distribution frequently occurs for life span data or data from strength of materials tests in engineering as well as for measurements of sustaining attention in psychology and more generally in problems from reliability theory.

## Matched Pairs

In particular, in clinical trials, one frequently collects observations $X_1, \ldots, X_n$ and $Y_1, \ldots, Y_n$ by sampling pairs $(X_1, Y_1), \ldots, (X_n, Y_n)$. The pairs $(X_i, Y_i)$ can, for example, be pairs of patients of approximately equal age, size and weight, or two measurements taken from the same person. The two samples $X_1, \ldots, X_n$ and $Y_1, \ldots, Y_n$ of such *matched pairs* $(X_i, Y_i)$ are typically not independent, but independence of the differences $D_i := X_i - Y_i$, $i = 1, \ldots, n$, can often be justified.

**2.3.8 Example** (Lead Data). Morton et al. (1982) studied lead in the blood of children whose parents worked in a factory where lead is used for making batteries. The authors were concerned that children were exposed to lead inadvertently brought home by their parents. They matched 33 such children from different families to 33 control children of the same age and neighborhood whose parents were employed in other industries not using lead. For a discussion of these data we refer to Rosenbaum (1993).

If we model the two samples of lead absorption by $X_1, \ldots, X_{33}$ and $Y_1, \ldots, Y_{33}$, then we obviously have matched pairs $(X_i, Y_i)$. Let us assume that the differences $D_i = X_i - Y_i$, $i = 1, \ldots, 33$, are independent and identically $N(\mu, \sigma^2)$ distributed random variables. The hypothesis $H_0 : \mu = 0$ or, equivalently, $E(X_i) = E(Y_i)$ that the parents' place of work is not related to the children's lead absorption can then be tested using the test statistic $t := \sqrt{33}\bar{D}_{33}/S_{33}$, which is by Theorem 2.2.1 (iii) $t_{32}$ distributed. By $\bar{D}_{33} := \sum_{i=1}^{33} D_i/33$ we denote the sample mean and by $S_{33}^2 := \sum_{i=1}^{33}(D_i - \bar{D}_{33})^2/32$ the sample variance of $D_1, \ldots, D_{33}$. The hypothesis $H_0 : \mu = 0$ is rejected if $|t|$ is significantly large or, equivalently, if the pertaining $p$–value is significantly small. This is the *one–sample $t$* test.

The UNIVARIATE Procedure
Variable: diff

Basic Statistical Measures

| Location | | Variability | |
|---|---|---|---|
| Mean | 15.96970 | Std Deviation | 15.86365 |
| Median | 15.00000 | Variance | 251.65530 |
| Mode | 25.00000 | Range | 69.00000 |
| | | Interquartile Range | 21.00000 |

Tests for Location: Mu0=0

| Test | -Statistic- | | -----p Value------ | |
|---|---|---|---|---|
| Student's t | t | 5.782966 | Pr > \|t\| | <.0001 |
| Sign | M | 12 | Pr >= \|M\| | <.0001 |
| Signed Rank | S | 235 | Pr >= \|S\| | <.0001 |

**Figure 2.3.3.** One–sample $t$ test of the lead data.

```
***    Program 2_3_3   ***;
TITLE1 'One-Sample t Test';
TITLE2 'Lead Data';
LIBNAME datalib 'c:\data';

DATA data1;
    SET datalib.lead;
    diff=withlead-nolead;

PROC UNIVARIATE DATA=data1;
    VAR diff;
RUN; QUIT;
```

The variable 'diff' is defined in the DATA step as the difference of the two variables 'withlead' and 'nolead', which contain the values of the matched pairs in this example. PROC UNIVARIATE with the statement 'VAR diff' computes among others the value of the one–sample $t$ test and the corresponding $p$–value for the hypothesis that the theoretical mean of the variable diff is zero.

We have in this example $\bar{D}_{33} = 15.9697, S_{33} = 15.86365, S_{33}/\sqrt{33} = 2.761507$ and $t = \bar{D}_{33}/(S_{33}/\sqrt{33}) = 5.782966$. The $p$–value is $P\{|t| > 5.782966\} = P\{t > 5.782966\} + P\{t < -5.782966\} < 0.0001$. This small value makes us doubt that the hypothesis $\mu = 0$ of identical means of lead absorption in the two samples is true. This result indicates that children actually absorb lead brought home by their parents.

## 2.4　A Nonparametric Alternative: The Wilcoxon Test

The use of *nonparametric* procedures requires no particular assumption about the distribution underlying a sample such as the normal distribution. We consider, therefore, in the sequel independent random variables $X, X_1, \ldots, X_n$ and $Y, Y_1, \ldots, Y_m$, where $X$ has some arbitrary common distribution function $F$ and $Y$ has some arbitrary common distribution function $G$. We focus on the problem, whether $F$ and $G$ coincide.

To test the hypothesis

$$H_0 : F = G$$

one can use the probability

$$q := P\{X < Y\}$$

of the event that the random variable $X$ is smaller than $Y$. Suppose that $F$ is continuous. If $F = G$, then we have $q = 1/2$. If, on the other hand, $q$ is different from $1/2$, then we have to reject the hypothesis $H_0$ that $F$ and $G$ coincide.

**2.4.1 Lemma.** *If $F$ is continuous and if $F = G$, then we have*

$$P\{X < Y\} = 1/2.$$

**Proof:** The assertion follows by exchanging $X$ and $Y$ so that

$$P\{X < Y\} = P\{Y < X\}$$

and by the fact that $P\{X = Y\} = 0$ for a continuous distribution function $F$, cf Exercise 33 in Chapter 1, which imply

$$1 = P\{X < Y\} + P\{Y < X\} + P\{X = Y\} = 2P\{X < Y\}. \qquad \square$$

## Ranks

We can write the probability $P\{X < Y\}$ by Fubini's Theorem as the double integral

$$q = P\{X < Y\} = E\left(1_{(-\infty,Y)}(X)\right) = \iint 1_{(-\infty,y)}(x)\, F(dx)\, G(dy).$$

A plausible estimator of $q$, which is based on $X_1, \ldots, X_n, Y_1, \ldots, Y_m$, is then defined by

$$\hat{q}_{n,m} := \iint 1_{(-\infty,y)}(x)\, F_n(dx)\, G_m(dy)$$

$$= \int \frac{1}{n} \sum_{i=1}^{n} 1_{(-\infty,y)}(X_i)\, G_m(dy) = \frac{1}{mn} \sum_{j=1}^{m} \sum_{i=1}^{n} 1_{(-\infty,Y_j)}(X_i),$$

see Exercise 4, Chapter 1. By $F_n$ we denote the empirical distribution function of $X_1, \ldots, X_n$, by $G_m$ that of $Y_1, \ldots, Y_m$ and by $1_A(t)$ the indicator function of a set $A$, i.e., $1_A(t) = 1$ if $t \in A$ and $0$ elsewhere, cf Section 1.1. We order the random observations $Y_1, \ldots, Y_m$ according to their size, thus obtaining the order statistics $Y_{1:m} \leq \cdots \leq Y_{m:m}$. If the values $Y_1, \ldots, Y_m, X_1, \ldots, X_n$ are different from each other, i.e., if there are no *ties*, then we obviously obtain

$$mn\hat{q}_{n,m} = \sum_{j=1}^{m} \sum_{i=1}^{n} 1_{(-\infty,Y_{j:m})}(X_i)$$

$$= \text{number of those } X_1, \ldots, X_n \text{ that are less than } Y_{1:m}$$
$$+ \text{number of those } X_1, \ldots, X_n \text{ that are less than } Y_{2:m}$$
$$\vdots$$
$$+ \text{number of those } X_1, \ldots, X_n \text{ that are less than } Y_{m:m}$$
$$= (R_{Y_{1:m}} - 1) + (R_{Y_{2:m}} - 2) + \cdots + (R_{Y_{m:m}} - m)$$
$$= \sum_{j=1}^{m} R_{Y_j} - \frac{m(m+1)}{2}.$$

Here, by

$$R_{Y_j} := \sum_{i=1}^{n} 1_{(-\infty, Y_j]}(X_i) + \sum_{k=1}^{m} 1_{(-\infty, Y_j]}(Y_k),$$

we denote the number of those observations in the combined sample $X_1, \ldots, X_n$, $Y_1, \ldots, Y_m$ that are less than $Y_j$. This is the *rank* of $Y_j$ in the combined sample. Note that the random variables $X_1, \ldots, X_n, Y_1, \ldots, Y_m$ are different from each other with probability 1 if the distribution functions $F$ and $G$ are continuous, see Exercise 22 in Chapter 1.

We expect the random observations $Y_1, \ldots, Y_m$ to be uniformly distributed in the combined sample if the hypothesis $H_0 : F = G$ is true. If, however, we actually have $F \neq G$, then this should be violated. Consider, for instance, the case $F < G$. The random observations $Y_1, \ldots, Y_m$ then have the tendency to attain smaller values than the $X_i$ in the combined sample and, thus, the sum $mn\hat{q}_{n,m}$ of the ranks of the $Y_j$ should be significantly small.

## A Linear Rank Statistic

If we test the hypothesis $H_0 : F = G$ with a continuous $F$ by means of $\hat{q}_{n,m}$, then we will, consequently, reject $H_0$ if the distance of $\hat{q}_{n,m}$ from $1/2$ is too large. Note that only a large distance contributes information about the stochastic background. A value $\hat{q}_{n,m}$ close to $1/2$, on the other hand, does *not* support the hypothesis $H_0$; it rather does not reject the hypothesis. Nevertheless, one can easily find examples $F \neq G$ such that $P\{X < Y\} = 1/2$ (Exercise 22).

**2.4.2 Theorem.** *Suppose that $F$ is continuous and that $F = G$. The linear rank statistic $U_{n,m} := mn\,\hat{q}_{n,m}$ then satisfies*

$$E(U_{n,m}) = \frac{mn}{2}, \quad Var(U_{n,m}) = \frac{mn(m+n+1)}{12}.$$

**Proof:** Put $Z_{ij} := 1_{(-\infty, Y_j]}(X_i)$. Lemma 2.4.1 then implies

$$E(U_{n,m}) = mn\, E(Z_{11}) = mn\, P\{X < Y\} = mn/2.$$

For the variance we obtain the representation

$$Var(U_{n,m}) = E\left(\left(\sum_{j=1}^{m}\sum_{i=1}^{n}(Z_{ij} - 1/2)\right)^2\right)$$

$$= \sum_{i=1}^{n}\sum_{j=1}^{m}\sum_{k=1}^{n}\sum_{l=1}^{m} E\left((Z_{ij} - 1/2)(Z_{kl} - 1/2)\right)$$

$$= \sum_{i=1}^{n}\sum_{j=1}^{m}\sum_{k=1}^{n}\sum_{l=1}^{m} \left(E(Z_{ij}\, Z_{kl}) - 1/4\right)$$

$$= \sum_{i=1}^{n}\sum_{j=1}^{m}\sum_{k=1}^{n}\sum_{l=1}^{m} \left(P\{X_i < Y_j,\ X_k < Y_l\} - 1/4\right)$$

since $Z_{ij}\, Z_{kl} = 1$ if $X_i < Y_j$ and $X_k < Y_l$, and $Z_{ij}\, Z_{kl} = 0$ elsewhere. We have further

$$P\{X_i < Y_j,\ X_k < Y_l\} = \begin{cases} 1/2 & \text{if } i = k,\ j = l, \\ 1/4 & \text{if } i \neq k,\ j \neq l, \\ 1/3 & \text{if } i = k,\ j \neq l \text{ or } i \neq k,\ j = l. \end{cases}$$

The first two cases are immediate from Lemma 2.4.1 and the independence of the two samples $X_1,\ldots,X_n$ and $Y_1,\ldots,Y_m$. The third case follows from symmetry arguments. As $X_i$ and $Y_j$ are independent and identically distributed, we have

$$P\{X_1 < Y_1,\ X_1 < Y_2\} = P\Big\{X_1 = \min(X_1, Y_1, Y_2)\Big\}$$

$$= P\Big\{Y_1 = \min(X_1, Y_1, Y_2)\Big\}$$

$$= P\Big\{Y_2 = \min(X_1, Y_1, Y_2)\Big\} = 1/3.$$

Note that $P\{Y_1 = X_1\} = 0$. These symmetry arguments apply to the final case $P\{X_1 < Y_1,\ X_2 < Y_1\} = P\{Y_1 = \max(X_1, X_2, Y_1)\} = 1/3$ as well. We, thus, obtain

$$E(Z_{ij}\, Z_{kl}) - 1/4 = \begin{cases} 1/4 & \text{if } i = k,\ j = l, \\ 0 & \text{if } i \neq k,\ j \neq l, \\ 1/12 & \text{if } i = k,\ j \neq l \text{ or } i \neq k,\ j = l. \end{cases}$$

This implies

$$Var(U_{n,m}) = \frac{nm}{4} + \frac{nm(m-1)}{12} + \frac{n(n-1)m}{12} = \frac{nm(m+n+1)}{12}. \qquad \square$$

The following result on the asymptotic normality of $U_{n,m}$ is a consequence of Example B in Section 5.5.1 of the monograph by Serfling (1980).

**2.4.3 Theorem.** *Suppose that $F$ is continuous and that $F = G$. Then for* $U_{n,m} = mn\,\hat{q}_{n,m}$,

$$P\left\{\frac{U_{n,m} - mn/2}{\sqrt{nm(m+n+1)/12}} \le x\right\} \xrightarrow[m,n\to\infty]{} \Phi(x), \qquad x \in \mathbb{R}.$$

Note that the asymptotic normality of

$$Z := \frac{U_{n,m} - mn/2}{(nm(m+n+1)/12)^{1/2}}$$

is not an immediate consequence of the central limit theorem for sums of independent random variables, since $U_{n,m}$ is not a sum of *independent* observations. Its asymptotic normality follows from corresponding results for $U$ *statistics*, as $U_{n,m}$ is actually of this type, cf Chapter 5 of the monograph by Serfling (1980).

## The Wilcoxon Test

In practice, one computes the *Wilcoxon rank sum test* or *Mann–Whitney U test* by computing the sum of ranks $\sum_{j=1}^{m} R_{Y_j}$ in the combined sample and then

$$Z = \frac{\sum_{j=1}^{m} R_{Y_j} - m(m+n+1)/2}{\sqrt{nm(m+n+1)/12}}.$$

The hypothesis $H_0 : F = G$ is rejected if $|Z|$ is too large, i.e., if the (approximate) $p$–value $1 - \Phi(|Z|) + \Phi(-|Z|) = 2(1 - \Phi(|Z|))$ is too small. The approximation of the distribution of $Z$ by the standard normal one is commonly considered to be sufficiently close for practical purposes if $n \ge 4$, $m \ge 4$ and $n + m \ge 20$; compare with Section 3 in Chapter 1 of Lehmann (1975). For smaller sizes one can easily compute the exact distribution of $Z$ under the hypothesis $F = G$ with a continuous $F$ by using combinatorial arguments, cf Exercise 23. For this, the crucial fact is that the distribution of $Z$ does not depend on $F$ whenever $F$ is continuous. Consider the representation $X_i = F^{-1}(V_i)$ and $Y_j = F^{-1}(W_j)$, where $V_1, \ldots, V_n, W_1, \ldots, W_m$ are independent, on $(0,1)$ uniformly distributed random variables, cf Corollary 1.6.4 (ii). From Lemma 1.6.3 and the continuity of $F$ we then obtain

$$U_{n,m} = \sum_{j=1}^{m}\sum_{i=1}^{n} 1_{(-\infty, F^{-1}(W_j))}(F^{-1}(V_i))$$

$$= \sum_{j=1}^{m}\sum_{i=1}^{n} 1_{(-\infty, W_j)}(F(F^{-1}(V_i))) = \sum_{j=1}^{m}\sum_{i=1}^{n} 1_{(-\infty, W_j)}(V_i).$$

Since this sum is independent of $F$, the test statistic $Z$ is *distribution free* under the hypothesis $F = G$ with $F$ continuous. Note, in addition, that the value of $Z$ is not altered if the $Y_j$ are replaced by the $X_i$.

As a generalization of the Wilcoxon test we will introduce in Chapter 5 the *Kruskal–Wallis test* in the framework of the analysis of variance. By means of this, one tests the hypothesis that the distribution of $k \geq 2$ independent samples coincide. For a thorough study of rank statistics we refer to the monograph by Lehmann (1975).

**2.4.4 Example** (Ice Data; Natrella (1963)). To determine the latent heat when ice is melting, two different methods $X$ and $Y$ were used. Two samples of sizes 13 and 8, respectively, were taken measuring the changes in total heat from ice at $-72°C$ to water at $0°C$ in calories per gram of mass. Do the two methods measure the same target value? For a discussion of the ice data we refer to Rice (1995), Section 11.2.

The two sample sizes 13 and 8 are quite small. It is, therefore, reasonable to test the hypothesis that both methods measure the same target value also by means of a nonparametric procedure if we cannot justify the assumption of normal data. Putting $m = 8$ and $n = 13$, we obtain for the sum of the ranks of the sample in the combined sample

$$\sum_{j=1}^{8} R_{Y_j} = 51; \ U_{n,m} - \frac{mn}{2} = \sum_{j=1}^{m} R_{Y_j} - \frac{m(m+n+1)}{2} = 51 - 88 = -37,$$
$$\sqrt{8 \cdot 13 \cdot (8+13+1)/12} \approx 13.81.$$

We have, consequently, $Z \approx -2.67$ with the pertaining $p$–value $P\{|Z| \geq 2.67\} = 1 - P\{Z \leq 2.67\} + P\{Z \leq -2.67\} \approx 2(1 - \Phi(2.67)) \approx 0.007$. This nonparametric test indicates, therefore, that the two methods do not measure the same target value; see also the discussion of ties and average ranks in the sequel.

N P A R 1 W A Y   P R O C E D U R E

Wilcoxon Scores (Rank Sums) for Variable CALORIES
Classified by Variable METHOD

| METHOD | N | Sum of Scores | Expected Under H0 | Std Dev Under H0 | Mean Score |
|---|---|---|---|---|---|
| x | 13 | 180.0 | 143.0 | 13.6503968 | 13.8461538 |
| y | 8 | 51.0 | 88.0 | 13.6503968 | 6.3750000 |

Average Scores Were Used for Ties

```
Wilcoxon 2-Sample Test (Normal Approximation)
(with Continuity Correction of .5)

S =  51.0000  Z = -2.67391  Prob > |Z| = 0.0075

T-Test Approx. Significance = 0.0146

Kruskal-Wallis Test (Chi-Square Approximation)
CHISQ =  7.3470  DF =  1  Prob > CHISQ = 0.0067
```

**Figure 2.4.1.** Wilcoxon test of the ice data.

```
***    Program 2_4_1    ***;
TITLE1 'Wilcoxon Test';
TITLE2 Ice Data';
LIBNAME datalib 'c:\data';

PROC NPAR1WAY DATA=datlib.ice WILCOXON;
    CLASS method;
    VAR calories;
RUN; QUIT;
```

The procedure NPAR1WAY with the option WILCOXON performs a Wilcoxon rank sum test for the variable defined in the VAR statement if the grouping variable named in the CLASS statement has exactly two levels. For any number of levels, this option performs a Kruskal–Wallis test, see Section 5.1. The syntax of the statements CLASS and VAR coincides with that for the TTEST procedure, see program 2_3_2.

## Ties and Average Ranks

The rank sum $\sum_{j=1}^{8} R_{Y_j}$ in this example is computed using *average ranks*. The reason is that there are *ties* among the observations in the combined sample. Theoretically, such ties do not occur if the underlying distribution function is continuous. In practice, however, they frequently occur, for example, caused by approximations in the number of digits in which the data are stored. The following table lists the combined and ordered sample of the ice data together with the ties:

$$
\begin{array}{lll}
1 & 79.94 & y \\
2 & 79.95 & y \\
3|4|5|6 & 79.97 \; x & yyy \\
7|8 & 79.98 \; x & y \\
9 & 80.00 \; x & \\
10|11|12|13 & 80.02 \; xxx & y \\
14|15|16|17 & 80.03 \; xxx & y \\
18|19|20 & 80.04 \; xxx & \\
21 & 80.05 \; x. &
\end{array}
$$

Take, for example, the value 79.97, which occurs three times in the $y$ sample and once in the $x$ sample. What are the ranks of the pertaining observations in the combined sample? The same problem occurs with 79.98|80.02|80.03. We solve this decision problem by computing for 79.97 the average

$$(3 + 4 + 5 + 6)/4 = 9/2$$

of the four pertaining possible ranks and by giving each of the three $y$ values 79.97 this average rank. Equally, we give 79.98 the average rank $(7 + 8)/2 = 15/2$, 80.02 the average rank $(10+11+12+13)/4 = 23/2$ and 80.03 the average rank $(14 + 15 + 16 + 17)/4 = 31/2$. The value 80.04, which occurs three times in the $x$ sample, is not used for the rank sum of the $Y_j$. We obtain, altogether,

$$\sum_{j=1}^{8} R_{Y_j} = 1 + 2 + 3 \cdot \frac{9}{2} + \frac{15}{2} + \frac{23}{2} + \frac{31}{2} = 51.$$

The occurrence of ties will generally reduce the theoretical variation of the ranks of the $Y_j$. The variance of the rank statistic $U_{n,m} = \sum_{j=1}^{m} R_{Y_j} - m(m + 1)/2$ should be corrected in this case. Under the hypothesis $F = G$ the variance $Var(U_{n,m}) = nm(m + n + 1)/12$ in the Wilcoxon test statistic $Z = (U_{n,m} - mn/2)/(Var(U_{n,m}))^{1/2}$ is, therefore, replaced by

$$\widetilde{Var}(U_{n,m}) := Var(U_{n,m}) - \frac{nm}{12(m + n)(m + n - 1)} \sum_{k=1}^{r}(b_k^3 - b_k).$$

We denote the number of distinct values in the combined sample $X_1, \ldots, X_n$, $Y_1, \ldots, Y_m$ by $r$ and by $b_k$ their frequencies, cf Example 3 in the Appendix of the monograph by Lehmann (1975). Here we have nine distinct values with the frequencies

$$b_1 = 1, \; b_2 = 1, \; b_3 = 4, \; b_4 = 2, \; b_5 = 1, \; b_6 = 4, \; b_7 = 4, \; b_8 = 3, \; b_9 = 1.$$

Thus, we obtain

$$\widetilde{Var}(U_{n,m}) = Var(U_{n,m}) - \frac{8 \cdot 13}{12 \cdot 21 \cdot 20}210 = 190.6666 - 4.3333 = 186.3333.$$

This leads to the standardization of the Wilcoxon statistic, which is also used by SAS,

$$\sqrt{\widetilde{Var}(U_{n,m})} = \sqrt{186.3333} = 13.6503968.$$

Note that $\widetilde{Var}(U_{n,m})$ coincides with $Var(U_{n,m})$ if there are no ties.

In addition, the value 0.5 is commonly added to $U_{n,m}$ to improve the normal approximation of its distribution in Theorem 2.4.3, cf Section 1.3 in Lehmann (1975). This addition of 0.5 is called *continuity correction*. The Wilcoxon test statistic used by SAS is, therefore, $Z = (U_{n,m} + 0.5 - mn/2)/(\widetilde{Var}(U_{n,m}))^{1/2}$. It attains the value $-2.67391$, see Figure 2.4.1. SAS uses that sample as the $Y$ sample, which has the smallest sample size.

## Exercises

**1.** Let $Y = (Y_1,\ldots,Y_m)^T$ be $m$-dimensional normal distributed with mean vector $\mu = (\mu_1,\ldots,\mu_m)^T$ and covariance matrix $\Sigma = AA^T$. Show that

$$E(Y) = \mu, \quad Cov(Y) = (E((Y_i - \mu_i)(Y_j - \mu_j)))_{1 \leq i,j \leq m} = \Sigma.$$

**2.** Show that every positive definite $n \times n$–matrix $\Sigma$ is the covariance matrix of an $n$–dimensional normal distributed random vector. (A symmetric $n \times n$–matrix $A$ is positive definite if $x^T A x > 0$, $x \in \mathbb{R}^n \setminus \{0\}$.) Hint: Apply the well–known decomposition of symmetric matrices as stated in (8.5) in Section 8.2.

**3.** Prove that if $X$ and $Y$ are independent normal distributed random variables with identical variances, then $X + Y$ and $X - Y$ are independent and normal distributed as well, cf. Exercise 4, Chapter 3. Hint: Corollary 2.1.4.

**4.** Give an example of normal distributed random variables $X_1$ and $X_2$, whose joint distribution $(X_1, X_2)$ is not bivariate normal. Hint: Consider, for example,

$$(X_1, X_2) := \begin{cases} (Z_1, |Z_2|), & \text{if } Z_1 \geq 0, \\ (Z_1, -|Z_2|), & \text{if } Z_1 < 0, \end{cases}$$

where $Z_1$ and $Z_2$ are $N(0,1)$ distributed.

**5.** (Gamma distribution) The distribution on $(0, \infty)$ with the density

$$g_{b,p}(x) = (b^p/\Gamma(p))x^{p-1}\exp(-bx), \quad x > 0,$$

is the *gamma distribution* with the parameters $b > 0, p > 0$, denoted by $\gamma_{b,p}$. With $b = 1/2$ and $p = n/2$ this is the chi–square distribution with $n$ degrees of freedom, and with $p = 1$ this is the exponential distribution with parameter $b$.

(i) Let $X_1, \ldots, X_n$ be independent random variables and let $X_i$ be $\gamma_{b,p_i}$ distributed, $i = 1, \ldots, n$. Prove that the sum $X_1 + \cdots + X_n$ is $\gamma_{b,p_1 + \cdots + p_n}$ distributed. This is the *convolution theorem* of the gamma distribution. Hint: Exercise 5, Chapter 1.

(ii) Establish that the convolution theorem of the gamma distribution implies the following well–known relation between the *beta function* $B(r, s) = \int_0^1 t^{r-1}(1 - t)^{s-1} \, dt$, $r, s > 0$, and the gamma function:

$$B(r, s) = \Gamma(r)\Gamma(s)/\Gamma(r + s), \quad r, s > 0.$$

**6.** (Expectation and variance of the chi–square distribution) Show that a $\chi_n^2$ distributed random variable $X$ satisfies $E(X) = n$, $Var(X) = 2n$. Hint: Exercise 20, Chapter 1.

**7.** Draw graphs of the densities of the chi–square distribution with $n = 30$, $50$ and $70$ degrees of freedom.

**8.** (Beta distribution) The distribution on $(0, 1)$ with the density

$$\frac{1}{B(r, s)} t^{r-1}(1 - t)^{s-1}, \quad 0 < t < 1,$$

is the *beta distribution*, where $B$ is the beta function (see Exercise 5). Prove that if $X$ is $F_{m,n}$ distributed, then $mX/(n + mX)$ is beta distributed with the parameters $r = m/2$ and $s = n/2$.

**9.** Show that if $X$ and $Y$ are independent and exponential distributed with parameter 1, then $X/Y$ is $F$ distributed. Find also the degrees of freedom.

**10.** Show that $F_{n,m}^{-1}(q) = 1/F_{m,n}^{-1}(1 - q)$ for $q \in (0, 1)$ and $m, n \in \mathbb{N}$.

**11.** Prove that if $X$ is $t_n$ distributed, then $X^2$ is $F_{1,n}$ distributed.

**12.** (Expectation and variance of the $t_n$ distribution) Let $X$ be a $t_n$ distributed random variable. Show that $E(X) = 0$ when $n > 1$ and $Var(X) = n/(n - 2)$ when $n > 2$. Prove that the expectation for $n = 1$, i.e., for the standard Cauchy distribution, does not exist. Hint to the variance: $\Gamma(x + 1) = x\Gamma(x)$, $x > 0$.

**13.** Let $X$ be uniformly distributed on $(-\pi/2, \pi/2)$ and let $f : \mathbb{R} \to \mathbb{R}$ be defined by $f(x) = \tan(x)$ for $|x| < \pi/2$ and 0 elsewhere. Show that the random variable $f(X)$ is Cauchy distributed. Remark: If a source of light, which is located at the point $(0,1)$ in the $\mathbb{R}^2$ plane, emits rays in each direction of the $x$ axis with the same intensity (uniform distribution of the ray angle), then the intensity along the $x$ axis follows a Cauchy distribution.

**14.** Let the random variable $X$ be Cauchy distributed. Plot the density of $(X - 2741)/954$.

**15.** Prove that the distribution function and the density of the $t_n$ distribution converge for $n \to \infty$ to their counterparts of the standard normal distribution. Hint: Dominated convergence theorem.

**16.** Show the existence of an orthogonal $n \times m$–matrix whose first row has constant entry $n^{-1/2}$, i.e., $(n^{-1/2}, \ldots, n^{-1/2})$.

**17.** Let $X_1, \ldots, X_n$ be independent and identical $N(\mu, \sigma^2)$ distributed random variables. Consider the case, where $\sigma^2$ is known and $\mu_0 \in \mathbb{R}$ is fixed. Define by means of the sample mean $\bar{X}_n$ a critical region of level $\alpha$ for testing $H_0 : \mu = \mu_0$ against $H_1 : \mu \neq \mu_0$. Minimize the type II error. What decision is made in the particular case $n = 5$, $\mu_0 = 2$, $\sigma^2 = 1$ and $\alpha = 0.05$ if the realizations are: $x_1 = 4.12$, $x_2 = 2.05$, $x_3 = 0.85$, $x_4 = 3.02$, $x_5 = 4.96$?

**18.** Physicists summarize observations $x = (x_1, \ldots, x_n)$ commonly as

$$\bar{x}_n \pm \frac{s_n}{\sqrt{n}},$$

where $\bar{x}_n$ is the mean and $s_n/\sqrt{n}$ is the standard error of the sample. ($s_n$ is the sample standard deviation.) Under the model assumption that the $x_i$ are realizations of independent $N(\mu, \sigma^2)$ distributed random variables, compute the probability of the event that the random interval

$$\left[\bar{x}_n - s_n/\sqrt{n}, \ \bar{x}_n + s_n/\sqrt{n}\right]$$

contains the parameter $\mu$. For the numerical computation suppose that $n$ is large and apply Exercise 15.

**19.** Let $X_1, \ldots, X_n$ be independent and $N(\mu_X, \sigma^2)$ distributed random variables and let $Y_1, \ldots, Y_n$ be independent and $N(\mu_Y, \sigma^2)$ distributed random variables. We assume in addition that the $X_i$ and $Y_j$ are independent as well. Fix $\sigma = 10$. How large should the sample size $n$ be chosen so that the confidence interval for $\mu_X - \mu_Y$ of level 0.95 has length 2? As $n$ will turn out to be quite large, use the normal approximation of the $t$ distribution.

**20.** (Ice data) Apply the $t$ test to the ice data.

**21.** Prove that a lognormal distributed random variable has the density

$$f(x) = \frac{1}{\sqrt{2\pi}\sigma x} \exp(-(\log(x) - \mu)^2/(2\sigma^2)), \quad x > 0,$$

and $f(x) = 0$ elsewhere.

**22.** Give an example of two random variables $X$ and $Y$ having different continuous distribution functions, but satisfying $P\{X < Y\} = 1/2$.

**23.** Compute the exact distribution of the rank statistic $U_{2,3}$ under the null hypothesis.

**24.** Let $X_1, \ldots, X_n$ be independent and identically distributed random variables with common continuous distribution function $F$. Denote by $X_{i:n}$ the $i$–th order statistic, by $R_{X_i} := \sum_{j=1}^{n} 1_{[0,\infty)} (X_i - X_j)$ the rank of $X_i$, $i = 1, \ldots, n$, and by $S(n)$ the set of all permutations of $(1, 2, \ldots, n)$. Show that

(i) $X_i = X_{R_{X_{i:n}}}$.

(ii) $P\{X_{\pi(1)} < \cdots < X_{\pi(n)}\} = 1/(n!)$, $\pi \in S(n)$.

(iii) $(R_{X_1}, \ldots, R_{X_n})$ is uniformly distributed on $S(n)$.

(iv) $R_{X_i}$ is uniformly distributed on $\{1, \ldots, n\}$ for $i = 1, \ldots, n$.

**25.** (Yoga data) To analyze the influence of a certain yoga exercise on blood pressure, the blood pressure of 14 individuals was measured in mmHg (systolic/diastolic) before and after the exercise. The following table lists the data. Import the data to SAS first. Does the variable *sex* have an influence on the blood pressure? Is the $t$ test applicable?

| No. | Sex | Age | Blood pressure before | afterwards |
|-----|-----|-----|-----------|------------|
| 1   | f   | 43  | 140/90    | 110/70     |
| 2   | f   | 39  | 100/80    | 120/70     |
| 3   | m   | 36  | 120/70    | 130/70     |
| 4   | m   | 76  | 130/100   | 190/130    |
| 5   | f   | 40  | 150/80    | 130/90     |
| 6   | f   | 49  | 115/75    | 120/80     |
| 7   | m   | 41  | 100/80    | 130/60     |
| 8   | f   | 27  | 140/80    | 120/70     |
| 9   | m   | 37  | 105/80    | 120/60     |
| 10  | f   | 21  | 105/80    | 110/70     |
| 11  | m   | 38  | 130/75    | 120/65     |
| 12  | f   | 52  | 120/90    | 110/85     |
| 13  | f   | 69  | 145/80    | 130/80     |
| 14  | m   | 32  | 115/85    | 125/65     |

# Chapter 3

# Regression Analysis

The comparison of two samples in the preceding chapter was based on the assumption that the underlying random variables $X$ and $Y$ were independent. In the following we will allow dependence. Regression analysis evaluates this dependence on the basis of a mean squared error and aims at the prediction of $Y$ based on the observation $X$. A popular model is the assumption that $Y$ is the sum of $X$ and an independent error variable. This is called the *linear model*.

## 3.1  Best Linear Approximation

In the sequel we will first compile various definitions and mathematical results concerning the dependence of random variables $X$ and $Y$. We will in particular derive constants $a^*, b^* \in I\!R$ such that the mean squared error $E((Y-(aX+b))^2)$ of the linear approximation of $Y$ by $aX + b$ is minimized.

### Covariance and Coefficient of Correlation

**3.1.1 Definition.** Let $X$ and $Y$ be square integrable random variables, i.e., $E(X^2) < \infty$, $E(Y^2) < \infty$. The number

$$Cov(X,Y) := E\big((X - E(X))(Y - E(Y))\big)$$
$$= E(XY) - E(X)E(Y)$$

is the *covariance* of $X$ and $Y$. If, in addition, $Var(X) =: \sigma^2(X) > 0$ and $Var(Y) = \sigma^2(Y) > 0$, then the number

$$\varrho(X,Y) := \frac{Cov(X,Y)}{\sigma(X)\sigma(Y)}$$

is the *coefficient of correlation* of $X$ and $Y$. The random variables $X$ and $Y$ are called *positive* or *negative correlated* or *uncorrelated* if $\varrho(X,Y) > 0$, $< 0$ or $= 0$, respectively. The covariance and the coefficient of correlation are obviously symmetric in $X, Y$.

The Cauchy–Schwarz inequality (see, for example, Section 5.1 in Fristedt and Gray (1997)) implies

$$
\begin{aligned}
|Cov(X,Y)| &= \left| E\big((X - E(X))(Y - E(Y))\big) \right| \\
&\leq E\big(|X - E(X)|\, |Y - E(Y)|\big) \\
&\leq \left(E(X - E(X))^2\right)^{1/2} \left(E(Y - E(Y))^2\right)^{1/2} \\
&= \sigma(X)\sigma(Y),
\end{aligned}
$$

and, thus,

$$
\varrho(X,Y) \in [-1,1].
$$

If $\varrho(X,Y) > 0$ or $< 0$, then $X - E(X)$ and $Y - E(Y)$ tend to have the same or different signs, respectively.

**3.1.2 Example.** Consider a two–dimensional normal distributed random vector $(X,Y)$. It has by Theorem 2.1.3 the density

$$
\varphi(x,y) = \frac{1}{2\pi \sigma_X \sigma_Y \sqrt{1 - \varrho^2}}
$$

$$
\times \exp\left\{ -\frac{1}{2(1 - \varrho^2)} \left( \frac{(x - \mu_X)^2}{\sigma_X^2} \right.\right.
$$

$$
\left.\left. +\frac{(y - \mu_Y)^2}{\sigma_Y^2} - \frac{2\varrho(x - \mu_X)(y - \mu_Y)}{\sigma_X \sigma_Y} \right) \right\},
$$

where $\mu_X, \mu_Y \in \mathbb{R}$, $\sigma_X, \sigma_Y \in (0,\infty)$ and $\varrho \in (-1,1)$. Then we have

$$
\mu_X = E(X), \;\; \mu_Y = E(Y), \;\; \sigma_X^2 = Var(X), \;\; \sigma_Y^2 = Var(Y),
$$

and $\varrho$ is the coefficient of correlation of $X$ and $Y$. Moreover, $X$ and $Y$ are independent if and only if $\varrho = 0$ (Exercise 9).

**3.1.3 Definition.** A *scatterplot* of the data $(x_1, y_1), \ldots, (x_n, y_n)$ is a simple two–dimensional graph in rectangular coordinates consisting of points, whose coordinates represent the $x, y$–values of these data. If $(x_1, y_1), \ldots, (x_n, y_n)$ are realizations of $n$ independent replicates of some random vector $(X, Y)$, then a large absolute value of the coefficient of correlation can be seen as a linearity in the scatterplot.

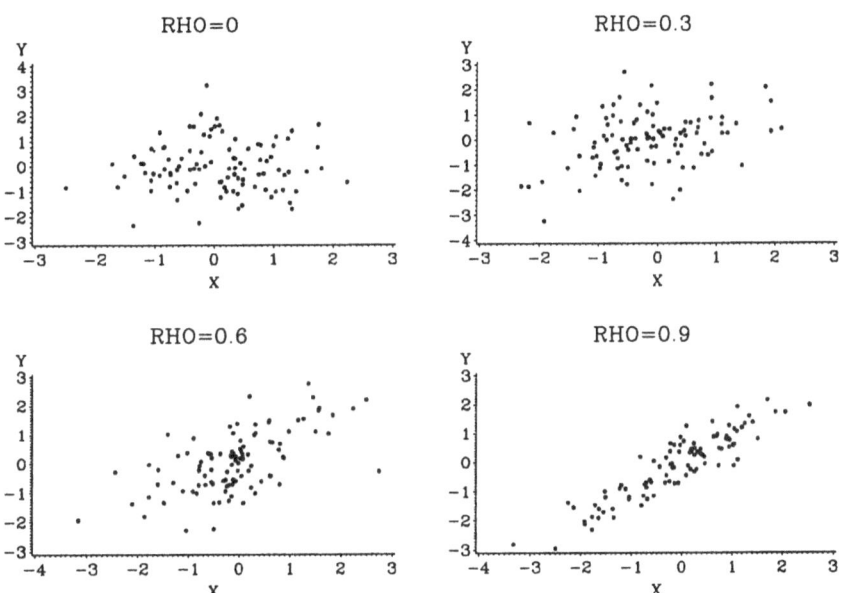

**Figure 3.1.1.** Scatterplots of 100 independent normal vectors; $\varrho = 0$, 0.3, 0.6, 0.9.

```
***    Program 3_1_1    ***;
TITLE1 'Scatterplots of Normal Vectors'
TITLE2 'Random Data'
LIBNAME datalib 'c:\data';

DATA datalib.corr1(KEEP=rho x y);
   DO rho=0, 0.3, 0.6, 0.9;
      DO i=1 TO 100;
         x=RANNOR(1);                 * generation of independent;
         z=RANNOR(1);                 * N(0,1) variables;
         y=rho*x + SQRT(1-rho**2)*z;  * see Exercise 11;
   OUTPUT; END; END;

                         ↓
```

```
                                      ↑
GOPTIONS NODISPLAY HBY=4 HTEXT=3;
SYMBOL1 V=DOT C=RED I=NONE;
TITLE1;
PROC GPLOT DATA=datalib.corr1 GOUT=corr;
    PLOT y*x;
    BY rho;
RUN; QUIT;

GOPTIONS DISPLAY;
PROC GREPLAY NOFS IGOUT=corr TC=SASHELP.TEMPLT;
    TEMPLATE=L2R2S;
    TREPLAY 1:GPLOT 2:GPLOT2 3:GPLOT1 4:GPLOT3;
RUN; DELETE _ALL_; QUIT;
```

In the DATA step 100 realizations of two normal distributed variables are created for 4 different values of $\varrho$ by means of the RANNOR function. The argument of this function (here 1) determines seed values used to calculate (pseudo) random numbers from a N(0,1) distribution.

A positive argument of the RANNOR function always returns the same stream of random numbers. A value 0 and negative values cause seed streams based on a computer clock observation.

One can use GREPLAY to display several graphics output at the same time. Each drawing is first stored to a graphics catalog by specifying the option 'GOUT=*catalog name*' (here 'GOUT=corr') in the GPLOT procedure. If this catalog does not yet exist, it is automatically created by SAS.

Then the procedure GREPLAY is invoked, which can be run in windows or line mode. The line mode is specified by the option 'NOFS'(=NO Full Screen). One can switch back and forth within a session. In addition to the input graphics catalog (here 'IGOUT=corr') one can specify a template–catalog. The template–catalog used here ('TC=SASHELP.TEMPLT') is contained in the SAS/GRAPH module. 'SASHELP.TEMPLT' contains numerous templates for designing the graphics output. The template 'L2R2S' specified in the present program displays the graphics by the panel '=Left 2, Right 2, with Space'.

The TREPLAY statement selects the catalog entries for replay in the template panel.

The options HBY=4 and HTEXT=3 in the GOPTIONS statement define the font height of the BY variable and the labels.

The following result justifies the use of the coefficient of correlation as a measure of the degree of linear dependence between two random variables.

**3.1.4 Theorem.** *Let $X$ and $Y$ be random variables with $0 < \sigma^2(X), \sigma^2(Y) < \infty$. Then we have*

(i) *$\varrho(X, Y) = 0$ if $X$ and $Y$ are independent.*

(ii) *If $\varrho(X, Y) = -1$ or 1, then there exist $a, b \in \mathbb{R}$ with $P\{Y = aX + b\} = 1$. The converse implication is also true.*

*(iii)  The mean squared deviation $E((Y - (aX + b))^2)$ of the linear approxima-
tion of $Y$ by $aX + b$ is minimized if and only if*

$$a^* = \frac{Cov(X, Y)}{\sigma^2(X)} \quad and \quad b^* = E(Y) - a^* E(X).$$

*In this case we have*

$$E\left((Y - (a^* X + b^*))^2\right) = (1 - \varrho^2(X, Y))\sigma^2(Y). \tag{3.1}$$

**Proof:** If $X$ and $Y$ are independent, then we have $Cov(X, Y) = E((X - E(X))(Y - E(Y))) = E(X - E(X))E(Y - E(Y)) = 0$ and, thus, $\varrho(X, Y) = 0$. This is part (i). To prove part (iii) we define for $a, b \in \mathbb{R}$ the function

$$\begin{aligned} p(a, b) &:= E\left((Y - aX - b)^2\right) \\ &= a^2 E(X^2) + b^2 + 2ab E(X) - 2a E(XY) - 2b E(Y) + E(Y^2). \end{aligned}$$

The function $p(a, b)$ is a polynomial of degree two of the variables $a, b$. Using partial derivatives it is easy to see that it has a unique minimum at $a^*, b^*$ (Exercise 12). Formula (3.1) follows from elementary computation. Finally, the first assertion of part (ii) is an immediate consequence of (3.1), since in this case

$$E\left((Y - (a^* X + b^*))^2\right) = 0,$$

which implies $P\{Y - (a^* X + b^*) = 0\} = 1$. The converse implication is elementary. $\qquad\square$

## The Linear Regression

The coefficient

$$a^* = \frac{Cov(X, Y)}{\sigma^2(X)} = \varrho(X, Y)\frac{\sigma(Y)}{\sigma(X)}$$

is a proper tool for the prediction of $Y$ given $X$, called the *regression* of $Y$ on $X$. This is due to the fact that by Theorem 3.1.4 the random variable

$$\hat{Y} := a^* X + b^* = E(Y) + a^* (X - E(X))$$

is the *best linear approximation* of $Y$ based on $X$ with respect to the mean squared error. The coefficient $a^*$ is the *regression coefficient* and the linear function

$$m(t) := a^* (t - E(X)) + E(Y)$$

is the *linear regression*. The error

$$Y - \hat{Y} = Y - m(X)$$

is the *residual*. The variable $Y$ is the *dependent* or *response variable* and $X$ is the *independent* or *predictor variable*.

Suppose that $X$ and $Y$ are random variables with $E(X) = E(Y) = 0$ and $Var(X) = Var(Y) = 1$. The best linear predictor $\hat{Y}$ of $Y$ given $X$ is then

$$\hat{Y} = a^* X = \varrho(X, Y)X.$$

Since the coefficient of correlation $\varrho(X, Y)$ of $X$ and $Y$ is in absolute value less than or equal to one, we obtain

$$|\hat{Y}| = |\varrho(X, Y)|\, |X| \leq |X|.$$

The predictor $\hat{Y}$ of $Y$ is, therefore, less than or equal to $|X|$. The name *regression analysis* for the analysis of the stochastic dependence of random variables by means of the mean squared error is due to this regression.

The empirical counterparts of the covariance $Cov(X, Y)$ and the coefficient of correlation $\varrho(X, Y)$ of two random variables $X$ and $Y$, based on a sample $(X_1, Y_1), \ldots, (X_n, Y_n)$ of independent replicates of $(X, Y)$, are the *sample covariance*

$$S_{X,Y,n} := (n-1)^{-1} \sum_{i=1}^{n} (X_i - \bar{X}_n)(Y_i - \bar{Y}_n) = (n-1)^{-1} \left( \sum_{i=1}^{n} X_i Y_i - n \bar{X}_n \bar{Y}_n \right)$$

and the *sample coefficient of correlation*

$$r_{X,Y,n} := \frac{S_{X,Y,n}}{S_{X,n} S_{Y,n}} = \frac{\sum\limits_{i=1}^{n} (X_i - \bar{X}_n)(Y_i - \bar{Y}_n)}{\left( \sum\limits_{i=1}^{n} (X_i - \bar{X}_n)^2 \sum\limits_{i=1}^{n} (Y_i - \bar{Y}_n)^2 \right)^{1/2}}.$$

By $\bar{X}_n := n^{-1} \sum_{i=1}^{n} X_i$, $\bar{Y}_n := n^{-1} \sum_{i=1}^{n} Y_i$ we denote the sample means of $X_1, \ldots, X_n$ and $Y_1, \ldots, Y_n$, respectively, and by $S_{X,n}^2 = (n-1)^{-1} \sum_{i=1}^{n} (X_i - \bar{X}_n)^2$, $S_{Y,n}^2 = (n-1)^{-1} \sum_{i=1}^{n} (Y_i - \bar{Y}_n)^2$ the corresponding sample variances.

**3.1.5 Theorem.** *Let $(X_1, Y_1), \ldots, (X_n, Y_n)$ be independent replicates of the random vector $(X, Y)$ with $0 < Var(X)$, $Var(Y) < \infty$.*

(i) *The sample covariance $S_{X,Y,n}$ is an unbiased estimator of $Cov(X, Y)$, i.e., we have for $n \in \mathbb{N}$*

$$E(S_{X,Y,n}) = Cov(X, Y).$$

(ii) *The strong law of large numbers (see, for example, Section 12.2 in Fristedt and Gray (1997)) implies*

$$S_{X,Y,n} \longrightarrow_{n \to \infty} Cov(X, Y) \qquad \text{almost surely}$$

*and*

$$r_{X,Y,n} \longrightarrow_{n\to\infty} \varrho(X,Y) \qquad \text{almost surely,}$$

i.e., $S_{X,Y,n}$ and $r_{X,Y,n}$ are strongly consistent estimators of $Cov(X,Y)$ and the coefficient of correlation $\varrho(X,Y)$, respectively.

**Proof:** Exercise 13.                                                                   $\square$

The numbers $S_{X,n}$, $S_{Y,n}$, $S_{X,Y,n}$ and $r_{X,Y,n}$ carry information about the shape of the set $\{(X_1,Y_1)^T,\ldots,(X_n,Y_n)^T\}$ in $\mathbb{R}^2$. A value of $r_{X,Y,n}$ close to 1 or $-1$ indicates a high linear dependence of $X$ and $Y$, in which case the best linear predictor

$$\hat{Y} = E(Y) + a^*(X - E(X))$$

$$= E(Y) + \frac{Cov(X,Y)}{Var(X)}(X - E(X)) = m(X)$$

of $Y$ given $X$ should yield a close approximation. Replacing the unknown parameters $E(Y)$, $E(X)$ and $Cov(X,Y)$, $Var(X)$ by their empirical counterparts, we obtain

$$\bar{Y}_n + \frac{S_{X,Y,n}}{S_{X,n}^2}(X - \bar{X}_n) = \frac{S_{X,Y,n}}{S_{X,n}^2}X + \bar{Y}_n - \frac{S_{X,Y,n}}{S_{X,n}^2}\bar{X}_n$$

$$=: \hat{\beta}_1 X + \hat{\beta}_0$$

as an empirical linear predictor of $Y$ given $X$. The linear function

$$\hat{m}(t) := \hat{\beta}_1 t + \hat{\beta}_0, \qquad t \in \mathbb{R},$$

with *slope* and *intercept* given by

$$\hat{\beta}_1 := \frac{S_{X,Y,n}}{S_{X,n}^2} \qquad \text{and} \qquad \hat{\beta}_0 := \bar{Y}_n - \hat{\beta}_1 \bar{X}_n$$

is the *empirical linear regression*, or *least squares line*, cf the following section. It approximates the linear regression line $m(t)$. The errors

$$Y_i - \hat{Y}_i := Y_i - \hat{m}(X_i), \qquad i = 1,\ldots,n,$$

are the empirical *residuals*.

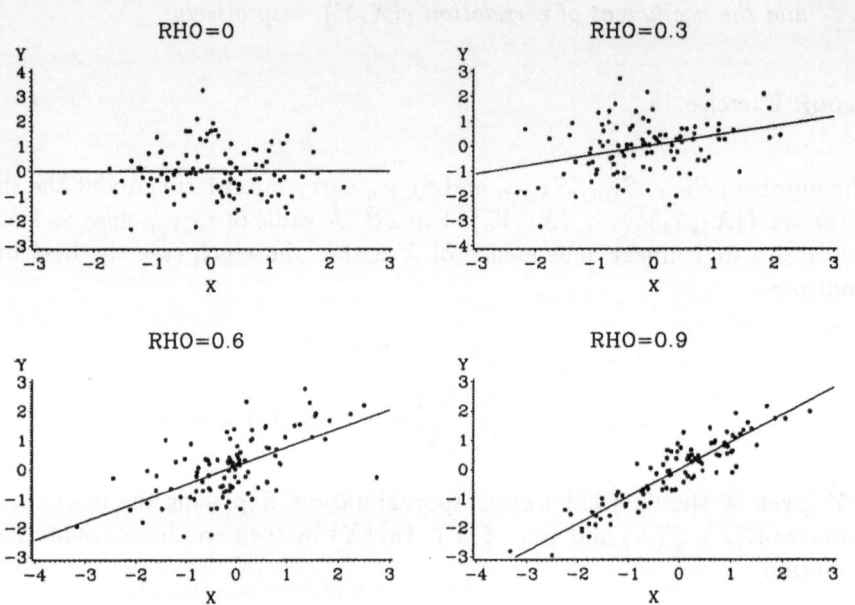

**Figure 3.1.2.** Scatterplots from Figure 3.1.1 with empirical linear regression lines.

| RHO | _DEPVAR_ | Intercept | X |
|-----|----------|-----------|---|
| 0.0 | Y | 0.00338 | 0.00298 |
| 0.3 | Y | 0.04857 | 0.38402 |
| 0.6 | Y | 0.12423 | 0.63432 |
| 0.9 | Y | -0.00419 | 0.93197 |

**Figure 3.1.3.** Intercept $\hat{\beta}_0$ and slope $\hat{\beta}_1$ of the regression lines in Figure 3.1.2.

```
***    Program 3_1_2   ***;
TITLE1 'Scatterplots with Regression Lines';
TITLE2 'Random Data';
LIBNAME datalib 'c:\data';

GOPTIONS NODISPLAY HBY=4 HTEXT=3;
SYMBOL1 V=DOT C=RED I=RL;
PROC GPLOT DATA=datalib.corr1 GOUT=corr2;
    PLOT y*x;
    BY rho;
RUN; QUIT;

GOPTIONS DISPLAY;
PROC GREPLAY NOFS IGOUT=corr2 TC=SASHELP.TEMPLT;
    TEMPLATE=L2R2S;
    TREPLAY 1:GPLOT 2:GPLOT2 3:GPLOT1 4:GPLOT3;
RUN; DELETE _ALL_; QUIT;

PROC REG DATA=datalib.corr1 OUTEST=regstats NOPRINT;
    MODEL y=x;
    BY rho;
PROC PRINT DATA=regstats;
    VAR _DEPVAR_ INTERCEPT x;
    ID rho;
RUN; QUIT;
```

This program uses the permanent file datalib.corr1 created in Program 3_1_1. The option 'I=RL' (=Regression Linear) in the SYMBOL statement draws a regression line, whose slope and intercept are automatically computed. The graphics output is created as in Program 3_1_1.

To print the parameters, PROC REG is additionally invoked. The results of this estimation are written to the temporary file 'regstats' by the option 'OUTEST=regstats'. The MODEL statement is required. It has the form *'dependent variable = independent variable'*.

PROC PRINT with the VAR statement then prints the correlation coefficient 'rho', the intercept and the slope of each of the four samples in Figure 3_1_1. These values are contained in the file regstat.

By the line $\hat{m}(t)$, which is computed from some *training sample* $(X_1, Y_1), \ldots,$ $(X_n, Y_n)$, we can predict *future* values of $Y$ given $X$. The performance of this prediction depends on the degree of the linear dependence of $X$ and $Y$, i.e., on their coefficient of correlation. If this coefficient is close to $-1$ or $1$, this prediction will perform well, whereas a value close to $0$ makes a linear prediction of $Y$ by means of $X$ less meaningful.

## 3.2   The Method of Least Squares

In the preceding section we derived the least squares line by mathematical arguments. Now we change our point of view by turning to a given set of data and deriving this least squares estimator by rational reasoning.

**3.2.1 Example** (Chestnut Data; Chapman and Demeritt (1936)). To investigate the relationship between the age of a tree and the diameter of its stem, measurements of the age in years and the diameter at breast height in foot were taken from 25 chestnut trees.

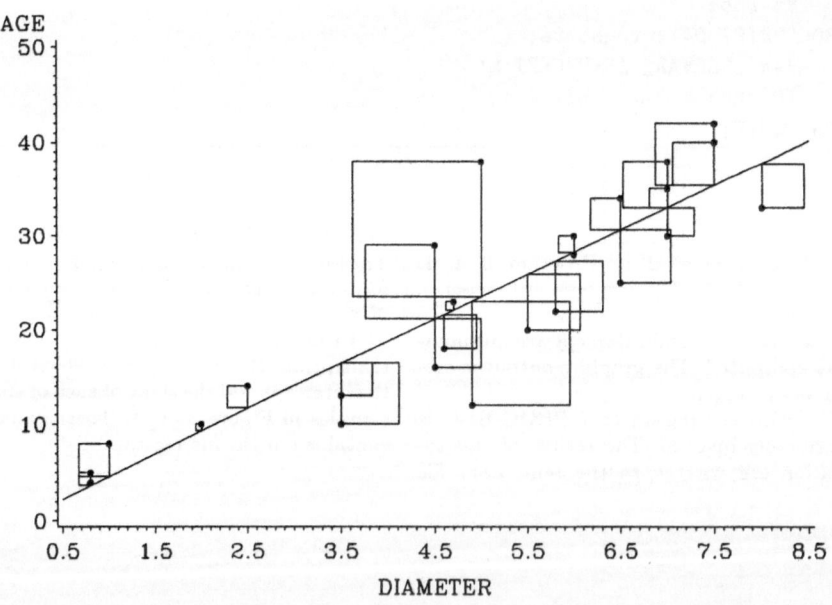

**Figure 3.2.1.** Scatterplot of chestnut data; method of least squares.

```
***   Program 3_2_1   ***;
TITLE1 'Least Squares';
TITLE2 'Chestnut Data';
LIBNAME datalib 'c:\data';

PROC REG DATA=datalib.chestnut NOPRINT;
    MODEL age=diameter;
    OUTPUT OUT=regdata P=predictd;
DATA chstnut2;
    SET regdata;
    XSYS='2'; YSYS='2'; LINE=1; SIZE=0.6;
    X=diameter; Y=age; FUNCTION='MOVE'; OUTPUT;
    X=diameter; Y=predictd; FUNCTION='DRAW'; OUTPUT;
    X=diameter+(predictd-age)/10.5; Y=predictd; FUNCTION='DRAW';
OUTPUT;
    X=diameter+(predictd-age)/10.5; Y=age; FUNCTION='DRAW'; OUTPUT;
    X=diameter; Y=age; FUNCTION='DRAW'; OUTPUT;

AXIS1 ORDER=(0.5 TO 8.5 BY 1);
SYMBOL1 V=DOT H=0.8 C=G I=RL;
PROC GPLOT DATA=datalib.chestnut ANNOTATE=chstnut2;
    PLOT age*diameter / HAXIS=AXIS1;
RUN; QUIT;
```

The results of PROC REG are written to the temporary file 'regdata' by the option 'OUT=regdata' in the OUTPUT statement. The option 'P=predictd' calculates the predicted values from the input data and the estimated model and stores them in the new variable 'predictd'.

The ANNOTATE facility enables the generation of a data set of graphics commands from which one can produce graphics output. One can use the data set to generate custom graphics or to enhance graphics output from a SAS/GRAPH procedure such as GPLOT; see SAS/GRAPH (1990) for details.

As we will use more and more complex applications later in this book, we explain some basic concepts of the ANNOTATE facility used in the above small program.

First, an ANNOTATE data set (here 'chstnut2') is composed of specially named variables. Each observation is interpreted as a graphics primitive to draw a graphics element or a programming function. Each observation contains variables that specify what graphics element to draw, where and how to draw it. The ANNOTATE graphics is then included in the SAS/GRAPH procedure GPLOT with the 'ANNOTATE=*data set*' option.

The SET statement in the DATA step loads the temporary file regdata. The observations XSYS='2' and YSYS='2' of the variables XSYS, YSYS specify that the coordinate system of the ANNOTATE graphics coincides with that of the GPLOT procedure. Both depend now on the data values. LINE controls the drawing of a line, SIZE determines the size of the graphics elements with which it is used, here the line thickness. The variable FUNCTION specifies what to draw. The function MOVE causes a move to the specified points without drawing a line. Here the specified points are the data with coordinates X=diameter, Y=age. The OUTPUT statement writes these results to the data set chstnut2.

The function DRAW draws a line from the actual point (diameter, age) to the pre-

dicted point (diameter, predictd). The result
is again written to chstnut2 by the OUTPUT
statement, etc. This is repeated for each observation in the file regdata.

To fit a straight line to a set $\{(x_i, y_i)^T,\ i = 1,\ldots,n\}$ of points in $\mathbb{R}^2$, we have to determine the slope $\beta_1$ and the intercept $\beta_0$ of the line $y = \beta_1 x + \beta_0$. The *method of least squares* suggests to take those parameters $\beta_0$ and $\beta_1$, which minimize the sum of the squares of the vertical distances of the points $(x_i, y_i)^T, i = 1,\ldots, n$, from the line:

$$R(\beta_0, \beta_1) := \sum_{i=1}^{n}(y_i - \beta_0 - \beta_1 x_i)^2 = \text{minimum}.$$

Using partial derivatives and equating them to zero, we obtain the following conditions, which necessarily have to be satisfied by minimizers of the function $R$:

$$\partial R(\beta_0, \beta_1)/\partial\beta_0 = -2\sum_{i=1}^{n}(y_i - \beta_0 - \beta_1 x_i) = 0,$$

$$\partial R(\beta_0, \beta_1)/\partial\beta_1 = -2\sum_{i=1}^{n}x_i(y_i - \beta_0 - \beta_1 x_i) = 0,$$

or, equivalently,

$$\sum_{i=1}^{n}y_i = n\beta_0 + \beta_1\sum_{i=1}^{n}x_i,$$

$$\sum_{i=1}^{n}x_i y_i = \beta_0\sum_{i=1}^{n}x_i + \beta_1\sum_{i=1}^{n}x_i^2.$$

The solutions $\hat{\beta}_0, \hat{\beta}_1$ of this system of equations are

$$\hat{\beta}_0 = \frac{(\sum_{i=1}^{n}x_i^2)(\sum_{i=1}^{n}y_i) - (\sum_{i=1}^{n}x_i)(\sum_{i=1}^{n}x_i y_i)}{n\sum_{i=1}^{n}x_i^2 - (\sum_{i=1}^{n}x_i)^2},$$

$$\hat{\beta}_1 = \frac{n\sum_{i=1}^{n}x_i y_i - (\sum_{i=1}^{n}x_i)(\sum_{i=1}^{n}y_i)}{n\sum_{i=1}^{n}x_i^2 - (\sum_{i=1}^{n}x_i)^2}, \tag{3.2}$$

provided the denominator is different from zero, cf Exercise 14. The vector $(\hat{\beta}_0, \hat{\beta}_1)$ is the *least squares estimator*. Under the model assumption of the preceding section where $(x_i, y_i)$ were realizations of $n$ independent replicates of a random vector $(X, Y)$, the vector $(\hat{\beta}_0, \hat{\beta}_1)$ estimates the parameters of the regression line $m(t)$. Without this model assumption, $(\hat{\beta}_0, \hat{\beta}_1)$ is just *one* suggestion for fitting a straight line to a set of points $\{(x_i, y_i)^T : i = 1, \ldots, n\}$, cf Exercise 4 in Chapter 8. The proof of the following result is Exercise 15.

**3.2.2 Theorem.** *With $\bar{x}_n := n^{-1} \sum_{i=1}^{n} x_i$ and $\bar{y}_n := n^{-1} \sum_{i=1}^{n} y_i$ we have*

$$\hat{\beta}_1 = \frac{\sum_{i=1}^{n}(x_i - \bar{x}_n)(y_i - \bar{y}_n)}{\sum_{i=1}^{n}(x_i - \bar{x}_n)^2}, \qquad \hat{\beta}_0 = \bar{y}_n - \hat{\beta}_1 \bar{x}_n.$$

For the chestnut data in Example 3.2.1 we obtain approximately $\hat{\beta}_1 = 4.73449$, $\hat{\beta}_0 = -0.13796$ and, thus, the estimated linear relationship

$$\text{age} \approx 4.73449 \times \text{diameter} - 0.13796$$

between the age of a chestnut tree and the diameter of its stem at breast height, measured in years and foot, respectively. By this formula we can now estimate the age of a chestnut tree without having to cut it and to count the annual rings. This second method would of course yield an exact result. The estimate of the age would not be affected essentially if we put the intercept $\hat{\beta}_0$ in the above formula equal to zero.

The REG Procedure
Model: MODEL1
Dependent Variable: AGE

Parameter Estimates

| Variable | DF | Parameter Estimate | Standard Error | t Value | Pr > |t| |
|---|---|---|---|---|---|
| Intercept | 1 | -0.13796 | 2.86033 | -0.05 | 0.9619 |
| DIAMETER | 1 | 4.73449 | 0.53535 | 8.84 | <.0001 |

**Figure 3.2.2.** Least squares estimates for the chestnut data.

```
***    Program 3_2_2    ***;
TITLE1 'Least Squares Estimates';
TITLE2 'Chestnut Data';
LIBNAME datalib 'c:\data';

PROC REG DATA=datalib.chestnut;
    MODEL age=diameter;
RUN; QUIT;
```

## Fixed and Random Design

$\hat{\beta}_0$ and $\hat{\beta}_1$ coincide obviously with those estimates of the slope and of the intercept of the regression line that we derived in the previous section. But this approach now covers the following two cases:

(a) The numbers $x_i$, $i = 1, \ldots, n$, are realizations of some random variables $X_1, \ldots, X_n$. This is the case of a *random design*.

(b) The numbers $x_i$, $i = 1, \ldots, n$ are nonrandom. This is the case of a *fixed design*.

## The Standard Model

We will derive in the sequel various statistical properties of the least squares estimator. Fitting a straight line to a set of points $\{(x_i, y_i)^T : i = 1, \ldots, n\}$ seems to be reasonable only if there is a certain linear relationship between $x_i$ and $y_i$. We will, therefore, consider in the following the simplest case, where the observation $y_i$ is a linear function of $x_i$ together with an additive random error $\varepsilon_i$. Precisely, our *standard model* is the assumption that $y_1, \ldots, y_n$ are realizations of some random variables $Y_1, \ldots, Y_n$ which satisfy

$$Y_i = \beta_0 + \beta_1 x_i + \varepsilon_i, \qquad i = 1, \ldots, n, \tag{3.3}$$

where $\varepsilon_1, \ldots, \varepsilon_n$ are independent random variables with $E(\varepsilon_i) = 0$ and $Var(\varepsilon_i) = \sigma^2$, $i = 1, \ldots, n$. The design points $x_1, \ldots, x_n$ are given numbers and we suppose that at least two of them differ.

**3.2.3 Theorem.** *Assume the standard model (3.3). The least squares estimates*

$$\hat{\beta}_1 = \frac{\sum_{i=1}^n (x_i - \bar{x}_n)(Y_i - \bar{Y}_n)}{\sum_{i=1}^n (x_i - \bar{x}_n)^2}, \qquad \hat{\beta}_0 = \bar{Y}_n - \hat{\beta}_1 \bar{x}_n$$

*of $\beta_0$ and $\beta_1$ are unbiased, i.e., $E(\hat{\beta}_j) = \beta_j$, $j = 0, 1$.*

**Proof:** From the model assumptions we obtain $E(Y_i) = \beta_0 + \beta_1 x_i$. Representation (3.2) now implies

$$E(\hat{\beta}_0) = \frac{(\sum_{i=1}^{n} x_i^2)(\sum_{i=1}^{n} E(Y_i)) - (\sum_{i=1}^{n} x_i)(\sum_{i=1}^{n} x_i E(Y_i))}{n \sum_{i=1}^{n} x_i^2 - (\sum_{i=1}^{n} x_i)^2}$$

$$= \frac{(\sum_{i=1}^{n} x_i^2)(n\beta_0 + \beta_1 \sum_{i=1}^{n} x_i) - (\sum_{i=1}^{n} x_i)(\beta_0 \sum_{i=1}^{n} x_i + \beta_1 \sum_{i=1}^{n} x_i^2)}{n \sum_{i=1}^{n} x_i^2 - (\sum_{i=1}^{n} x_i)^2} = \beta_0.$$

The proof for $\hat{\beta}_1$ uses similar arguments. $\qquad\square$

In the above proof we only made use of the additivity of the error terms $\varepsilon_i$ in the standard model (3.3) and of the assumption $E(\varepsilon_i) = 0$, $i = 1, \ldots, n$. We did not use their independence or the assumption of equal variances. The above result remains valid without these model assumptions.

In the standard model we have $Var(Y_i) = \sigma^2$ and $Cov(Y_i, Y_j) = 0$, $i \neq j$. The variances of $\hat{\beta}_0$ and $\hat{\beta}_1$ as well as their covariance can, therefore, be computed by elementary arguments, see the next theorem. For this result it would obviously be sufficient to require only that the $\varepsilon_i$ are uncorrelated instead of being independent random variables, i.e., that $Cov(\varepsilon_i, \varepsilon_j) = 0$, $i \neq j$.

**3.2.4 Theorem.** *In the standard model (3.3) we have*

$$Var(\hat{\beta}_0) = \frac{\sigma^2 \sum_{i=1}^{n} x_i^2}{n \sum_{i=1}^{n} x_i^2 - (\sum_{i=1}^{n} x_i)^2} = \frac{\sigma^2}{n} \frac{\frac{1}{n}\sum_{i=1}^{n} x_i^2}{\frac{1}{n}\sum_{i=1}^{n} (x_i - \bar{x}_n)^2},$$

$$Var(\hat{\beta}_1) = \frac{n\sigma^2}{n \sum_{i=1}^{n} x_i^2 - (\sum_{i=1}^{n} x_i)^2} = \frac{\sigma^2}{n} \frac{1}{\frac{1}{n}\sum_{i=1}^{n} (x_i - \bar{x}_n)^2},$$

$$Cov(\hat{\beta}_0, \hat{\beta}_1) = \frac{-\sigma^2 \sum_{i=1}^{n} x_i}{n \sum_{i=1}^{n} x_i^2 - (\sum_{i=1}^{n} x_i)^2} = -\frac{\sigma^2}{n} \frac{\frac{1}{n}\sum_{i=1}^{n} x_i}{\frac{1}{n}\sum_{i=1}^{n} (x_i - \bar{x}_n)^2}.$$

**Proof:** The representation

$$\hat{\beta}_1 = \frac{\sum_{i=1}^{n}(x_i - \bar{x}_n)(Y_i - \bar{Y}_n)}{\sum_{i=1}^{n}(x_i - \bar{x}_n)^2} = \frac{\sum_{i=1}^{n}(x_i - \bar{x}_n)Y_i}{\sum_{i=1}^{n}(x_i - \bar{x}_n)^2}$$

implies

$$Var(\hat{\beta}_1) = \frac{\sum_{i=1}^{n}(x_i - \bar{x}_n)^2 \, Var(Y_i)}{(\sum_{i=1}^{n}(x_i - \bar{x}_n)^2)^2} = \frac{\sigma^2}{\sum_{i=1}^{n}(x_i - \bar{x}_n)^2}.$$

The other equations are obtained by similar arguments.                              □

## Residual Sum of Squares

The relation

$$\sigma^2 = E(\varepsilon_i^2) = E((Y_i - \beta_0 - \beta_1 x_i)^2), \qquad i = 1, \ldots, n,$$

suggests an estimator of the commonly unknown model parameter $\sigma^2$ using the
*residual sum of squares*

$$RSS := R(\hat{\beta}_0, \hat{\beta}_1) = \sum_{i=1}^{n}(Y_i - \hat{\beta}_0 - \hat{\beta}_1 x_i)^2,$$

where the empirical residuals $\hat{\varepsilon}_i := Y_i - \hat{\beta}_0 - \hat{\beta}_1 x_i$ approximate the errors
$\varepsilon_i = Y_i - \beta_0 - \beta_1 x_i$. By the law of large numbers one expects the approximation

$$\frac{RSS}{n} = \frac{1}{n}\sum_{i=1}^{n}(Y_i - \hat{\beta}_0 - \hat{\beta}_1 x_i)^2$$

$$\approx \frac{1}{n}\sum_{i=1}^{n}(Y_i - \beta_0 - \beta_1 x_i)^2 = \frac{1}{n}\sum_{i=1}^{n}\varepsilon_i^2 \approx E(\varepsilon_1^2) = \sigma^2.$$

The proof of the following result is Exercise 17, see also Lemma 3.3.20.

**3.2.5 Theorem.** *The ratio*

$$S^2 := \frac{RSS}{n-2}$$

*is an unbiased estimator of $\sigma^2$.*

Let us particularly assume that the errors $\varepsilon_i$ in the standard model are
normal distributed. The convolution theorem of the normal distribution 2.1.5
together with (3.2) then implies that $\hat{\beta}_0$ and $\hat{\beta}_1$ are normal distributed as well.
Replacing $\sigma^2$ by $S^2$ in the formulas of Theorem 3.2.4, we obtain estimators of
their variances. We denote these by $S_{\hat{\beta}_1}^2$ and $S_{\hat{\beta}_0}^2$. In analogy to Theorem 2.2.1
it can be shown that under the normality assumption for the errors, the ratio
$(\hat{\beta}_j - \beta_j)/S_{\hat{\beta}_j}$ is $t_{n-2}$ distributed, $j = 0, 1$, and $RSS/\sigma^2$ is $\chi_{n-2}^2$ distributed, cf
Theorem 3.3.23 for a general result.

Without the normality assumption for the $\varepsilon_i$, it can be shown by a suitable
version of the central limit theorem that under proper conditions on the design

points $x_i$, the ratio $(\hat{\beta}_j - \beta_j)/S_{\hat{\beta}_j}$ is approximately standard normal distributed for a large sample size $n$, cf Section 7.2 in Sen and Singer (1993). We can, therefore, use the $t_{n-2}$ distribution in this case as the approximating distribution as well. This enables us to derive confidence intervals for $\beta_j$ that are based on $\hat{\beta}_j$ and $S_{\hat{\beta}_j}$ and which have approximatively the desired level for a large $n$.

One can now use $|\hat{\beta}_j|/\hat{S}_{\hat{\beta}_j}$ for testing the hypothesis $H_0 : \beta_j = 0$. If the resulting $p$–value is significantly small, then one will reject this hypothesis. For $j = 0$ this rejection would mean that the intercept $\beta_0$ in the standard model is different from zero, for $j = 1$ the rejection would indicate a linear relationship between the independent variable $Y_i$ and the explanatory one $x_i$, see Figure 3.2.2.

**3.2.6 Example** (Geo Data). The following table lists the numbers $y_i$ of students of geography at the Catholic University of Eichstaett during the winter semesters 1986/87 up to 1994/95:

| Year | 86 | 87 | 88 | 89 | 90 | 91 | 92 | 93 | 94 |
|---|---|---|---|---|---|---|---|---|---|
| Number | 5 | 14 | 24 | 64 | 100 | 129 | 200 | 176 | 177 |

The linear relationship between the numbers $y_i$ of students and the years $x_i$ is described by the estimated standard model

$$y_i = -2287.7222 + 26.5167 x_i + \hat{\varepsilon}_i, \qquad i = 1, \ldots, 9,$$

with $\bar{y}_9 = 98.78$. The estimated standard deviation of $\varepsilon_i$, $\hat{\beta}_0$ and $\hat{\beta}_1$ are

$$S = 23.4141, \quad \hat{S}_{\hat{\beta}_0} = 272.1591, \quad \hat{S}_{\hat{\beta}_1} = 3.0227.$$

The $p$–values are less than 0.0001 for the test statistics

$$\frac{|\hat{\beta}_0|}{\hat{S}_{\hat{\beta}_0}} = 8.41 \quad \text{and} \quad \frac{|\hat{\beta}_1|}{\hat{S}_{\hat{\beta}_1}} = 8.77.$$

If (3.3) actually describes the relation between $y_i$ and $x_i$ then we ought to be pretty sure that $\beta_0 \neq 0$ and $\beta_1 \neq 0$.

Often a linear relationship between variables is approximately given only after a transformation such as logarithm or square root, which both spread out small data and condense large ones. If we replace $y_i$ for the geo data, for example, by $\sqrt{y_i}$, then we obtain the estimated standard model

$$\sqrt{y_i} = -133.0422 + 1.5782 x_i + \hat{\varepsilon}_i, \qquad i = 1, \ldots, 9,$$

with the considerably smaller estimated standard errors

$$S = 1.2822, \quad \hat{S}_{\hat{\beta}_0} = 14.9044, \quad \hat{S}_{\hat{\beta}_1} = 0.1655.$$

The $R^2$ values 0.9166 and 0.9285 as measures of the goodness of fit of the estimated standard model to the data will be discussed in the next section in the general framework of multiple linear regression. Values close to 1 indicate a good fit. The better fit of the square root model is also visible in the residual plots, where the empirical residuals $\hat{\varepsilon}_i$, i.e., the errors in the prediction of the dependent variables by $\hat{\beta}_0 + \hat{\beta}_1 x_i$, are plotted against $x_i$, see also Example 3.3.26.

The REG Procedure
Model: MODEL1
Dependent Variable: NUMBER

```
            Root MSE              23.41409    R-Square      0.9166
            Dependent Mean        98.77778    Adj R-Sq      0.9047
            Coeff Var             23.70381
```

Parameter Estimates

| Variable | DF | Parameter Estimate | Standard Error | t Value | Pr > \|t\| |
|---|---|---|---|---|---|
| Intercept | 1 | -2287.72222 | 272.15913 | -8.41 | <.0001 |
| YEAR | 1 | 26.51667 | 3.02275 | 8.77 | <.0001 |

-----------------------------------------------------------------

Model: MODEL2
Dependent Variable: tnumber

```
            Root MSE               1.28224    R-Square      0.9285
            Dependent Mean         8.99414    Adj R-Sq      0.9183
            Coeff Var             14.25637
```

Parameter Estimates

| Variable | DF | Parameter Estimate | Standard Error | t Value | Pr > \|t\| |
|---|---|---|---|---|---|
| Intercept | 1 | -133.04224 | 14.90439 | -8.93 | <.0001 |
| YEAR | 1 | 1.57818 | 0.16554 | 9.53 | <.0001 |

**Figure 3.2.3.** Least squares estimates for the geo data, initial data (*year, number*) and transformed data (*year, $\sqrt{number}$*).

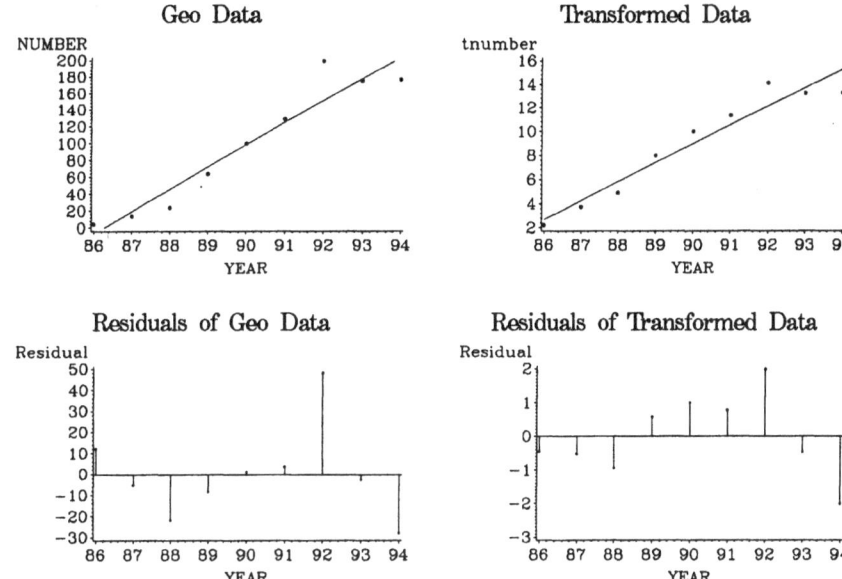

**Figure 3.2.4.** Scatterplots with regression lines and residual plots of the geo
data, initial data (*year, number*) and transformed data (*year, $\sqrt{number}$*).

```
***    Program 3_2_3    ***;
TITLE1 'Linear Regression';
TITLE2 'Geo Data';
LIBNAME datalib 'c:\data';

DATA data1;
    SET datalib.geo;
    tnumber=SQRT(number);

PROC REG DATA=data1;
    MODEL number=year;
    OUTPUT OUT=regdat_o R=res_o;
    MODEL tnumber=year;
    OUTPUT OUT=regdat_t R=res_t;
RUN;

GOPTIONS NODISPLAY HTEXT=3;
SYMBOL1 V=DOT C=R I=RL;
SYMBOL2 V=DOT C=R I=NEEDLE H=0.5;
PROC GPLOT DATA=regdat_o GOUT=fig3_2_4;
    TITLE1 'Geo Data'; TITLE2;
    PLOT number*year=1 / NAME='origdata';              RUN;
    TITLE1 'Residuals of Geo Data';
    PLOT res_o*year=2 / NAME='origres' VREF=0;         RUN;
                        ↓
```

```
                                           ↑
PROC GPLOT DATA=regdat_t GOUT=fig3_2_4;
    TITLE1 'Transformed Data'; TITLE2;
    PLOT tnumber*year=1 / NAME='tdata';                    RUN;
    TITLE1 'Residuals of Transformed Data';
    PLOT res_t*year=2 / NAME='tres' VREF=0;                RUN;

GOPTIONS DISPLAY;
PROC GREPLAY NOFS IGOUT=fig3_2_4 TC=SASHELP.TEMPLT;
    TEMPLATE=L2R2S;
        TREPLAY 1:origdata 2:origres 3:tdat 4:tres;
RUN; DELETE _ALL_; QUIT;
```

In the DATA step the observations of the original variable 'number' are transformed to their square roots by the function SQRT. These values are assigned to the variable 't-number' and written to the temporary file 'data1'.

The subsequent REG procedure shows that several MODEL statements can be included. The result of each regression is written to the file specified by the 'OUT=*dataset*'

option in the OUTPUT statement. Due to the option 'R=*name*', these OUTPUT files contain the residuals in the named variable.

The option I=NEEDLE in the SYMBOL statement draws a vertical line from each data point to a horizontal line at the 0 value on the vertical axis or at the minimum value on the vertical axis if it is greater than 0.

The RUN statements in PROC GPLOT display the specified titles.

We assumed in our standard model (3.3) that the errors $\varepsilon_i$ have identical variances $\sigma^2$. Such errors are called *homoscedastic*. Otherwise, they are called *heteroscedastic*. If the assumption of homoscedasticity is actually violated, then the use of estimators, tests or confidence intervals, which are built on this assumption, might lead to erroneous conclusions.

The least squares estimates $(\hat{\beta}_0, \hat{\beta}_1)$ are obviously quite sensitive to outliers, each has breakdown point $1/n$. For this reason, the interest in robust alternatives such as the *least median squares estimator* $(\hat{b}_0, \hat{b}_1)$ of $(\beta_0, \beta_1)$ has been increasing. The pair $(\hat{b}_0, \hat{b}_1)$ minimizes the median of the residual squares $(y_i - b_0 - b_1 x_i)^2$, $i = 1, \ldots, n$. Its breakdown point in the model (3.3) is $1/2$ for an even sample size and $1/2 - 1/(2n)$ for an odd $n$, cf Theorem 2 in Section 4, Chapter 3 in Rousseeuw and Leroy (1987). The least median estimator is, therefore, quite robust against outliers. It can, however, be quite sensitive to small changes in the central part of the data, cf Hettmansperger and Sheather (1992).

## 3.3 Multiple Linear Regression

In more complex situations the variable $Y$ may depend on several variables $x_1, \ldots, x_{p-1}$. In this case it is advisable to use the matrix notation for the description of the *multiple linear regression* model and the pertaining least squares estimators.

We consider, therefore, in the following one dependent variable and $p-1$ explanatory ones, $p \geq 2$, i.e., we consider the data

$$(y_i, x_{i1}, \ldots, x_{ip-1})^T \in I\!\!R^p, \qquad i = 1, \ldots, n.$$

**3.3.1 Example** (Economy Data; Institut der deutschen Wirtschaft (1993), No. 147–158). The percentages of the unemployed (*unempld*) in twenty industrial nations is to be explained by a linear model of the following variables: The increase of the gross national product within one year (*gnp*), the inflation (*inflatn*), the public investment rate (*invest*), the tax rate (*tax*) in percentage of the gross national product, the population size (*populatn*) in millions, the average labor cost (*labcost*) per hour, the number of days on strike (*strike*) per 1000 employed and the number of running nuclear power stations (*nukes*), i.e.,

$$unempld = \beta_0 + \beta_1 \ gnp + \beta_2 \ inflatn + \beta_3 \ invest + \beta_4 \ tax + \beta_5 \ populatn$$
$$+\beta_6 \ labcost + \beta_7 \ strike + \beta_8 \ nukes.$$

If not explicitly mentioned, all variables are given in percent and concern the year 1990. Missing observations in the data were estimated using multiple linear regression as described in Example 3.3.4.

If we assume a linear relationship between the $y_i$ and $x_{i1}, \ldots, x_{ip-1}$, then we can fit a hyperplane to the set of data in $I\!\!R^p$. This generalizes the fit of a straight line to a set of points in $I\!\!R^2$. We will require in the sequel that the number $n$ of the data is larger than the number $p-1$ of explanatory variables, i.e., we will require $p \leq n$. This is obviously a plausible and very weak condition. But it also has a strong mathematical impact, cf Lemma 3.3.3.

We want to determine, therefore, those real numbers $\beta_0, \beta_1, \ldots, \beta_{p-1}$ such that in the ideal case, the data points in $I\!\!R^p$ satisfy the equations

$$y_i = \beta_0 + \beta_1 x_{i1} + \ldots + \beta_{p-1} x_{ip-1}, \qquad i = 1, \ldots, n.$$

Put

$$\boldsymbol{y} := \begin{pmatrix} y_1 \\ \vdots \\ y_n \end{pmatrix}, \quad \boldsymbol{\beta} := \begin{pmatrix} \beta_0 \\ \vdots \\ \beta_{p-1} \end{pmatrix}$$

and

$$X := \begin{pmatrix} 1 & x_{11} & \ldots & x_{1p-1} \\ \vdots & \vdots & & \vdots \\ 1 & x_{n1} & \ldots & x_{np-1} \end{pmatrix}.$$

$y$ and $\beta$ are vectors of dimension $n$ and $p$, respectively, and $X$ is a $n \times p$-matrix, called *design matrix*. By using matrix notation, the above $n$ equations can be written in closed form as $y = X\beta$.

## The Normal Equations

The method of least squares aims at finding that hyperplane or, equivalently, that vector $\beta$ such that the residual sum of squares is minimized:

$$R(\beta) := \sum_{i=1}^{n}(y_i - \beta_0 - \beta_1 x_{i1} - \cdots - \beta_{p-1} x_{ip-1})^2$$

$$= \|y - X\beta\|^2 = \text{ minimum.}$$

By

$$\|z\| := (z^T z)^{1/2} = \sqrt{z_1^2 + \cdots + z_n^2}$$

we denote the usual length or *Euclidian norm* of an arbitrary vector $z = (z_1, \ldots, z_n)^T \in \mathbb{R}^n$.

**3.3.2 Example.** In the case $p = 2$, where we have to fit a straight line to the set of points $\{(x_i, y_i)^T, i = 1, \ldots, n\}$ in $\mathbb{R}^2$, we obtain

$$y = \begin{pmatrix} y_1 \\ \vdots \\ y_n \end{pmatrix}, \quad \hat{\beta} := \begin{pmatrix} \hat{\beta}_0 \\ \hat{\beta}_1 \end{pmatrix}, \quad X = \begin{pmatrix} 1 & x_1 \\ \vdots & \vdots \\ 1 & x_n \end{pmatrix}$$

and for the vector of the empirical residuals

$$\hat{\varepsilon} := y - \hat{y} := y - X\hat{\beta} = \begin{pmatrix} y_1 - \hat{\beta}_0 - \hat{\beta}_1 x_1 \\ \vdots \\ y_n - \hat{\beta}_0 - \hat{\beta}_1 x_n \end{pmatrix}.$$

Using partial derivatives and equating them to zero as in (3.2), we obtain that a minimizing vector $\beta$ of the function $R$ is a solution of the equations

$$n\beta_0 + \beta_1 \sum_{i=1}^{n} x_{i1} + \cdots + \beta_{p-1} \sum_{i=1}^{n} x_{ip-1} = \sum_{i=1}^{n} y_i,$$

$$\beta_0 \sum_{i=1}^{n} x_{ik} + \beta_1 \sum_{i=1}^{n} x_{i1} x_{ik} + \cdots + \beta_{p-1} \sum_{i=1}^{n} x_{ip-1} x_{ik} = \sum_{i=1}^{n} y_i x_{ik}, \quad (3.4)$$

$k = 1, \ldots, p - 1$. Using matrix notation, we can write these equations as

$$X^T X \beta = X^T y. \tag{3.5}$$

The equations (3.4) or, equivalently, (3.5) are called *normal equations*. If the $p \times p$–matrix $X^T X$ is invertible, then

$$\hat{\beta} := (X^T X)^{-1} X^T y \tag{3.6}$$

is obviously the unique solution of (3.5). The vector $\hat{\beta} \in \mathbb{R}^p$ is again called *least squares estimate*.

**3.3.3 Lemma.** *The rank of $X^T X$ coincides with the rank of $X$. The normal equations (3.5) have a unique solution if and only if $X^T X$ is invertible, i.e., if it has full rank $p$, which is equivalent to the condition that the $p$ columns of $X$ are linearly independent. Here we assume that $p \leq n$.*

**Proof:** The linear functions, which are defined by the two matrices $X$ and $X^T X$, have the same kernels. Recall that the kernel of a matrix or of a linear function is the subset of the elements that are mapped onto 0. This can easily be seen as follows: $Xu = 0$ implies $X^T Xu = 0$ for $u \in \mathbb{R}^p$; the equation $X^T Xu = 0$ implies on the other hand $\|Xu\|^2 = u^T X^T Xu = 0$, i.e., $Xu = 0$. Therefore, the kernels coincide.

Recall now that the rank of an arbitrary linear function $L : \mathbb{R}^p \to \mathbb{R}^s$ satisfies the equation

$$\text{rank } L = p - \text{dimension of the kernel of } L,$$

cf Section 3.5 in Kwak and Hong (1997). The fact that the kernels of $X$ and $X^T X$ coincide implies, therefore,

$$\text{rank } X = \text{rank } (X^T X). \qquad \square$$

**3.3.4 Example** (Air Pollution Data). Monthly averages of the following aerial toxic agents were measured at various locations in Bavaria for the two months July 1993 and April 1994: sulfur dioxide (*so2*), carbon monoxide (*co*), nitrogen oxide (*no*), nitrogen dioxide (*no2*), ozone (*o3*) and floating particles (*dust*). The first five variables were measured in milligram per cubic meter of air (mg/m$^3$), *dust* was measured in microgram per cubic meter of air ($\mu$g/m$^3$). Some of the 48 sets of data contain *missing values*, only 26 are complete. The following *scatterplot matrix* shows scatterplots of pairwise variables. There seems to be, for example, a certain linearity between ozone and nitrogen oxide.

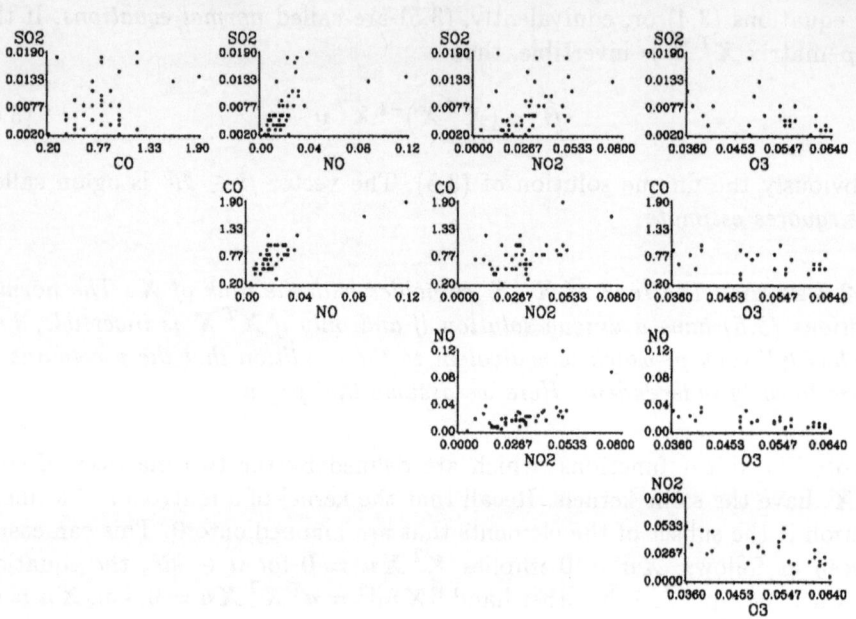

**Figure 3.3.1.** Scatterplot matrix of air pollution data.

```
***    Program 3_3_1    ***;
TITLE1 'Scatterplot Matrix';
TITLE2 'Air Pollution Data';
LIBNAME datalib 'c:\data';

GOPTIONS NODISPLAY;
AXIS1 MAJOR=(N=4) LABEL=(H=5.5) VALUE=(H=4);
SYMBOL1 V=DOT C=G H=1 I=NONE;
PROC GPLOT DATA=datalib.air GOUT=fig3_3_1;

    PLOT so2*( co no no2 o3)    /HAXIS=AXIS1 VAXIS=AXIS1;
    PLOT co *(    no no2 o3)    /HAXIS=AXIS1 VAXIS=AXIS1;
    PLOT no *(       no2 o3)    /HAXIS=AXIS1 VAXIS=AXIS1;
    PLOT no2*          o3       /HAXIS=AXIS1 VAXIS=AXIS1;
RUN; QUIT;
                                   ↓
```

```
                                    ↑
GOPTIONS DISPLAY;
%mkfields (4,4);
PROC GREPLAY IGOUT=fig3_3_1 TC=tempcat TEMPLATE=newtemp NOFS;
     TREPLAY 1:1   2:2    3:3    4:4
                   6:5    7:6    8:7
                          11:8   12:9
                                 16:10;
RUN; DELETE _ALL_; QUIT;
```

This program shows that the procedure GPLOT may contain several PLOT statements, and that several drawings can be plotted by just one PLOT call. The statement 'PLOT so2*(co no no2 o3)', for instance, draws four plots, where so2 is plotted against each of the variables in parentheses.

PROC GREPLAY is preceded by the statement %mkfields(4,4). Statements beginning with a '%' are interpreted as a macro, i.e., as a reusable program. The macro 'mkfields($m, n$)' is individually programmed. It generates a template with $m$ rows and $n$ columns. The name 'newtemp' is assigned to this template and it is stored in the temporary catalog 'tempcat'.

The subsequent procedure GREPLAY then makes use of this template.

The two ozone values are missing in the data measured at the city of Augsburg. We take o3 now as the dependent variable and so2, co, no, no2 and *dust* as explanatory variables. Thus, we can estimate the missing values by means of the least squares estimator $\hat{\beta}$ that is computed from the $n = 26$ complete sets of data:

$$o3 \approx \hat{\beta}_0 + \hat{\beta}_1 \, so2 + \hat{\beta}_2 \, no + \hat{\beta}_3 \, no2 + \hat{\beta}_4 \, co + \hat{\beta}_5 \, dust.$$

We obtain the rounded values

$$\hat{\beta}^T = (0.0693, \, 0.2443, \, -0.5797, \, -0.1082, \, 0.01, \, -0.0004)$$

and, thus, the two estimates

$$\widehat{o3} \, (1993) = -0.0082 \text{ and } \widehat{o3} \, (1994) = 0.0079$$

of the average measurements of ozone at Augsburg in July 1993 and April 1994. $\widehat{o3}$ (1993) should reasonably be modified to zero as negative values of ozone are impossible. This result, however, reminds us of the fact that we compute only *estimates*. The missing ozone values can be estimated for those 18 sets of data that completely contain the remaining variables. The following program computes $\hat{\beta}$, prints the ozone values and computes the estimates of these missing values. If o3 is not a missing observation in a set of data, then the program puts $\widehat{o3} = o3$.

| Intercept | SO2 | NO | NO2 | CO | DUST |
|-----------|-----|-----|------|-----|------|
| 0.069280 | 0.24429 | -0.57969 | -0.10819 | .009960620 | -.000361752 |

| NAME | O3 | O3HAT |
|------|-----|-------|
| ANSBACH | 0.057 | 0.057000 |
| ANSBACH | 0.061 | 0.061000 |
| ASCHAFFENBURG | . | . |
| ASCHAFFENBURG | . | 0.052702 |
| AUGSBURG | . | 0.007923 |
| AUGSBURG | . | -0.008198 |
| BAD REICHENHALL | 0.064 | 0.064000 |
| BAYREUTH | . | 0.046161 |
| BAYREUTH | . | 0.048551 |
| BURGHAUSEN | 0.061 | 0.061000 |
| BURGHAUSEN | 0.063 | 0.063000 |
| ERLANGEN | . | 0.049908 |
| ERLANGEN | . | 0.046546 |
| AIRPORT MUC | . | 0.058706 |
| FUERTH | . | . |
| FUERTH | . | 0.055701 |
| HOF | 0.048 | 0.048000 |
| HOF | 0.056 | 0.056000 |
| INGOLSTADT | . | 0.052492 |
| INGOLSTADT | . | 0.053304 |
| KELHEIM | . | 0.049829 |
| KELHEIM | . | 0.048712 |
| KEMPTEN | 0.063 | 0.063000 |
| KEMPTEN | 0.062 | 0.062000 |
| KULMBACH | 0.050 | 0.050000 |
| KULMBACH | 0.057 | 0.057000 |
| LANDSHUT | . | 0.053890 |
| LANDSHUT | . | 0.048897 |
| MUNICH | 0.055 | 0.055000 |
| MUNICH | 0.055 | 0.055000 |
| NEU-ULM | 0.054 | 0.054000 |
| NEU-ULM | 0.048 | 0.048000 |
| NUREMBERG | 0.041 | 0.041000 |
| NUREMBERG | 0.051 | 0.051000 |
| OBERAUDORF | 0.056 | 0.056000 |
| OBERAUDORF | 0.048 | 0.048000 |

| | | |
|---|---|---|
| PASSAU | 0.037 | 0.037000 |
| PASSAU | 0.040 | 0.040000 |
| REGENSBURG | 0.036 | 0.036000 |
| REGENSBURG | 0.039 | 0.039000 |
| SCHWEINFURT | 0.049 | 0.049000 |
| SCHWEINFURT | 0.041 | 0.041000 |
| TROSTBERG | 0.062 | 0.062000 |
| TROSTBERG | 0.060 | 0.060000 |
| WEIDEN | . | 0.043009 |
| WEIDEN | . | 0.049877 |
| WUERZBURG | . | 0.047014 |
| WUERZBURG | 0.054 | 0.054000 |

**Figure 3.3.2.** Least squares estimates for the air pollution data and estimates of the missing values of ozone.

```
***   Program 3_3_2   ***;
TITLE1 'Estimates of Missing Ozone Values';
TITLE2 'Air Pollution Data';
LIBNAME datalib 'c:\data';

PROC REG DATA=datalib.air OUTEST=regdata NOPRINT;
    o3hat: MODEL o3=so2 no no2 co dust;
PROC PRINT DATA=regdata NOOBS;
    VAR INTERCEPT so2 no no2 co dust;
PROC SCORE DATA=datalib.air SCORE=regdata TYPE=PARMS OUT=air1;
    VAR so2 no no2 co dust;

DATA datalib.air1;
    SET air1;
    IF o3 ^=. THEN o3hat=o3;   * ' ^=' means 'unequal';
PROC SORT;
    BY name;
PROC PRINT DATA=datalib.air1;
    VAR o3 o3hat;
    ID name;
RUN; QUIT;
```

This program shows how missing values in a data set can be substituted by linear regression estimates.

The results of PROC REG are written to the temporary file 'regdata' by the option 'OUTEST=regdata'. By '*label*: MODEL ...' one can name the resulting model (here: 'o3hat: MODEL ...'). Note that a regression analysis is computed only for those data sets, which contain no missing observations of those variables specified in the MODEL statement.

PROC PRINT prints the estimated parameters.

PROC SCORE multiplies the values of a variable in a raw data set (here 'DATA=datalib.air') by the matching coefficients in the data set of scoring coeffi-

cients (here 'SCORE=regdata'). The resulting products are then summed up to produce
the value of the new score variable. This new
variable automatically is the label 'o3hat' of
the preceding regression MODEL statement.
If the scoring coefficients were computed by
PROC REG, the option 'TYPE=PARMS'
would have to be specified. Note that PROC
SCORE computes estimates of o3 also in

those data sets, where o3 is a missing value.
The option 'OUT=air1' writes the original
variable and the new one to the temporary
file 'air1'.

The IF statement in the subsequent DATA
step replaces the missing values of the variable o3 by the corresponding values of the
variable o3hat.

## Mean Vector, Cross–Covariance Matrix

**3.3.5 Definition.** Consider arbitrary integrable random variables $Y_1, \ldots, Y_n$
and put $\boldsymbol{Y} = (Y_1, \ldots, Y_n)^T$. The vector

$$\boldsymbol{\mu_Y} := E(\boldsymbol{Y}) := \begin{pmatrix} E(Y_1) \\ \vdots \\ E(Y_n) \end{pmatrix}$$

is the mean vector of $\boldsymbol{Y}$. If $Y_1, \ldots, Y_n$ are in addition square integrable then
the $n \times n$–matrix

$$\boldsymbol{\Sigma_Y} := (Cov(Y_i, Y_j))$$

is the covariance matrix of $\boldsymbol{Y}$.

**3.3.6 Theorem.** *Let* $\boldsymbol{Y} = (Y_1, \ldots, Y_n)^T$ *be a random vector,* $c \in \mathbb{R}^m$ *and* $\boldsymbol{A}$
*an arbitrary* $m \times n$*–matrix. Then we have for* $\boldsymbol{Z} := c + \boldsymbol{AY}$

$$E(\boldsymbol{Z}) = c + \boldsymbol{A}\, E(\boldsymbol{Y})$$

*if* $E(Y_i)$ *exists for* $i = 1, \ldots, n$.

**Proof:** With $\boldsymbol{A} = (a_{ij})$ and $c = (c_1, \ldots, c_m)^T$, the $i$–th component of $\boldsymbol{Z}$ is

$$Z_i = c_i + \sum_{j=1}^{n} a_{ij} Y_j.$$

This implies

$$E(Z_i) = c_i + \sum_{j=1}^{n} a_{ij}\, E(Y_j),$$

and, thus, the assertion.                                                                    □

**3.3.7 Theorem.** *Suppose, in addition to the assumptions of Theorem 3.3.6, that $E(Y_i^2) < \infty$, $i = 1, \ldots, n$. Then the covariance matrix $\Sigma_Z$ of $Z = c + AY$ is given by*

$$\Sigma_Z = A\,\Sigma_Y\,A^T.$$

**Proof:** We have for $i, j = 1, \ldots, n$

$$Cov(Z_i, Z_j) = Cov\left(\sum_{k=1}^{n} a_{ik}Y_k, \sum_{l=1}^{n} a_{jl}Y_l\right) = \sum_{k=1}^{n}\sum_{l=1}^{n} a_{ik}\,Cov(Y_k, Y_l)a_{jl},$$

i.e., $\Sigma_Z = A\,\Sigma_Y\,A^T$. □

**3.3.8 Example.** Assume that the random vector $Y = (Y_1, \ldots, Y_n)^T$ has the covariance matrix $\Sigma_Y = \sigma^2 I_n$, where $I_n$ is the $n \times n$ identity matrix and $\sigma^2 > 0$. We choose an integer $k \in \{0, 1, \ldots, n-1\}$ and put $Z_i := (Y_i + Y_{i+1} + \cdots + Y_{i+k})/(k+1)$. Then $Z_1, \ldots, Z_{n-k}$ are a *moving average*. For $Z = (Z_1, \ldots, Z_{n-k})^T$ we obtain the representation

$$Z = AY,$$

where

$$A = \frac{1}{k+1}\begin{pmatrix} 1 & \cdots & 1 & 0 & \cdots & & 0 \\ 0 & 1 & \cdots & 1 & 0 & \cdots & 0 \\ \vdots & & & \ddots & & & \vdots \\ 0 & \cdots & & 0 & \underbrace{1 \quad \cdots \quad 1}_{k+1} \end{pmatrix}$$

is a $(n-k) \times n$–matrix. From Theorem 3.3.7 we obtain for the $(n-k) \times (n-k)$–covariance matrix $\Sigma_Z = \sigma^2 A I_n A^T$ of $Z$:

$$\Sigma_Z = \frac{\sigma^2}{(k+1)^2}\begin{pmatrix} k+1 & k & k-1 & \cdots & 1 & 0 & 0 & \cdots & 0 \\ k & k+1 & k & \cdots & 2 & 1 & 0 & \cdots & 0 \\ \vdots & & \ddots & & & & & & \vdots \\ 1 & 2 & 3 & \cdots & k+1 & k & k-1 & \cdots & 0 \\ 0 & 1 & 2 & \cdots & k & k+1 & k & \cdots & 0 \\ \vdots & & & & & & \ddots & & \vdots \\ 0 & & & \cdots & & & \cdots & k+1 & k \\ 0 & 0 & 0 & \cdots & & & \cdots & k & k+1 \end{pmatrix}.$$

**3.3.9 Definition.** Let $A = (a_{ij})$ be an arbitrary $n \times n$–matrix. The function

$$\mathbb{R}^n \ni x = (x_1, \ldots, x_n)^T \longmapsto x^T A x = \sum_{i=1}^{n}\sum_{j=1}^{n} x_i\,a_{ij}\,x_j$$

is a *quadratic form*.

In the next theorem we compute the expectation of a *random* quadratic form $\boldsymbol{Z}^T \boldsymbol{A} \boldsymbol{Z}$, where $\boldsymbol{Z} = (Z_1, \ldots, Z_n)^T$ is an $n$ dimensional random vector.

**3.3.10 Theorem.** *Let $\boldsymbol{Z} = (Z_1, \ldots, Z_n)^T$ be a random vector with mean vector $\boldsymbol{\mu_Z} = (\mu_1, \ldots, \mu_n)^T$ and $n \times n$–covariance matrix $\boldsymbol{\Sigma_Z} = (\sigma_{ij})$. Then we have for an arbitrary $n \times n$–matrix $\boldsymbol{A}$*

$$E(\boldsymbol{Z}^T \boldsymbol{A} \boldsymbol{Z}) = \; tr\left(\boldsymbol{A} \boldsymbol{\Sigma_Z}\right) + \boldsymbol{\mu_Z}^T \boldsymbol{A} \boldsymbol{\mu_Z}.$$

**Proof:** The trace of a square matrix is the sum of the entries in its leading diagonal. From the representation

$$E(Z_i Z_j) = \sigma_{ij} + \mu_i \, \mu_j$$

we obtain

$$E(\boldsymbol{Z}^T \boldsymbol{A} \boldsymbol{Z}) = E\left( \sum_{i=1}^n \sum_{j=1}^n Z_i \, a_{ij} \, Z_j \right)$$

$$= \sum_{i=1}^n \sum_{j=1}^n a_{ij} \, \sigma_{ij} + \sum_{i=1}^n \sum_{j=1}^n \mu_i \, \mu_j \, a_{ij} = \; tr\left(\boldsymbol{A} \boldsymbol{\Sigma_Z}\right) + \boldsymbol{\mu_Z}^T \boldsymbol{A} \boldsymbol{\mu_Z}.$$

Note that the matrix $\boldsymbol{\Sigma_Z}$ is symmetric, i.e., we have $\sigma_{ij} = \sigma_{ji}$, $i, j = 1, \ldots, n$.
$\square$

**3.3.11 Example.** An arbitrary random vector $\boldsymbol{Z} = (Z_1, \ldots, Z_n)^T$ satisfies

$$\bar{Z}_n := n^{-1} \sum_{i=1}^n Z_i = n^{-1} \boldsymbol{1}^T \boldsymbol{Z},$$

where $\boldsymbol{1} = (1, \ldots, 1)^T \in I\!\!R^n$. Putting

$$\boldsymbol{A} := \boldsymbol{I}_n - n^{-1} \boldsymbol{1} \boldsymbol{1}^T = \begin{pmatrix} 1 & & 0 \\ & \ddots & \\ 0 & & 1 \end{pmatrix} - n^{-1} \begin{pmatrix} 1 & \cdots & 1 \\ \vdots & & \vdots \\ 1 & \cdots & 1 \end{pmatrix},$$

we obtain

$$\sum_{i=1}^n (Z_i - \bar{Z}_n)^2 = \|\boldsymbol{A} \boldsymbol{Z}\|^2 = \boldsymbol{Z}^T \boldsymbol{A}^T \boldsymbol{A} \boldsymbol{Z}.$$

The matrix $\boldsymbol{A}$ is symmetric and idempotent, i.e., $\boldsymbol{A}^T \boldsymbol{A} = \boldsymbol{A} \boldsymbol{A} = \boldsymbol{A}$. This implies

$$\sum_{i=1}^n (Z_i - \bar{Z}_n)^2 = \boldsymbol{Z}^T \boldsymbol{A} \boldsymbol{Z}.$$

The sample covariance is, consequently, a random quadratic form. Suppose that the random variables $Z_1, \ldots, Z_n$ are uncorrelated with identical means $\mu$ and variances $\sigma^2$. Theorem 3.3.10 then implies

$$E\left(\sum_{i=1}^{n}(Z_i - \bar{Z}_n)^2\right) = E(\mathbf{Z}^T \mathbf{A} \mathbf{Z}) = \sigma^2 \, tr(\mathbf{A}) + \boldsymbol{\mu_Z}^T \mathbf{A} \boldsymbol{\mu_Z} = \sigma^2(n-1),$$

since $tr(\mathbf{A}) = n - 1$ and $\boldsymbol{\mu_Z}^T \mathbf{A} \boldsymbol{\mu_Z} = \mu^2 \mathbf{1}^T \mathbf{A} \mathbf{1} = 0$. In this case the sample covariance $(n-1)^{-1} \sum_{i=1}^{n} (Z_i - \bar{Z}_n)^2$ is an unbiased estimator of $\sigma^2$, cf Exercise 7.

**3.3.12 Definition.** Let $\mathbf{Y} = (Y_1, \ldots, Y_p)^T$ and $\mathbf{W} = (W_1, \ldots, W_m)^T$ be random vectors. We assume that $E(Y_i^2) < \infty$ and $E(W_j^2) < \infty$ for $i = 1, \ldots, p$ and $j = 1, \ldots, m$. The $p \times m$–matrix

$$\Sigma_{YW} := (Cov(Y_i, W_j))$$

is the *cross–covariance matrix of* $\mathbf{Y}$ *and* $\mathbf{W}$.

**3.3.13 Theorem.** *Consider a random vector* $\mathbf{Z} = (Z_1, \ldots, Z_n)^T$ *with covariance matrix* $\Sigma_Z$. *Let* $\mathbf{A}$ *be an arbitrary* $p \times n$–*matrix and let* $\mathbf{B}$ *be an arbitrary* $m \times n$–*matrix. The cross–covariance matrix* $\Sigma_{YW}$ *of* $\mathbf{Y} := \mathbf{A}\mathbf{Z}$ *and* $\mathbf{W} := \mathbf{B}\mathbf{Z}$ *is*

$$\Sigma_{YW} = \mathbf{A}\Sigma_Z \mathbf{B}^T.$$

**Proof:** Exercise 24. □

**3.3.14 Example.** Let $\mathbf{Z} = (Z_1, \ldots, Z_n)^T$ be a random vector with $E(\mathbf{Z}) = \mu \mathbf{1}$ and $\Sigma_Z = \sigma^2 \mathbf{I}_n$, where $\mu \in \mathbb{R}$ and $\sigma^2 > 0$. The random variables $Z_1, \ldots, Z_n$ have identical means and variances, and they are uncorrelated. By putting $\mathbf{A} := \mathbf{I}_n - n^{-1} \mathbf{1} \mathbf{1}^T$ and $\mathbf{B} := n^{-1} \mathbf{1}^T$ we obtain from Example 3.3.11

$$\mathbf{Y} := \mathbf{A}\mathbf{Z} = (Z_1 - \bar{Z}_n, \ldots, Z_n - \bar{Z}_n)^T, \qquad \mathbf{W} := \mathbf{B}\mathbf{Z} = \bar{Z}_n.$$

Theorem 3.3.13 implies that the cross–covariance matrix $\Sigma_{YW}$ of dimension $n \times 1$ equals

$$\Sigma_{YW} = (\mathbf{I}_n - n^{-1} \mathbf{1} \mathbf{1}^T)(\sigma^2 \mathbf{I}_n) n^{-1} \mathbf{1} = 0.$$

Under the above conditions on $\mathbf{Z}$, the sample mean $\bar{Z}_n$ and the deviations $Z_i - \bar{Z}_n$ from the mean are, therefore, uncorrelated for each $i = 1, \ldots, n$.

## The Multiple Standard Model

In the sequel we will derive several statistical properties of the least squares estimator in the case of multiple linear regression, thus generalizing results of the preceding section. We will assume the following *standard model of multiple linear regression*:

$$Y_i = \beta_0 + \sum_{j=1}^{p-1} \beta_j \, x_{ij} + \varepsilon_i, \qquad i = 1, \ldots, n,$$

or, in matrix notation

$$Y = X\beta + \varepsilon. \tag{3.7}$$

The error terms $\varepsilon_i$ are assumed to be uncorrelated random variables with identical means zero and identical variances viz.,

$$E(\varepsilon_i) = 0, \quad Var(\varepsilon_i) = \sigma^2, \qquad i = 1, \ldots, n,$$
$$Cov(\varepsilon_i, \varepsilon_j) = 0, \qquad i \neq j,$$

or, in matrix notation

$$E(\varepsilon) = 0, \; \Sigma_\varepsilon = \sigma^2 I_n.$$

Note that we require the errors $\varepsilon_i$ to be uncorrelated but not necessarily independent. We assume further that the design matrix

$$X = \begin{pmatrix} 1 & x_{11} & \cdots & x_{1p-1} \\ \vdots & \vdots & & \vdots \\ 1 & x_{n1} & \cdots & x_{np-1} \end{pmatrix}$$

has rank $p$. The columns of $X$ are in this case linearly independent vectors in $\mathbb{R}^n$ and, thus, the vector $\beta$ in (3.5) is uniquely determined.

## Moments of the Least Squares Estimator

**3.3.15 Theorem.** *In model (3.7), the least squares estimator*

$$\hat{\beta} = (X^T X)^{-1} X^T Y$$

*of $\beta$ is unbiased, i.e., $E(\hat{\beta}) = \beta$.*

**Proof:** The model $Y = X\beta + \varepsilon$ implies

$$\hat{\beta} = (X^T X)^{-1} X^T (X\beta + \varepsilon) = (X^T X)^{-1} X^T X\beta + (X^T X)^{-1} X^T \varepsilon$$
$$= \beta + (X^T X)^{-1} X^T \varepsilon.$$

The assertion is now immediate from Theorem 3.3.6.                            □

**3.3.16 Theorem.** *The covariance matrix $\Sigma_{\hat{\beta}}$ of $\hat{\beta}$ in the standard model (3.7) is*

$$\Sigma_{\hat{\beta}} = \sigma^2 (X^T X)^{-1}.$$

**Proof:** Since $\hat{\beta} = \beta + (X^T X)^{-1} X^T \varepsilon$, Theorem 3.3.7 implies that the covariance matrix $\Sigma_{\hat{\beta}}$ is given by

$$
\begin{aligned}
\Sigma_{\hat{\beta}} &= (X^T X)^{-1} X^T \Sigma_\varepsilon ((X^T X)^{-1} X^T)^T \\
&= (X^T X)^{-1} X^T \Sigma_\varepsilon X (X^T X)^{-1} \\
&= (X^T X)^{-1} X^T (\sigma^2 I_n) X (X^T X)^{-1} = \sigma^2 (X^T X)^{-1}.
\end{aligned}
$$

Note that $((X^T X)^{-1})^T = ((X^T X)^T)^{-1} = (X^T X)^{-1}$. $\qquad\square$

The assumption that the errors $\varepsilon_i$ are uncorrelated also yielded the computation of the covariance matrix of the least squares estimator $(\hat{\beta}_0, \hat{\beta}_1)^T$ in the preceding section, cf Theorem 3.2.4.

**3.3.17 Example.** In the case $p = 2$ of an univariate linear regression

$$
Y_i = \beta_0 + \beta_1 x_i + \varepsilon_i, \qquad i = 1, \ldots, n,
$$

we have

$$
X = \begin{pmatrix} 1 & x_1 \\ \vdots & \vdots \\ 1 & x_n \end{pmatrix},
$$

and, hence,

$$
X^T X = \begin{pmatrix} 1 & \cdots & 1 \\ x_1 & \cdots & x_n \end{pmatrix} \begin{pmatrix} 1 & x_1 \\ \vdots & \vdots \\ 1 & x_n \end{pmatrix} = \begin{pmatrix} n & \sum_{i=1}^{n} x_i \\ \sum_{i=1}^{n} x_i & \sum_{i=1}^{n} x_i^2 \end{pmatrix}.
$$

By Theorem 3.3.16, the least squares estimator $\hat{\beta} = (\hat{\beta}_0, \hat{\beta}_1)^T$ has the covariance matrix

$$
\sigma^2 (X^T X)^{-1} = \frac{\sigma^2}{n \sum_{i=1}^{n} x_i^2 - (\sum_{i=1}^{n} x_i)^2} \begin{pmatrix} \sum_{i=1}^{n} x_i^2 & -\sum_{i=1}^{n} x_i \\ -\sum_{i=1}^{n} x_i & n \end{pmatrix}.
$$

And by (3.6), $\hat{\beta}$ has the representation

$$
\hat{\beta} = (X^T X)^{-1} X^T Y = (X^T X)^{-1} \begin{pmatrix} \sum_{i=1}^{n} Y_i \\ \sum_{i=1}^{n} x_i Y_i \end{pmatrix}.
$$

$$= \frac{1}{n \sum_{i=1}^{n} x_i^2 - (\sum_{i=1}^{n} x_i)^2} \left( \begin{array}{c} (\sum_{i=1}^{n} x_i^2)(\sum_{i=1}^{n} Y_i) - (\sum_{i=1}^{n} x_i)(\sum_{i=1}^{n} x_i Y_i) \\ n \sum_{i=1}^{n} x_i Y_i - (\sum_{i=1}^{n} x_i)(\sum_{i=1}^{n} Y_i) \end{array} \right).$$

This coincides with (3.2).

## The Gauss–Markov Theorem

In the standard model (3.7), the least squares estimator $\hat{\beta} = (X^T X)^{-1} X^T Y$ has the least variance among all linear and unbiased estimators of $\beta$, i.e., $\hat{\beta}$ is in this sense a *best linear unbiased estimator* or *BLUE*, for short. This is the content of the following *Gauss–Markov Theorem*.

**3.3.18 Theorem.** *In the standard model (3.7), any unbiased estimator $\tilde{\beta} = (\tilde{\beta}_0, \ldots, \tilde{\beta}_{p-1})^T$ of $\beta$, of the form $\tilde{\beta} = AY$, where $A$ is a $p \times n$-matrix, satisfies the inequality*

$$Var(\hat{\beta}_j) \leq Var(\tilde{\beta}_j), \qquad j = 0, 1, \ldots, p - 1.$$

**Proof:** Theorem 3.3.6 and the unbiasedness of $\tilde{\beta}$ imply that for any underlying parameter $\beta \in \mathbb{R}^p$

$$\beta = E(\tilde{\beta}) = AE(Y) = AX\beta,$$

i.e., $(I_p - AX)\beta = 0$. Since this equation holds for *any* $\beta \in \mathbb{R}^p$, we obtain $I_p = AX$. Let $v$ be an arbitrary vector in $\mathbb{R}^p$. We will prove the inequality

$$Var(v^T \hat{\beta}) \leq Var(v^T \tilde{\beta}).$$

Suitable choices of $v$ then imply the assertion. By Theorem 3.3.7 and Theorem 3.3.16 we obtain

$$Var(v^T \hat{\beta}) = \sigma^2 v^T (X^T X)^{-1} v$$

and

$$Var(v^T \tilde{\beta}) = Var(v^T AY) = Var(v^T A\varepsilon) = \sigma^2 v^T AA^T v.$$

The above inequality is, consequently, established if

$$v^T (AA^T - (X^T X)^{-1}) v \geq 0.$$

The equation $AX = I_p$ together with the symmetry of $X^T X$ implies

$$AA^T - (X^T X)^{-1} = (A - (X^T X)^{-1} X^T)(A - (X^T X)^{-1} X^T)^T.$$

This equation implies immediately that the matrix $\boldsymbol{A}\boldsymbol{A}^T - (\boldsymbol{X}^T\boldsymbol{X})^{-1}$ is positive semidefinite and, thus, yields the required inequality: Let $\boldsymbol{B}$ be an arbitrary $p \times n$–matrix. Then we have with $\boldsymbol{w} := \boldsymbol{B}^T\boldsymbol{v} \in \mathbb{R}^n$

$$v^T \boldsymbol{B}\boldsymbol{B}^T v = w^T w \geq 0. \qquad \square$$

## The Hat Matrix

The unknown parameter $\sigma^2$ in our standard model (3.7) is the expectation of each $\varepsilon_i^2$. Since the empirical residuals are estimates of the errors $\varepsilon_i$, the obvious choice is to take the residual sum of squares as an estimator of $\sigma^2$, just like we did in Theorem 3.2.5 in the univariate case. But the multiple regression model offers additional insight into the mathematical properties of the least squares estimator as we will see below. The vector of the empirical residuals is

$$\hat{\varepsilon} = \boldsymbol{Y} - \boldsymbol{X}\hat{\beta} = \boldsymbol{Y} - \boldsymbol{X}(\boldsymbol{X}^T\boldsymbol{X})^{-1}\boldsymbol{X}^T\boldsymbol{Y} = \boldsymbol{Y} - \boldsymbol{P}\boldsymbol{Y},$$

where $\boldsymbol{P} := \boldsymbol{X}(\boldsymbol{X}^T\boldsymbol{X})^{-1}\boldsymbol{X}^T$ is a $n \times n$–matrix.

It turns out that the matrix $\boldsymbol{P}$ is actually a *projection matrix*. Such a matrix is defined by the conditions that it is symmetric and idempotent, cf the next lemma and Exercises 27 and 28. By the equation $\boldsymbol{P}\boldsymbol{Y} = \boldsymbol{X}\hat{\beta}$ it projects the vector $\boldsymbol{Y}$ into that $p$–dimensional linear subspace of $\mathbb{R}^n$ which is spanned by the columns of the matrix $\boldsymbol{X}$. Recall that we assume that these $p$ columns are linearly independent vectors in $\mathbb{R}^n$. By the common notation $\hat{\boldsymbol{Y}} := \boldsymbol{P}\boldsymbol{Y} = \boldsymbol{X}\hat{\beta}$ for the prediction of $\boldsymbol{Y}$ by $\boldsymbol{X}\hat{\beta}$ one usually calls $\boldsymbol{P}$ the *hat matrix*: The multiplication of $\boldsymbol{Y}$ by the matrix $\boldsymbol{P}$ puts a hat on $\boldsymbol{Y}$. One will, therefore, expect in particular that $\boldsymbol{P}\hat{\boldsymbol{Y}} = \hat{\boldsymbol{Y}}$. The following lemma shows that this is actually true.

**3.3.19 Theorem.** *The hat matrix* $\boldsymbol{P} = \boldsymbol{X}(\boldsymbol{X}^T\boldsymbol{X})^{-1}\boldsymbol{X}^T$ *is a projection matrix, i.e., it satisfies* $\boldsymbol{P} = \boldsymbol{P}^T = \boldsymbol{P}^2$.

**Proof:** Exercise 25. $\qquad \square$

## Residual Sum of Squares

For the residual sum of squares $RSS$, from Lemma 3.3.19 with $\hat{\boldsymbol{Y}} = \boldsymbol{X}\hat{\beta} = (\hat{Y}_i)_{1 \leq i \leq n} = \boldsymbol{P}\boldsymbol{Y}$, we obtain the representation

$$RSS := \sum_{i=1}^{n}(Y_i - \hat{Y}_i)^2 = \|\boldsymbol{Y} - \boldsymbol{P}\boldsymbol{Y}\|^2 = \|(\boldsymbol{I}_n - \boldsymbol{P})\boldsymbol{Y}\|^2$$

$$= \boldsymbol{Y}^T(\boldsymbol{I}_n - \boldsymbol{P})^T(\boldsymbol{I}_n - \boldsymbol{P})\boldsymbol{Y} = \boldsymbol{Y}^T(\boldsymbol{I}_n - \boldsymbol{P})\boldsymbol{Y}.$$

$RSS$ is, therefore, a random quadratic form. By Theorem 3.3.10 we can now compute the expectation of $RSS$ in a simple manner.

**3.3.20 Lemma.** *In the standard model (3.7) we have*

$$E(RSS) = E(||\boldsymbol{Y} - \boldsymbol{PY}||^2) = (n - p)\sigma^2.$$

**Proof:** From Theorem 3.3.10 we obtain

$$\begin{aligned} E(RSS) &= E(\boldsymbol{Y}^T(\boldsymbol{I}_n - \boldsymbol{P})\boldsymbol{Y}) \\ &= \sigma^2 \, tr(\boldsymbol{I}_n - \boldsymbol{P}) + E(\boldsymbol{Y})^T(\boldsymbol{I}_n - \boldsymbol{P})E(\boldsymbol{Y}), \end{aligned}$$

where $E(\boldsymbol{Y}) = \boldsymbol{X\beta}$ and, thus,

$$(\boldsymbol{I}_n - \boldsymbol{P})E(\boldsymbol{Y}) = (\boldsymbol{I}_n - \boldsymbol{X}(\boldsymbol{X}^T\boldsymbol{X})^{-1}\boldsymbol{X}^T)\boldsymbol{X\beta} = (\boldsymbol{X} - \boldsymbol{X})\boldsymbol{\beta} = \boldsymbol{0}.$$

It remains to show that $tr(\boldsymbol{I}_n - \boldsymbol{P}) = n - p$. We have $tr(\boldsymbol{I}_n - \boldsymbol{P}) = tr(\boldsymbol{I}_n) - tr(\boldsymbol{P}) = n - tr(\boldsymbol{P})$. Note that in general $tr(\boldsymbol{AB}) = tr(\boldsymbol{BA})$ for an arbitrary $m \times n$–matrix $\boldsymbol{A}$ and a $n \times m$ matrix $\boldsymbol{B}$, see Exercise 26. This implies that

$$tr(\boldsymbol{P}) = tr\left(\underbrace{\boldsymbol{X}(\boldsymbol{X}^T\boldsymbol{X})^{-1}}_{n \times p}\underbrace{\boldsymbol{X}^T}_{p \times n}\right) = tr(\boldsymbol{X}^T\boldsymbol{X}(\boldsymbol{X}^T\boldsymbol{X})^{-1})$$

$$= tr(\boldsymbol{I}_p) = p$$

which completes the proof.                                                                     $\square$

Lemma 3.3.20 immediately implies that in the case $n > p$

$$S^2 := \frac{RSS}{n - p} = \frac{||\boldsymbol{Y} - \boldsymbol{PY}||^2}{n - p} = \frac{||\boldsymbol{Y} - \hat{\boldsymbol{Y}}||^2}{n - p} = \frac{||\boldsymbol{Y} - \boldsymbol{X}\hat{\boldsymbol{\beta}}||^2}{n - p}$$

is an unbiased estimator of $\sigma^2$ in the standard model (3.7).

**3.3.21 Corollary.** *In the standard model (3.7), we have, for $n > p$*

$$E(S^2) = \sigma^2.$$

## The Variance Decomposition

The derivation of a measure of fit of $\boldsymbol{Y}$ by $\hat{\boldsymbol{Y}}$ is based on the *variance decomposition*

$$\sum_{i=1}^{n}(Y_i - \bar{Y}_n)^2 = \sum_{i=1}^{n}\left((Y_i - \hat{Y}_i) + (\hat{Y}_i - \bar{Y}_n)\right)^2$$

$$= \sum_{i=1}^{n}\hat{\varepsilon}_i^2 + \sum_{i=1}^{n}(\hat{Y}_i - \bar{Y}_n)^2 = RSS + \sum_{i=1}^{n}(\hat{Y}_i - \bar{Y}_n)^2,$$

where $\bar{Y}_n := n^{-1} \sum_{i=1}^{n} Y_i$ is the arithmetic mean of the observations $Y_1, \ldots, Y_n$ and $(\hat{Y}_1, \ldots, \hat{Y}_n)^T = \hat{Y} = X\hat{\beta}$ is the predictor of $Y$. Thus, we have

$$total\ variation = error + model\ variation,$$

since the term $\sum_{i=1}^{n}(Y_i - \hat{Y}_i)(\hat{Y}_i - \bar{Y}_n)$ vanishes which can be seen as follows. The normal equations (3.5) immediately imply

$$\sum_{i=1}^{n}(Y_i - \hat{Y}_i) = 0 \quad \text{and} \quad \sum_{i=1}^{n}(Y_i - \hat{Y}_i)x_{ik} = 0, \qquad k = 1, \ldots, p-1.$$

This yields $\sum_{i=1}^{n}(Y_i - \hat{Y}_i)\bar{Y}_n = \bar{Y}_n \sum_{i=1}^{n}(Y_i - \hat{Y}_i) = 0$ and

$$\sum_{i=1}^{n}(Y_i - \hat{Y}_i)\hat{Y}_i = \sum_{i=1}^{n}(Y_i - \hat{Y}_i)\left(\sum_{k=0}^{p-1}x_{ik}\hat{\beta}_k\right) = \sum_{k=0}^{p-1}\hat{\beta}_k\left(\sum_{i=1}^{n}(Y_i - \hat{Y}_i)x_{ik}\right) = 0.$$

## The $R$–Square Value

The ratio

$$R^2 := \frac{\sum_{i=1}^{n}(\hat{Y}_i - \bar{Y}_n)^2}{\sum_{i=1}^{n}(Y_i - \bar{Y}_n)^2} \qquad \in [0, 1]$$

of the model variation and the total variation is used as a *measure of the model fit*. If it is close to 1, then the ratio error/total variation is close to 0 and the approximation of $Y$ by $\hat{Y}$ performs well. We have

$$R^2 = 1 - \frac{RSS}{\sum_{i=1}^{n}(Y_i - \bar{Y}_n)^2}$$

and, hence,

$$R^2 = 1 \iff RSS = 0$$

as well as

$$R^2 = 0 \iff RSS = \sum_{i=1}^{n}(Y_i - \bar{Y}_n)^2.$$

A value of $R^2$ close to 0 indicates only a minor performance of $\hat{Y}$ as an approximation of $Y$, since in this case the residual sum of squares $RSS$ equals approximately the total variation $\sum_{i=1}^{n}(Y_i - \bar{Y}_n)^2$. Finally we have

$$\sum_{i=1}^{n}(\hat{Y}_i - \bar{Y}_n)(Y_i - \bar{Y}_n) = \sum_{i=1}^{n}(\hat{Y}_i - \bar{Y}_n)\left((Y_i - \hat{Y}_i) + (\hat{Y}_i - \bar{Y}_n)\right) = \sum_{i=1}^{n}(\hat{Y}_i - \bar{Y}_n)^2,$$

since $\sum_{i=1}^{n}(\hat{Y}_i - \bar{Y}_n)(Y_i - \hat{Y}_i) = 0$ as shown above. This implies the representation

$$\frac{\sum_{i=1}^{n}(\hat{Y}_i - \bar{Y}_n)(Y_i - \bar{Y}_n)}{(\sum_{i=1}^{n}(\hat{Y}_i - \bar{Y}_n)^2)^{1/2}(\sum_{i=1}^{n}(Y_i - \bar{Y}_n)^2)^{1/2}} = \frac{(\sum_{i=1}^{n}(\hat{Y}_i - \bar{Y}_n)^2)^{1/2}}{(\sum_{i=1}^{n}(Y_i - \bar{Y}_n)^2)^{1/2}} = R.$$

Note that the arithmetic means $\bar{Y}_n = n^{-1}\sum_{i=1}^{n} Y_i$ and $n^{-1}\sum_{i=1}^{n} \hat{Y}_i$ coincide. This is due to $\sum_{i=1}^{n}(Y_i - \hat{Y}_i) = 0$. The above representation shows, therefore, that $R$ is actually the sample coefficient of correlation of $\boldsymbol{Y}$ and $\hat{\boldsymbol{Y}} = \boldsymbol{PY} = \boldsymbol{X\hat{\beta}}$. For this reason $R^2$ is also called *squared multiple coefficient of correlation*.

The residual sum of squares $RSS$ will in general decrease if the number of explanatory variables is increased. This, in turn, increases the value of $R^2$. The *adjusted* $R^2$ aims at diminishing this dependence of the number of explanatory variables. It is defined by

$$R_a^2 := 1 - \frac{n-1}{n-p}(1 - R^2).$$

Note that this value can be negative. For the prediction of ozone using the explanatory variables *so2, no, no2, co* and *dust* from the air pollution data in Example 3.3.4, we obtain from the $n = 26$ complete sets of data the values

$$\bar{y}_{26} = 0.0518,\ S = \sqrt{RSS/20} = 0.0057,\ CV = S/\bar{y}_{26} = 11.07\%,$$
$$R^2 = 0.6442,\ R_a^2 = 0.5553.$$

```
                         The REG Procedure
                          Model: MODEL1
                      Dependent Variable: O3


        Root MSE                 0.00573    R-Square      0.6442
        Dependent Mean           0.05181    Adj R-Sq      0.5553
        Coeff Var               11.06685
```

**Figure 3.3.3.** Mean, $S$ and $R^2$ values for the prediction of ozone in Example 3.3.4.

```
***    Program 3_3_3   ***;
TITLE1 'Multiple Regression';
TITLE2 'Air Pollution Data';
LIBNAME datalib 'c:\data';

PROC REG DATA=datalib.air;
    MODEL o3=so2 no no2 co dust;
    OUTPUT OUT=datalib.regdata R=res STUDENT=studres;
RUN; QUIT;
```

The option 'STUDENT=studres' in the OUTPUT statement adds the new variable 'studres' to the file 'regdata'. This variable contains the values of the studentized residuals. It is used in the subsequent Program 3_3_4.

One frequently encounters the strange effect that a linear regression analysis *without* an intercept, i.e., with predetermined $\beta_0 = 0$, has a larger $R^2$ value than the same analysis *with* an intercept, i.e., where $\beta_0 \in \mathbb{R}$ is estimated from the data. This is due to the fact that the regression hyperplane $y = x^T \tilde{\beta}$, $x \in \mathbb{R}^{p-1}$, with resulting least squares estimator $\tilde{\beta} \in \mathbb{R}^{p-1}$ without intercept, is forced to contain the origin in $\mathbb{R}^p$. The hyperplane will, therefore, usually have a steeper slope than the regression plane $y = (1, x^T)\hat{\beta}$ with intercept. This steeper slope now causes an increase of the total deviation $\sum_{i=1}^{n} (\tilde{Y}_i - \bar{Y}_n)^2$ of the projections $\tilde{Y}_i = x_i^T \tilde{\beta}$, of the $Y_i$ onto the hyperplane. The numerator of $R^2$, thus, increases. For a related discussion of $R^2$ and further references we refer to Anderson–Sprecher (1994). The evaluation of $R^2$ has to be done carefully and requires some experience.

### Correlations of the Residuals

In the standard model (3.7) the covariance matrix $\Sigma_{\hat{\varepsilon}}$ of the residual vector $\hat{\varepsilon} = Y - \hat{Y} = (I_n - P)Y$ is by Theorem 3.3.7 and Lemma 3.3.19 equal to

$$\Sigma_{\hat{\varepsilon}} = (I_n - P)(\sigma^2 I_n)(I_n - P)^T = \sigma^2(I_n - P).$$

The residuals $\hat{\varepsilon}_i$ have by Theorem 3.3.6 zero means, i.e., $E(\hat{\varepsilon}) = (I_n - P)E(Y) = 0$. Their variances $E(\hat{\varepsilon}_i^2) = \sigma^2(1 - p_{ii})$, however, will usually differ, where $p_{ii}$, $i = 1, \ldots, n$, are the entries in the main diagonal of the $n \times n$ projection matrix $P = (p_{ij})$. Only the standardized residuals $\hat{\varepsilon}_i/(\sigma(1 - p_{ii})^{1/2})$, $i = 1, \ldots, n$, have identical variances equal to 1 in the model (3.7). One can now test for homoscedasticity of the errors $\varepsilon_i$ by means of an analysis of the studentized residuals $\hat{\varepsilon}_i/(S(1 - p_{ii})^{1/2})$. The following plot shows, for example, that no outliers violate the 2– or 3–$\sigma$–rule among the residuals for the ozone prediction in Example 3.3.4.

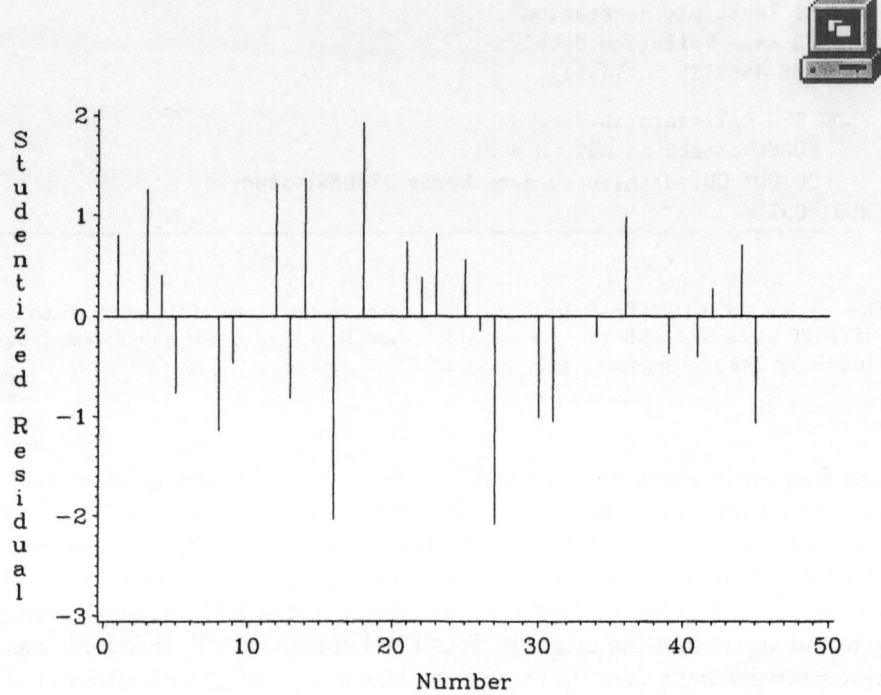

**Figure 3.3.4.** Plot of the studentized residuals for the ozone prediction in Example 3.3.4.

                          Univariate Procedure

Variable=STUDRES              Studentized Residual

                               Extremes

                  Lowest       ID            Highest      ID
              -2.08219(OBERAUDORF)      0.980768(ANSBACH    )
              -2.03549(NUREMBERG )      1.263521(MUNICH     )
              -1.13978(PASSAU    )      1.32015 (HOF        )
               -1.0663(NEU-ULM   )      1.373226(ANSBACH    )
              -1.05845(REGENSBURG)      1.930866(SCHWEINFURT)

**Figure 3.3.5.** Extrema of the studentized residuals in Example 3.3.4.

```
***    Program 3_3_4    ***;
TITLE1 'Plot of Studentized Residuals';
TITLE2 'Air Pollution Data';
LIBNAME datalib 'c:\data';

DATA regdata1;
    SET datalib.regdata;
    number =_N_;
    LABEL number ='Number';

SYMBOL1 V=NONE C=RED I=NEEDLE;
PROC GPLOT DATA=regdata1;
    PLOT studres*number;

PROC UNIVARIATE DATA=regdata1 PLOT;
    VAR studres;
    ID name;
RUN; QUIT;
```

This program uses the permanent file 'reg-data' that was created in Program 3_3_3.

Figure 3.3.5 displays parts of the output of PROC UNIVARIATE.

**3.3.22 Theorem.** *The residuals $\hat{\varepsilon}$ and the predictors $\hat{Y}$ are uncorrelated in the standard model (3.7), i.e., their cross-covariance matrix $\Sigma_{\hat{\varepsilon}\hat{Y}}$ vanishes.*

**Proof:** Writing $\hat{\varepsilon} = (I_n - P)Y$ and $\hat{Y} = PY$ we obtain from Theorem 3.3.13 and Lemma 3.3.19 that

$$\Sigma_{\hat{\varepsilon}\hat{Y}} = (I_n - P)(\sigma^2 I_n)P^T = \sigma^2(P^T - PP^T) = 0. \qquad \square$$

## The Distribution of the Least Squares Estimator

Suppose that the errors $\varepsilon_i$ in the standard model (3.7) are independent and identically normal distributed. The convolution theorem of the normal distribution 2.1.5 together with Theorems 3.3.15 and 3.3.16 then immediately imply that each component $\hat{\beta}_{i-1}$ of the least squares estimator $\hat{\beta}$ is normal distributed with mean $\beta_{i-1}$ and variance $\sigma^2 c_{ii}$, $i = 1, \ldots, p$, where $(X^T X)^{-1} = (c_{ij})$. If we estimate $\sigma^2$ by $S^2 = RSS/(n-p) = \|Y - \hat{Y}\|^2/(n-p) = \|Y - PY\|^2/(n-p)$, we will expect that $(\hat{\beta}_{i-1} - \beta_{i-1})/(S\sqrt{c_{ii}})$ is approximately standard normal distributed. Actually, we have the following result.

**3.3.23 Theorem.** *Suppose that the errors $\varepsilon_i$ in the standard model (3.7) are independent and identically $N(0, \sigma^2)$ distributed random variables. Then we have*

*(i) $\hat{\beta}$ is $N(\beta, \sigma^2(X^T X)^{-1})$ distributed,*

*(ii) $\hat{\beta}$ and $\hat{Y} = X\hat{\beta}$ are independent of RSS,*

*(iii) $RSS/\sigma^2$ is $\chi^2_{n-p}$ distributed,*

*(iv) $(\hat{\beta}_{i-1} - \beta_{i-1})/(S\sqrt{c_{ii}})$ is $t_{n-p}$ distributed, $i = 1, \ldots, p$.*

A typical application of Theorem 3.3.23 is to test the hypothesis $H_0 : \beta_i = 0$ that the explanatory variable $x_i$ in model (3.7) has no influence on the dependent variable and can, therefore, be dropped from the model. This hypothesis will be rejected if $|\hat{\beta}_i|/(S\sqrt{c_{i+1i+1}})$ is significantly large or, equivalently, its $p$-value $2(1 - t_{n-p}(|\hat{\beta}_i|/(S\sqrt{c_{i+1i+1}})))$ is significantly small. Under the null hypothesis, by Theorem 3.3.23, note that $|\hat{\beta}_i|/(S\sqrt{c_{i+1i+1}})$ has the distribution function

$$P\{|\hat{\beta}_i|/(S\sqrt{c_{i+1i+1}}) \leq x\} = t_{n-p}(x) - t_{n-p}(-x) = 2t_{n-p}(x) - 1, \quad x \geq 0.$$

This yields the above $p$-value. The following lemma will be crucial for the proof of Theorem 3.3.23.

**3.3.24 Lemma.** *Let $P$ be an arbitrary $n \times n$-projection matrix with rank $p \in \{1, \ldots, n-1\}$ and let $Z$ be an $n$-dimensional standard normal random vector. Then $PZ$ and $Z - PZ$ are independent. Further $(Z - PZ)^T(Z - PZ)$ is $\chi^2_{n-p}$ distributed and $(PZ)^T PZ$ is $\chi^2_p$ distributed.*

**Proof:** Choose orthogonal vectors $a_1, \ldots, a_n \in \mathbb{R}^n$ of length 1, i.e.,

$$a_i^T a_j = \begin{cases} 1 & \text{if} \quad i = j, \\ 0 & \text{if} \quad i \neq j, \end{cases}$$

such that $a_1, \ldots, a_p$ span the range $\{Px : x \in \mathbb{R}^n\}$ of $P$. The $n \times n$-matrix $A$, whose columns are given by the vectors $a_1, \ldots, a_n$ is an orthogonal matrix, i.e., $A^T A = I_n = AA^T$. Corollary 2.1.4, thus, implies that $A\eta$ is $n$-dimensional standard normal distributed if $\eta = (\eta_1, \ldots, \eta_n)^T$ is an $n$-dimensional standard normal distributed random vector. Without loss of generality we can assume, therefore, the particular representation $Z = A\eta$. The fact that $P$ is a projection matrix implies that

$$PZ = P(a_1\eta_1 + \cdots + a_n\eta_n) = a_1\eta_1 + \cdots + a_p\eta_p,$$

cf Exercise 28 and, hence,

$$Z - PZ = a_{p+1}\eta_{p+1} + \cdots + a_n\eta_n.$$

The independence of the $\eta_1, \ldots, \eta_n$ obviously yields the independence of $PZ$ and $Z - PZ$. Since

$$(Z - PZ)^T(Z - PZ)$$
$$= (a_{p+1}\eta_{p+1} + \cdots + a_n\eta_n)^T(a_{p+1}\eta_{p+1} + \cdots + a_n\eta_n) = \eta_{p+1}^2 + \cdots + \eta_n^2,$$

we obtain that $(Z - PZ)^T(Z - PZ)$ is $\chi_{n-p}^2$ distributed by Definition 2.1.6. By using the same arguments we obtain that $(PZ)^T PZ$ is $\chi_p^2$ distributed. $\quad\square$

**Proof of Theorem 3.2.23:** The vector $\varepsilon^* := \sigma^{-1}\varepsilon$ is $n$–dimensional standard normal distributed. From Definition 2.1.2 we immediately obtain, therefore, that

$$\hat{\beta} = (X^T X)^{-1}X^T Y = (X^T X)^{-1}X^T(X\beta + \sigma\varepsilon^*) = \beta + \sigma(X^T X)^{-1}X^T\varepsilon^*$$

is $N(\beta, \sigma^2(X^T X)^{-1})$ distributed. This is part (i).

Since $\hat{\beta} = (X^T X)^{-1}X^T\hat{Y}$ and $RSS = (Y - \hat{Y})^T(Y - \hat{Y}) = (Y - PY)^T(Y - PY)$, part (ii) follows if $\hat{Y} = PY$ and $Y - PY$ are independent. But this is a consequence of Lemma 3.3.24. This auxiliary result implies also that $RSS/\sigma^2 = \sigma^{-2}(Y - PY)^T(Y - PY) = (\varepsilon^* - P\varepsilon^*)^T(\varepsilon^* - P\varepsilon^*)$ is $\chi_{n-p}^2$ distributed, which is (iii). Part (iv) is a consequence of (i)–(iii) and Definition 2.1.11 of the $t$ distribution. $\quad\square$

**3.3.25 Corollary.** *Under the conditions of Theorem 3.3.23 we have:*

(i) *The model variation $\sum_{i=1}^n(\hat{Y}_i - \bar{Y}_n)^2$ and the error $RSS = \sum_{i=1}^n(Y_i - \hat{Y}_i)^2$ are independent.*

(ii) *If, in addition, $\beta_1 = \beta_2 = \cdots = \beta_{p-1} = 0$, then $\sigma^{-2}\sum_{i=1}^n(\hat{Y}_i - \bar{Y}_n)^2$ is $\chi_{p-1}^2$ distributed.*

**Proof:** The normal equations (3.4) imply that $\bar{Y}_n = n^{-1}\sum_{i=1}^n \hat{Y}_i = n^{-1}1^T\hat{Y}$, where $1 = (1, \ldots, 1)^T \in \mathbb{R}^n$, and, hence,

$$\sum_{i=1}^n(\hat{Y}_i - \bar{Y}_n)^2 = (\hat{Y} - n^{-1}1^T\hat{Y}1)^T(\hat{Y} - n^{-1}1^T\hat{Y}1) = \|X\hat{\beta} - n^{-1}1^T X\hat{\beta}1\|^2.$$

Theorem 3.3.23 (ii) now implies part (i).

If $\beta_1 = \beta_2 = \cdots = \beta_{p-1} = 0$, then $Y^* := \sigma^{-1}(Y - \beta_0 1)$ is $n$-dimensional standard normal distributed. Further, the line $G := \{s1 : s \in \mathbb{R}\} \subset \mathbb{R}^n$ is a subset of the range $V := \{Px : x \in \mathbb{R}^n\}$ of $P$: With $x := X(s, 0, \ldots, 0)^T$

we have $Px = X(s, 0, \ldots, 0)^T = (s, \ldots, s)^T = s\mathbf{1}$, i.e., $G \subset V$. This implies $Ps\mathbf{1} = s\mathbf{1}$ and, thus,

$$
\sigma^{-2} \sum_{i=1}^{n} (\hat{Y}_i - \bar{Y}_n)^2 = \sigma^{-2} \sum_{i=1}^{n} \left( \hat{Y}_i - n^{-1} \sum_{j=1}^{n} \hat{Y}_j \right)^2
$$

$$
= \sum_{i=1}^{n} \left( \hat{Y}_i^* - n^{-1} \sum_{j=1}^{n} \hat{Y}_j^* \right)^2
$$

$$
= \sum_{i=1}^{n} \hat{Y}_i^{*2} - n \left( n^{-1} \sum_{j=1}^{n} \hat{Y}_j^* \right)^2
$$

$$
= \| PY^* \|^2 - \| n^{-1} \mathbf{1}\,\mathbf{1}^T PY^* \|^2.
$$

Note that $n^{-1} \mathbf{1}\,\mathbf{1}^T$ is an $n \times n$ projection matrix whose range is $G$. We choose now orthogonal vectors $a_1, \ldots, a_n \in \mathbb{R}^n$ of length 1 such that $a_1$ spans the line $G$ and $a_1, \ldots, a_p$ span the linear space $V$. As in the proof of Lemma 3.3.24 we can assume the representation $Y^* = a_1 \eta_1 + \cdots + a_n \eta_n$, where $\eta = (\eta_1, \ldots, \eta_n)^T$ is $n$–dimensional standard normal distributed. Then we obtain

$$
PY^* = a_1 \eta_1 + \cdots + a_p \eta_p, \quad n^{-1} \mathbf{1}\,\mathbf{1}^T PY^* = a_1 \eta_1,
$$

cf Exercise 28 and, thus,

$$
\| PY^* \|^2 - \| n^{-1} \mathbf{1}\,\mathbf{1}^T PY^* \|^2 = \sum_{i=1}^{p} \eta_i^2 - \eta_1^2 = \sum_{i=2}^{p} \eta_i^2.
$$

Part (ii) is now an immediate consequence of Definition 2.1.6 of the chi–square distribution.                                                                                    $\square$

By means of Theorem 3.3.23, part (iii), and Corollary 3.3.25 we can define the *global F test* for the hypothesis $H_0 : \beta_1 = \cdots = \beta_{p-1} = 0$. The resulting test statistic

$$
F := \frac{\text{model variation}/(p-1)}{\text{error}/(n-p)} = \frac{\sum_{i=1}^{n} (\hat{Y}_i - \bar{Y}_n)^2/(p-1)}{\sum_{i=1}^{n} (Y_i - \hat{Y}_i)^2/(n-p)}
$$

is by Definition 2.1.9 $F$ distributed with $(p-1, n-p)$ degrees of freedom. If $F$ is significantly large or, equivalently, the resulting $p$–value $(1 - F_{p-1,n-p}(F))$ is significantly small, then $H_0$ will be rejected.

## Polynomial Regression

By means of the multiple linear model we can in particular test for a polynomial regression. If we put in the standard model (3.7)

$$x_{ij} = x_i^j, \qquad j = 1, \ldots, p-1, \qquad i = 1, \ldots, n,$$

then we obtain a *polynomial regression* model

$$Y_i = \beta_0 + \sum_{j=1}^{p-1} \beta_j x_i^j + \varepsilon_i, \qquad i = 1, \ldots, n.$$

The degree $p-1$ of this polynomial can be checked by testing the hypothesis $H_0 : \beta_{p-1} = 0$. It might be decreased as well as increased.

**3.3.26 Example.** We consider a polynomial model of degree 3 for the geo data in Example 3.2.6, i.e., we assume that the number of students $Y$ is a cubic polynomial of the explanatory variable *year*. We obtain the following result, where $p = 4$.

The $R^2$ value is 0.9760, the adjusted value $R_a^2$ is 0.9616 and, thus, both are larger than those of the transformed model $\sqrt{Y} = year + \varepsilon$ in Figure 3.2.3. These values are rather close to 1 so that the model

$$Y = 971\,829 - 32\,539\,year + 362.71198\,year^2 - 1.34597\,year^3 + \varepsilon$$

seems to fit the data quite well. The following Figure 3.3.7 uses, however, the OUTEST file generated in Program 3_3_6, where the parameter estimates are stored with up to 9 decimal digits. The number of digits printed in the output in Figure 3.3.6 is not sufficiently large to give the displayed polynomial. The global $F$ statistic has the rounded value $F = 14\,974/220.78 = 67.82$ with (3,5) degrees of freedom and a rounded $p$–value $1 - F_{3,5}(67.82) = 0.0002$. The global hypothesis $\beta_1 = \beta_2 = \beta_3 = 0$ will be rejected under the assumption of normal distributed errors. The single hypotheses $\beta_{i-1} = 0$ will be rejected as well, due to the $p$–values of roughly 0.02, of the corresponding test statistics $|\hat{\beta}_{i-1}|/(S\sqrt{c_{ii}})$, $i = 1, \ldots, 4$. These results are, therefore, in favor of the above *complete* model.

The REG Procedure
Model: MODEL1
Dependent Variable: NUMBER

Analysis of Variance

| Source | DF | Sum of Squares | Mean Square | F Value | Pr > F |
|---|---|---|---|---|---|
| Model | 3 | 44922 | 14974 | 67.82 | 0.0002 |
| Error | 5 | 1103.87626 | 220.77525 | | |
| Corrected Total | 8 | 46026 | | | |

| | | | | |
|---|---|---|---|---|
| Root MSE | 14.85851 | R-Square | 0.9760 |
| Dependent Mean | 98.77778 | Adj R-Sq | 0.9616 |
| Coeff Var | 15.04236 | | |

Parameter Estimates

| Variable | DF | Parameter Estimate | Standard Error | t Value | Pr > |t| |
|---|---|---|---|---|---|
| Intercept | 1 | 971829 | 286548 | 3.39 | 0.0194 |
| year1 | 1 | -32539 | 9559.35849 | -3.40 | 0.0192 |
| year2 | 1 | 362.71198 | 106.25656 | 3.41 | 0.0190 |
| year3 | 1 | -1.34597 | 0.39353 | -3.42 | 0.0188 |

**Figure 3.3.6.** Polynomial regression of the geo data.

```
***    Program 3_3_6    ***;
TITLE1 'Polynomial Regression';
TITLE2 'Geo Data';
LIBNAME datalib 'c:\data';

DATA geo1;
    SET datalib.geo;
    year1=year;
    year2=year**2;
    year3=year**3;

PROC REG DATA=geo1 OUTEST=estimate SINGULAR=1E-12;
    MODEL number=year1 year2 year3;
RUN; QUIT;
```

A polynomial regression can easily be computed by defining new variables in a DATA step (here 'year1', 'year2', 'year3'), which contain powers of the original variable (here 'year'). These new variables are included in the MODEL statement.

PROC REG checks for singularity of $X^T X$ using the following mechanism: If a pivot is less than SINGULAR*RSS, then a singularity is declared. The default value of SINGULAR is $10^{-8}$. In the above program it is set to $10^{-12}$. Otherwise the matrix $X^T X$ would be declared to be singular.

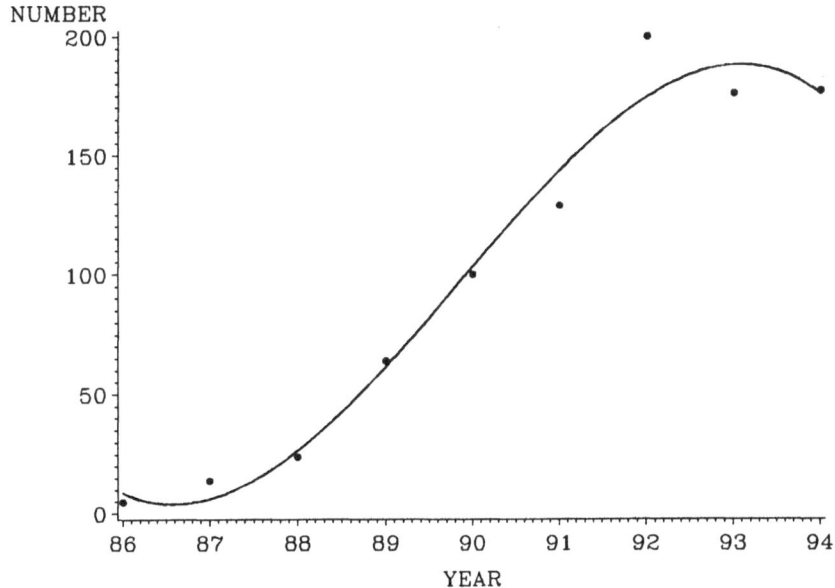

**Figure 3.3.7.** Regression polynomial of degree 3 of the geo data.

```
***    Program 3_3_7   ***;
TITLE1 'Polynomial Regression';
TITLE2 'Geo Data';
LIBNAME datalib 'c:\data';

DATA geo2;
    SET estimate(KEEP=INTERCEPT YEAR1 YEAR2 YEAR3);
    DO t=86 TO 94 BY 0.1;
        predictd=INTERCEPT+YEAR1*t+YEAR2*t**2+YEAR3*t**3;
    OUTPUT; END;
DATA geo3;
    MERGE datalib.geo geo2;
                            ↓
```

```
                              ↑
AXIS1 MAJOR=(N=5);
SYMBOL1 V=DOT C=G I=NONE;
SYMBOL2 V=NONE C=G I=JOIN W=2;
PROC GPLOT DATA=geo3;
    PLOT number*year=1 predictd*t=2 / OVERLAY VAXIS=AXIS1;
RUN; QUIT;
```

In the first DATA step the regression polynomial is computed at a grid of points, cf Program 2_1_1. The file 'estimate' was created in Program 3_3_6. It contains among others the estimated coefficients of the regression, i.e., of INTERCEPT, YEAR1, YEAR2, YEAR3.

The KEEP data set option specifies the variables that will be included in the output data set.

The MERGE statement in the second DATA step combines the two data sets 'datalib.geo' and 'geo2' and writes the results to the file 'geo3'.

## Exercises

**1.** Show that square integrable random variables $X$ and $Y$ satisfy

(i)  $Cov(X + a, Y + b) = Cov(X, Y)$, $a, b \in I\!R$.

(ii)  $Cov(cX, dY) = cd\, Cov(X, Y)$, $c, d \in I\!R$.

(iii)  $Cov(X, Y + Z) = Cov(X, Y) + Cov(X, Z)$.

**2.** Let $X_1, \ldots, X_n$ be independent and identically distributed random variables. Denote by $F$ the common distribution function and by $F_n$ the empirical distribution function of $X_1, \ldots, X_n$. Establish the relation

$$Cov(F_n(u), F_n(v)) = n^{-1}(F(\min(u, v)) - F(u)F(v)).$$

Hint: Exercise 1.

**3.** Show that $Var(X \pm Y) = Var(X) + Var(Y) \pm 2\, Cov(X, Y)$.

**4.** Let $X$ and $Y$ be arbitrary square integrable random variables. Show that $X + Y$ and $X - Y$ are uncorrelated if and only if the variances of $X$ and $Y$ coincide.

**5.** Consider square integrable random variables $X_1, X_2$ with distribution functions $F_1$ and $F_2$. Prove the following relation for the covariance:

$$Cov(X_1, X_2) = \int (F(u, v) - F_1(u)\, F_2(v))\, du\, dv,$$

where $F(u, v) := P\{X_1 \leq u,\ X_2 \leq v\}$, $u, v \in I\!R$, is the joint distribution function of $X_1, X_2$. Hint: Exercise 13, Chapter 1 and Fubini's Theorem.

**13.** Prove Theorem 3.1.5 (i).

**14.** Put $R(\beta_0, \beta_1) = \sum_{i=1}^{n}(y_i - \beta_0 - \beta_1 x_i)^2$ and suppose that $\sum_{i=1}^{n}(x_i - \bar{x}_n)^2 > 0$. Show that the condition $(\partial R/\partial \beta_0)(b_0, b_1) = 0 = (\partial R/\partial \beta_1)(b_0, b_1)$ is necessary as well as sufficient for $R$ to have a minimum at $(b_0, b_1)$.

**15.** Prove Theorem 3.2.2.

**16.** Consider the standard model $Y_i = \beta_0 + \beta_1 x_i + \varepsilon_i$, $i = 1, \ldots, n$, where $\varepsilon_1, \ldots, \varepsilon_n$ are independent and have identical variance $\sigma^2$. We suppose $x_i \in [-1, 1]$, $i = 1, \ldots, n$. For which $n$ points $x_i$ is the variance of the estimator $\hat{\beta}_1$ a minimum?

**17.** Prove Theorem 3.2.5.

**18.** (Yoga data) Check whether the effect of a yoga exercise (= difference of the systolic blood pressure before and after the exercise) is associated with the systolic pressure before:

- Draw a scatterplot,

- Compute a regression,

- Check the residuals for normality,

- Draw a plot of the residuals.

If outliers can be identified, drop them from the data set, repeat the analysis and compare the results.

**19.** Consider square integrable random variables $X_1, \ldots, X_n, Y_1, \ldots, Y_n$. Suppose that $X = (X_1, \ldots, X_n)^T$ and $Y = (Y_1, \ldots, Y_n)^T$ are uncorrelated, i.e., $X_i, Y_j$ are uncorrelated for $i, j = 1, \ldots, n$. Prove that $Cov(X + Y) = Cov(X) + Cov(Y)$.

**20.** Suppose that $X_1, \ldots, X_n$ are square integrable random variables and put $X := (X_1, \ldots, X_n)^T$. Show that the covariance matrix $\Sigma_X$ is positive semidefinite, i.e., the symmetric matrix $\Sigma_X$ satisfies $x^T \Sigma_X x \geq 0$ for $x \in \mathbb{R}^n$.

**21.** Let $A$ be an arbitrary $n \times n$-matrix and $Q_A(x) := x^T A x$, $x \in \mathbb{R}^n$, the pertaining quadratic form. Show that

(i) $\left(\dfrac{\partial Q_A}{\partial x_1}(x), \ldots, \dfrac{\partial Q_A}{\partial x_n}(x)\right)^T = 2Ax$.

(ii) $Q_A(\lambda x) = \lambda^2 Q_A(x)$, $\lambda \in \mathbb{R}$ (homogeneity).

(iii) $|Q_A(x)| \leq \|A\| \|x\|^2$, where $\|x\|$ is the Euclidean norm or length of $x$ and $\|A\| := \sup\{\|Ax\| : \|x\| \leq 1\}$.

**6.** Let $X_1, \ldots, X_n$ be random variables with $E(X_i) = \mu$, $Var(X_i) = \sigma^2$, $i$ and $\varrho(X_i, X_j) = \varrho$ for $1 \leq i \neq j \leq n$. Compute $Var(\bar{X}_n)$ and deduce $-(n-1)^{-1} \leq \varrho \leq 1$.

**7.** Consider random variables $X_1, \ldots, X_n$ with $E(X_i) = \mu$, $Var(X_i) = 1, \ldots, n$, and $Cov(X_i, X_j) = \varrho\sigma^2$, $\varrho \in (-1, 1)$, $1 \leq i \neq j \leq n$. Show that $(1 -$ is an unbiased estimator of $\sigma^2$. (For $\varrho = 0$ this implies the unbiasedness of the variance $S_n^2$ also for uncorrelated random variables, which generalizes Exerc Chapter 1.)

**8.** Let $X_1, \ldots, X_n$ be square integrable random variables with $E(X_i) = \mu$, $1, \ldots, n$, and $Cov(X_i, X_j) = 0$ for $j > i + 1$. Show that with

$$Q_1 = \sum_{i=1}^{n}(X_i - \bar{X}_n)^2$$

and

$$Q_2 = (X_n - X_1)^2 + \sum_{i=1}^{n-1}(X_i - X_{i+1})^2$$

we have

$$E\left(\frac{3Q_2 - Q_1}{n(n-3)}\right) = Var(\bar{X}_n), \quad n > 3.$$

**9.** Let $(X, Y)$ be bivariate normal. Show that

(i) the vector $(X, Y)^T$ has the density specified in Example 3.1.2. Hint: Exercise 1, Chapter 2.

(ii) $X$ and $Y$ are independent $\Leftrightarrow \varrho = 0$.

**10.** Show by a counterexample that the following implication is *not* true in general:

$X, Y$ are uncorrelated and each is normal distributed $\Rightarrow X, Y$ independent.

Hint: Let $X$ be $N(0, 1)$ distributed. Put $Y := ZX$, where $X, Z$ are independent and $P\{Z = -1\} = 1/2 = P\{Z = 1\}$, or consider the continuous function $f(c) = Cov(X, Y_c)$, where $Y_c := X$ if $|X| \leq c$ and $Y_c := -X$ elsewhere, $c \geq 0$.

**11.** Consider independent and standard normal distributed random variables $X$ and $Y$. Show that with $\varrho \in (-1, 1)$ the random vector $(X, \varrho X + \sqrt{1 - \varrho^2}Y)^T$ is bivariate normal distributed with coefficient of correlation $\varrho$.

**12.** Let $X$ and $Y$ be square integrable random variables and denote by $p : \mathbb{R}^2 \rightarrow \mathbb{R}$ the polynominal

$$p(a, b) = E((Y - aX - b)^2).$$

Show that $p$ has a minimum in

$$(a^*, b^*) = (Cov(X, Y)/\sigma^2(X), E(Y) - a^*E(X)).$$

**13.** Prove Theorem 3.1.5 (i).

**14.** Put $R(\beta_0, \beta_1) = \sum_{i=1}^{n} (y_i - \beta_0 - \beta_1 x_i)^2$ and suppose that $\sum_{i=1}^{n} (x_i - \bar{x}_n)^2 > 0$. Show that the condition $(\partial R/\partial\beta_0)(b_0, b_1) = 0 = (\partial R/\partial\beta_1)(b_0, b_1)$ is necessary as well as sufficient for $R$ to have a minimum at $(b_0, b_1)$.

**15.** Prove Theorem 3.2.2.

**16.** Consider the standard model $Y_i = \beta_0 + \beta_1 x_i + \varepsilon_i$, $i = 1, \ldots, n$, where $\varepsilon_1, \ldots, \varepsilon_n$ are independent and have identical variance $\sigma^2$. We suppose $x_i \in [-1, 1]$, $i = 1, \ldots, n$. For which $n$ points $x_i$ is the variance of the estimator $\hat{\beta}_1$ a minimum?

**17.** Prove Theorem 3.2.5.

**18.** (Yoga data) Check whether the effect of a yoga exercise ($=$ difference of the systolic blood pressure before and after the exercise) is associated with the systolic pressure before:

- Draw a scatterplot,
- Compute a regression,
- Check the residuals for normality,
- Draw a plot of the residuals.

If outliers can be identified, drop them from the data set, repeat the analysis and compare the results.

**19.** Consider square integrable random variables $X_1, \ldots, X_n$, $Y_1, \ldots, Y_n$. Suppose that $X = (X_1, \ldots, X_n)^T$ and $Y = (Y_1, \ldots, Y_n)^T$ are uncorrelated, i.e., $X_i$, $Y_j$ are uncorrelated for $i, j = 1, \ldots, n$. Prove that $Cov(X + Y) = Cov(X) + Cov(Y)$.

**20.** Suppose that $X_1, \ldots, X_n$ are square integrable random variables and put $X := (X_1, \ldots, X_n)^T$. Show that the covariance matrix $\Sigma_X$ is positive semidefinite, i.e., the symmetric matrix $\Sigma_X$ satisfies $x^T \Sigma_X x \geq 0$ for $x \in \mathbb{R}^n$.

**21.** Let $A$ be an arbitrary $n \times n$–matrix and $Q_A(x) := x^T A x$, $x \in \mathbb{R}^n$, the pertaining quadratic form. Show that

(i) $\left( \dfrac{\partial Q_A}{\partial x_1}(x), \ldots, \dfrac{\partial Q_A}{\partial x_n}(x) \right)^T = 2Ax$.

(ii) $Q_A(\lambda x) = \lambda^2 Q_A(x)$, $\lambda \in \mathbb{R}$ (homogeneity).

(iii) $|Q_A(x)| \leq \|A\| \|x\|^2$, where $\|x\|$ is the Euclidean norm or length of $x$ and $\|A\| := \sup\{\|Ax\| : \|x\| \leq 1\}$.

**6.** Let $X_1, \ldots, X_n$ be random variables with $E(X_i) = \mu$, $Var(X_i) = \sigma^2$, $i = 1, \ldots, n$, and $\varrho(X_i, X_j) = \varrho$ for $1 \leq i \neq j \leq n$. Compute $Var(\bar{X}_n)$ and deduce the bound $-(n-1)^{-1} \leq \varrho \leq 1$.

**7.** Consider random variables $X_1, \ldots, X_n$ with $E(X_i) = \mu$, $Var(X_i) = \sigma^2$, $i = 1, \ldots, n$, and $Cov(X_i, X_j) = \varrho\sigma^2$, $\varrho \in (-1, 1)$, $1 \leq i \neq j \leq n$. Show that $(1-\varrho)^{-1}S_n^2$ is an unbiased estimator of $\sigma^2$. (For $\varrho = 0$ this implies the unbiasedness of the sample variance $S_n^2$ also for uncorrelated random variables, which generalizes Exercise 16, Chapter 1.)

**8.** Let $X_1, \ldots, X_n$ be square integrable random variables with $E(X_i) = \mu$, $i = 1, \ldots, n$, and $Cov(X_i, X_j) = 0$ for $j > i + 1$. Show that with

$$Q_1 = \sum_{i=1}^{n}(X_i - \bar{X}_n)^2$$

and

$$Q_2 = (X_n - X_1)^2 + \sum_{i=1}^{n-1}(X_i - X_{i+1})^2$$

we have

$$E\left(\frac{3Q_2 - Q_1}{n(n-3)}\right) = Var(\bar{X}_n), \quad n > 3.$$

**9.** Let $(X, Y)$ be bivariate normal. Show that

(i) the vector $(X, Y)^T$ has the density specified in Example 3.1.2. Hint: Exercise 1, Chapter 2.

(ii) $X$ and $Y$ are independent $\Leftrightarrow \varrho = 0$.

**10.** Show by a counterexample that the following implication is *not* true in general:

$X, Y$ are uncorrelated and each is normal distributed $\Rightarrow X, Y$ independent.

Hint: Let $X$ be $N(0, 1)$ distributed. Put $Y := ZX$, where $X, Z$ are independent and $P\{Z = -1\} = 1/2 = P\{Z = 1\}$, or consider the continuous function $f(c) = Cov(X, Y_c)$, where $Y_c := X$ if $|X| \leq c$ and $Y_c := -X$ elsewhere, $c \geq 0$.

**11.** Consider independent and standard normal distributed random variables $X$ and $Y$. Show that with $\varrho \in (-1, 1)$ the random vector $(X, \varrho X + \sqrt{1-\varrho^2}Y)^T$ is bivariate normal distributed with coefficient of correlation $\varrho$.

**12.** Let $X$ and $Y$ be square integrable random variables and denote by $p : \mathbb{R}^2 \to \mathbb{R}$ the polynominal

$$p(a, b) = E((Y - aX - b)^2).$$

Show that $p$ has a minimum in

$$(a^*, b^*) = (Cov(X, Y)/\sigma^2(X), E(Y) - a^*E(X)).$$

**22.** Let $A$ be an arbitrary symmetric $n \times n$–matrix. Prove that $A$ is idempotent and has rank $r \in \{0, \ldots, n\}$ if and only if $r$ eigenvalues of $A$ are equal to 1 and $n - r$ eigenvalues of $A$ are equal to 0.

**23.** Consider an $n$–dimensional standard normal random vector $X = (X_1, \ldots, X_n)^T$ and let $A$ be a symmetric and idempotent $n \times n$–matrix with rank $p$. Prove that the random quadratic form $X^T A X$ is chi–square distributed with $p$ degrees of freedom.

**24.** Prove Theorem 3.3.13.

**25.** Prove Theorem 3.3.19.

**26.** Let $A$ be an $m \times n$–matrix and let $B$ be an $n \times m$–matrix. Show that $tr(AB) = tr(BA)$. This implies in particular $tr(\Lambda) = tr(R^T \Lambda R)$ for a diagonal matrix $\Lambda$ and an orthogonal matrix $R$.

**27.** Consider an arbitrary $n \times n$–projection matrix $P$, i.e., $P$ is symmetric and idempotent. Prove that

  (i)  $tr(P) = \text{rank}(P)$. Hint: Exercises 22 and 26,

  (ii)  $I_n - P$ is a projection matrix,

  (iii)  $P$ is positive semidefinite.

**28.** Let $a_1, \ldots, a_n$ be orthogonal vectors in $\mathbb{R}^n$ and let $P$ be a projection matrix with rank $p \in \{1, \ldots, n - 1\}$. Suppose that the range of $P$ is spanned by the vectors $a_1, \ldots, a_p$. Show the implication

$$x = \sum_{i=1}^{n} c_i a_i \Rightarrow P x = \sum_{i=1}^{p} c_i a_i.$$

**29.** (Economy data; see Example 3.3.1) (i) Which linear model explains the unemployment best? (ii) Compute a polynomial regression of degree 3, 4 and 5 for the inflation (*inflatn*) depending of the increase of the gross national product (*gnp*). Evaluate the goodness–of–fit of each polynomial model and the influence of each coefficient. Plot one of the polynomials joint with the data.

**30.** A sequence $Z_0, Z_1, \ldots, Z_n$ of random variables with $E(Z_i) = \mu$, $i = 0, 1, \ldots, n$, is called an *autoregressive process of order* $p$, denoted by AR($p$) process if there are constants $\beta_1, \ldots, \beta_p$ such that for $t = p, p + 1, \ldots, n$

$$Z_t - \mu = \beta_1(Z_{t-1} - \mu) + \beta_2(Z_{t-2} - \mu) + \cdots + \beta_p(Z_{t-p} - \mu) + \varepsilon_t.$$

$\varepsilon_p, \ldots, \varepsilon_n$ are assumed to be uncorrelated random variables with identical variances. Such processes are quite common for modeling time series, see Example 1.5.1.

   If we view the above model as a multiple linear regression model with a random design, then we can estimate the unknown vector of parameters $\beta = (\beta_1, \ldots, \beta_p)^T$ and $\mu$ by means of the least squares estimator $\hat{\beta} = (\hat{\beta}_0, \hat{\beta}_1, \ldots, \hat{\beta}_p)^T$. Do this for the sunspot data in Example 1.5.1 by modeling the yearly averages of the daily sunspots as an AR(10)–process. Hint: The yearly averages can simply be computed from the monthly averages by using the SAS procedure PROC MEANS together with the statement BY *year*, which generates subgroups. The backshift operation $Z_t \rightarrow Z_{t-1}$ can be done by using the SAS function LAG .

# Chapter 4

# Categorical Data Analysis

In this chapter we will mainly be concerned with random variables whose outcomes are not ordinary numbers, but elements of several possible categories, classes or groups. The hair color or the eye color of a newborn baby are typical examples. A common problem is the question, whether there is a relationship between these categorical outcomes, for example, between the hair color and the eye color. This question is investigated by means of contingency tables. If there is a relationship, then we will try to describe it first by a linear model in the framework of categorical regression. A nonlinear relationship will be described by a generalized linear model. Logit or probit models are particular examples. Finally, we will fit a simple neural network.

## 4.1 Contingency Tables

Suppose, for example, that we want to investigate the question whether there is any relationship between the income of an individual and urbanity. We consider the two classes 'high' and 'low' of income and the two classes 'urban' and 'rural' of urbanity. We collect data for the analysis of this question by a survey, where we query a certain number of randomly chosen individuals. A $2 \times 2$ table of urbanity by income displays the frequencies of each of the four possible combinations of rural | urban with high | low. An analysis of the frequencies in the cells of this table will help to answer the question whether there is a relationship between urbanity and income, see Example 4.1.2. In this section we will introduce some tools for the analysis of such *contingency tables* such as Fisher's exact test or Pearson's chi–square test.

### Fisher's Exact Test for Independence

For Fisher's exact test we will need the following result on probability distributions in certain urn models.

**4.1.1 Lemma.** *Consider an urn, which contains $K$ balls. Among these are $W$ white ones and $K - W$ black ones. Suppose we pick at random $m$ balls out of* *this urn without replacement. The probability $P\{N = k\}$, $k \in \{0, 1, \ldots, m\}$, of the event that the random number $N$ of white balls in the sample is $k$, equals*

$$P\{N = k\} = \binom{W}{k}\binom{K - W}{m - k} \bigg/ \binom{K}{m} =: H(K, W, m)(\{k\}).$$

The probability distribution $H(K, W, m)$ is the *hypergeometric distribution* with the parameters $K, W$ and $m$.

**Proof:** The result follows from the fact that a set with $s$ elements has $\binom{s}{r} = s!/(r!(s - r)!)$ subsets with $r$ elements, cf Section II.4 in Feller (1968).  □

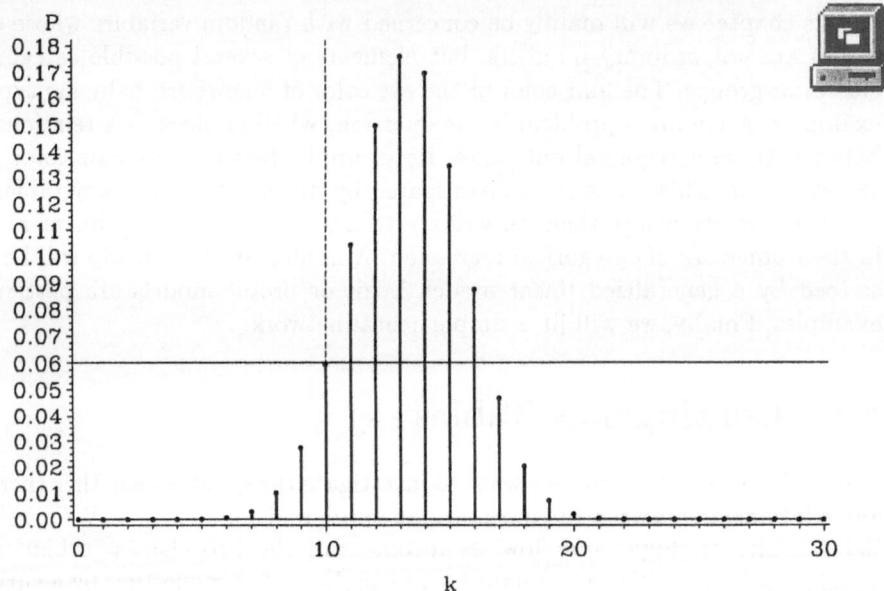

**Figure 4.1.1.** Hypergeometric distribution with the parameters $K = 90, W = 40, m = 30$; lines correspond to the $p$-value of $N = 10$, see below.

```
***    Program 4_1_1    ***;
TITLE1 'Hypergeometric Distribution';
TITLE2 'with K=90, W=40 and m=30';
LIBNAME datalib 'c:\data';

DATA hypergeo;
    K=90; W=40; m=30;
    DO j=0 TO 30;
    IF j=0 THEN p=PROBHYPR(K,W,m,j);
        ELSE p=PROBHYPR(K,W,m,j) - PROBHYPR(K,W,m,j-1);
        OUTPUT;
    END;
```

↓

```
                                    ↑
AXIS1 LABEL=('k');
SYMBOL1 V=DOT H=0.6 WIDTH=2 I=NEEDLE;
PROC GPLOT DATA=hypergeo;
    PLOT p*j / HREF=10 LHREF=5 VREF=0.06 LVREF=1 HAXIS=AXIS1;
RUN; QUIT;
```

The probabilities of the outcomes k=0,...,30 of the hypergeometric distribution are computed as the differences of the values of the distribution function at k and k-1. The lines, which correspond to the $p$–value of N=10, are vertically and horizontally drawn by the options 'VREF=' and 'HREF=' in the PLOT statement. The definition of the $p$–value is given below. The line types are specified by 'LHREF=' and 'LVREF='.

The following example introduces Fisher's exact test.

**4.1.2 Example** (NHANES II Data). The following table displays data from the Second National Health and Nutrition Examination Survey (NHANES II) of the National Center for Health Statistics. This survey was designed to collect data for the analysis of the U.S. civilian noninstitutionalized population aged 6 months to 74 years. NHANES II took four years (1976 – 1980). The table contains the results of $n = 16\,547$ queries; the second lines are percentages. The table is analyzed in Little and Wu (1991).

```
              Table of urbanity by income

        urbanity                income


        Frequency |
        Percent   |high    |low     |  Total

        ----------+--------+--------+
        rural     |  2548  |  3235  |   5783
                  | 15.40  | 19.55  |  34.95
        ----------+--------+--------+
        urban     |  5295  |  5469  |  10764
                  | 32.00  | 33.05  |  65.05
        ----------+--------+--------+
        Total        7843     8704    16547
                     47.40    52.60   100.00
```

**Figure 4.1.2.** 2 × 2 table of urbanity by income.

```
***     Program 4_1_2    ***;
TITLE1 '2 x 2 table';
TITLE2 'NHANES II Data';

DATA table;
    INPUT urbanity $ income $ freqncy;
    CARDS;
urban   low     5469
urban   high    5295
rural   low     3235
rural   high    2548
;

PROC FREQ DATA=table;
    TABLES urbanity*income / NOCOL NOROW;
    WEIGHT freqncy;
RUN; QUIT;
```

Contingency tables can be computed by means of various procedures such as FREQ, TABULATE and SQL.

In this program we use PROC FREQ. A table request is put in the TABLES statement for each wanted cross tabulation table. These are generated by two variables joined with an asterisk. The first variable specifies the rows and the second one the columns of the table.

PROC FREQ prints by default these items per cell: The number of subjects that have the indicated values of the two variables, the percentage of the total frequency count represented by that cell, the row percentage (the percentage of the total frequency count for that row represented by the cell) and the column percentage (the percentage of the total frequency count for that column represented by the cell). Row and column percentages are suppressed by the options NOROW and NO-COL.

Normally, each observation contributes a value of 1 to the frequency count. If, however, the data are frequency counts as in this program, the WEIGHT statement must be used. Then each observation contributes the weighting variable's value for that observation.

47.40 % of the members of the complete sample have a high income; 2548/5783 = 44.06 % of the rural members have a high income. This indicates that the income might be independent of the urbanity. But in order to be able to evaluate this result, we compute its probability of occurrence under the assumption that income and urbanity are independent. We obtain, thus, the following model: Among the total of 16 547 there are 7 843 members with a high income. If we now take at random 5 783 members without replacement out of the total of 16 547, then the number $N$ of members with high income in this subsample is hypergeometric $H(16\,547, 7\,843, 5\,783)$ distributed, i.e.,

$$P\{N = k\} = \frac{\binom{7\,843}{k}\binom{8\,704}{5\,783 - k}}{\binom{16\,547}{5\,783}}, \qquad k = 0, 1, \ldots, 5\,783.$$

If urbanity and income are actually independent, then we can query in particular the $5\,783$ rural members about their income. This *randomization* by the urbanity does not affect the distribution of $N$ under the hypothesis $H_0$: 'urbanity and income are independent', i.e., $N$ is, under this randomization, $H(16\,547, 7\,843, 5\,783)$ distributed as well. We will reject, therefore, this hypothesis if $N$ is significantly small or large. This is the 2–tail case. In our example we have $N = 2\,548$ with the corresponding $p$–value $2.915 \cdot 10^{-10}$, which leads to the rejection of $H_0$. For the definition of the $p$–value in this case see below.

```
                        The FREQ Procedure


              Statistics for Table of urbanity by income


        Statistic                      DF       Value        Prob
        ------------------------------------------------------------
        Chi-Square                      1      39.7325      <.0001
        Likelihood Ratio Chi-Square     1      39.7979      <.0001
        Continuity Adj. Chi-Square      1      39.5270      <.0001
        Mantel-Haenszel Chi-Square      1      39.7301      <.0001
        Phi Coefficient                        -0.0490
        Contingency Coefficient                 0.0489
        Cramer's V                             -0.0490


                        Fisher's Exact Test
                   ----------------------------------
                   Cell (1,1) Frequency (F)      2548
                   Left-sided Pr <= F       1.566E-10
                   Right-sided Pr >= F         1.0000

                   Table Probability (P)    2.984E-11
                   Two-sided Pr <= P        2.915E-10


                   Sample Size = 16547
```

**Figure 4.1.3.** Fisher's exact test for the NHANES II data in Example 4.1.2.

```
***    Program 4_1_3   ***;
TITLE1 'Fisher's Exact Test';
TITLE2 'NHANES II Data';

DATA table;
    INPUT urbanity $ income $ freqncy;
    CARDS;
urban  low    5469
urban  high   5295
rural  low    3235
rural  high   2548
;

PROC FREQ DATA=table;
    TABLES urbanity*income / NOPRINT EXACT;
    WEIGHT freqncy;
RUN; QUIT;
```

Program 4_1_3 is almost identical with Program 4_1_2; the differences are as follows: The option NOPRINT in the TABLES statement suppresses the printout of the table. The option EXACT performs Fisher's exact test as well as a list of further tests for the null hypothesis of no association between the row variable and the column variable.

The test for independence of categorical data in the example above is called *Fisher's exact test*. It is performed in general as follows. Suppose we have independent realizations $(X_1, Y_1), \ldots, (X_n, Y_n)$ of two categorical variables $X$ and $Y$, where each has two categories $I_1, I_2$ and $J_1, J_2$, respectively. The frequency counts $n_{ij}$ for each class combination $I_i \times J_j$ are displayed by the following $2 \times 2$ table of $X$ by $Y$:

|  |  | Y in class | |  |
|---|---|---|---|---|
|  |  | $J_1$ | $J_2$ |  |
| X in class | $I_1$ | $n_{11}$ | $n_{12}$ | $\Sigma = n_{1.}$ |
|  | $I_2$ | $n_{21}$ | $n_{22}$ | $\Sigma - n_{2.}$ |
|  |  | $\Sigma = n_{.1}$ | $\Sigma = n_{.2}$ | $\Sigma = n$ |

(4.1)

The hypothesis $H_0$: '$X$ and $Y$ are independent' is rejected if $n_{11}$ is significantly large or small, i.e., the corresponding $p$–value, derived from the hypergeometric distribution, is significantly small. In this 2–tail case the $p$–value is defined by

$$p := \sum_{\substack{k \in \{0, 1, \ldots, n\}: \\ P\{N=k\} \leq P\{N=n_{11}\}}} P\{N = k\},$$

where $N$ is hypergeometric $H(n, n_{.1}, n_{1.})$ distributed under $H_0$, see the following theorem. The $p$-value is, therefore, the sum of all those possibilities $P\{N = k\}$ that are less than or equal to $P\{N = n_{11}\}$. Since the function $g(k) := P\{N = k\}$, $k = 0, \ldots, n$, first increases and then decreases, cf Exercise 1, a small $p$-value indicates an extreme value of $N$. This is illustrated in Figure 4.1.1 by the additional lines for the realization $N = 10$.

**4.1.3 Theorem.** *Suppose $X$ and $Y$ are independent. Conditional on the marginal frequencies $\sum_{i=1}^{n} 1_{I_1}(X_i) = n_{1.}$, $\sum_{i=1}^{n} 1_{J_1}(Y_i) = n_{.1}$, the cell frequency $N_{11} = \sum_{i=1}^{n} 1_{I_1}(X_i) 1_{J_1}(Y_i)$ is hypergeometric distributed with the parameters $n, n_{.1}, n_{1.}$, i.e.,*

$$P\left(N_{11} = k \,\Big|\, \sum_{i=1}^{n} 1_{I_1}(X_i) = n_{1.}, \sum_{i=1}^{n} 1_{J_1}(Y_i) = n_{.1}\right)$$

$$= H(n, n_{.1}, n_{1.})(\{k\}) := \frac{\binom{n_{.1}}{k}\binom{n - n_{.1}}{n_{1.} - k}}{\binom{n}{n_{1.}}}, \qquad k = 0, 1, \ldots$$

**Proof:** By the assumed independence we have

$$P\left(N_{11} = k \,\Big|\, \sum_{i=1}^{n} 1_{I_1}(X_i) = n_{1.}, \sum_{i=1}^{n} 1_{J_1}(Y_i) = n_{.1}\right)$$

$$= \frac{P\{\sum_{i=1}^{n} 1_{I_1}(X_i) 1_{J_1}(Y_i) = k, \sum_{i=1}^{n} 1_{I_1}(X_i) = n_{1.}, \sum_{i=1}^{n} 1_{J_1}(Y_i) = n_{.1}\}}{P\{\sum_{i=1}^{n} 1_{I_1}(X_i) = n_{1.}\}P\{\sum_{i=1}^{n} 1_{J_1}(Y_i) = n_{.1}\}}$$

$$= \frac{P\{\sum_{i=1}^{n_{.1}} 1_{I_1}(X_i) = k, \sum_{i=1}^{n} 1_{I_1}(X_i) = n_{1.}\}}{P\{\sum_{i=1}^{n} 1_{I_1}(X_i) = n_{1.}\}}$$

$$= \frac{P\{\sum_{i=1}^{n_{.1}} 1_{I_1}(X_i) = k, \sum_{j=n_{.1}+1}^{n} 1_{I_1}(X_j) = n_{1.} - k\}}{P\{\sum_{i=1}^{n} 1_{I_1}(X_i) = n_{1.}\}}$$

$$= \frac{P\{\sum_{i=1}^{n_{.1}} 1_{I_1}(X_i) = k\} \, P\{\sum_{j=n_{.1}+1}^{n} 1_{I_1}(X_j) = n_{1.} - k\}}{P\{\sum_{i=1}^{n} 1_{I_1}(X_i) = n_{1.}\}}$$

$$= \frac{B(n_{.1}, q)(\{k\}) B(n - n_{.1}, q)(\{n_{1.} - k\})}{B(n, q)(\{n_{1.}\})}$$

$$= \frac{\binom{n_{.1}}{k}\binom{n - n_{.1}}{n_{1.} - k}}{\binom{n}{n_{1.}}}, \qquad k = 0, 1, \ldots$$

where $q = P\{X \in I_1\}$ and $B(m, q)(\{k\}) = \binom{m}{k} q^k (1 - q)^{m-k}$, $k = 0, 1, \ldots, m$, is the binomial distribution with the parameters $m \in \mathbb{N}$ and $q \in [0, 1]$, cf Exercise 23 in Chapter 1). $\qquad \square$

## Fisher's Exact Test for Homogeneity

We can use Fisher's exact test for independence also for the comparison of probabilities. A typical problem is the question, whether a new drug $M_1$ comes with higher probability of healing $p_1$ than the one $M_2$ that is already on the market and which has probability of success $p_2$. To decide this question, two groups of patients were considered. 18 patients in group 1 got the drug $M_1$ and 13 patients in group 2 got $M_2$. The following $2 \times 2$ table of healing by drug lists the results.

```
                The FREQ Procedure

              Table of healing by drug

         healing      drug

         Frequency |
         Percent   |M1       |M2       |  Total
         ----------+---------+---------+
         healing   |      6  |      4  |     10
                   |  19.35  |  12.90  |  32.26
         ----------+---------+---------+
         no healing|     12  |      9  |     21
                   |  38.71  |  29.03  |  67.74
         ----------+---------+---------+
         Total           18        13        31
                      58.06     41.94    100.00
```

```
          Statistics for Table of healing by drug

  Statistic                      DF     Value      Prob
  -------------------------------------------------------
  Chi-Square                      1     0.0227    0.8802
  Likelihood Ratio Chi-Square     1     0.0228    0.8801
  Continuity Adj. Chi-Square      1     0.0000    1.0000
  Mantel-Haenszel Chi-Square      1     0.0220    0.8821
  Phi Coefficient                       0.0271
  Contingency Coefficient               0.0271
  Cramer's V                            0.0271

    WARNING: 25% of the cells have expected counts less
         than 5. Chi-Square may not be a valid test.
```

```
                        Fisher's Exact Test
              ------------------------------------
              Cell (1,1) Frequency (F)              6
              Left-sided Pr <= F              0.7026
              Right-sided Pr >= F             0.5967

              Table Probability (P)          0.2993
              Two-sided Pr <= P              1.0000

                        Sample Size = 31
```

**Figure 4.1.4.** Fisher's exact test for homogeneity; healing | drug.

Program 4_1_4 underlying Figure 4.1.4 is a combination of Programs 4_1_2 and 4_1_3. The options in the TABLES statement are NOROW NOCOL EXACT.

The warning at the end of the output refers to the chi–square tests that are automatically performed if the option EXACT for Fisher's exact test is specified.

18 among the total of 31 patients belong to group 1. If we select at random 10 out of the 31 patients, then the random number $N$ of members from group 1 in this sample is hypergeometric $H(31, 18, 10)$ distributed. Suppose the probabilities of success of the two drugs $M_1$ and $M_2$ coincide, i.e., $p_1 = p_2$. Then a patient's successful recovery does not depend on the group he or she belongs to. We can, therefore, randomize by a successful recovery and select for the sample particularly the 10 healed patients. Their healing is independent of their group membership. The distribution $H(31, 18, 10)$ of $N$ is, consequently, not affected by this selection. We will now reject $H_0 : p_1 = p_2$ and in particular $H_0 : p_1 \leq p_2$ if the entry $N$ in the cell healing $| M_1$ of the $2 \times 2$ contingency table of healing by drug is significantly large. Here we have $N = 6$ and $p = \sum_{k=6}^{10} H(31, 18, 10)(\{k\}) = 0.5967$, which is the right–tail case. This value is so large that we will not reject the hypothesis $H_0 : p_1 \leq p_2$.

The above reasoning about the independence of healing and group membership in the case of identical probabilities of success $p_1, p_2$ in the two groups is mathematically settled in the following result by putting $X =$ 'successful recovery' and $Y =$ 'group'.

**4.1.4 Lemma.** *Let $X$ and $Y$ be random variables such that $X$ takes values $a, b$ and $Y$ takes values $c, d$. If the conditional probabilities satisfy the equation*

$$P(X = a|Y = c) = P(X = a|Y = d),$$

*then $X$ and $Y$ are independent.*

**Proof:** Put $p := P(X = a | Y = c)$. We have

$$
\begin{aligned}
P\{X = a\} &= P(X = a | Y = c) P\{Y = c\} + P(X = a | Y = d) P\{Y = d\} \\
&= p(P\{Y = c\} + P\{Y = d\}) = p
\end{aligned}
$$

and, hence,

$$
\begin{aligned}
P\{X = a, Y = c\} &= P(X = a | Y = c) P\{Y = c\} \\
&= p P\{Y = c\} = P\{X = a\} P\{Y = c\}.
\end{aligned}
$$

This implies the assertion, cf Exercise 5.                                              □

The following result settles Fisher's test for homogeneity mathematically. It is an immediate consequence of the above lemma and Theorem 4.1.3. We consider again the $2 \times 2$ contingency table (4.1).

**4.1.5 Theorem.** *The random variables $X$ and $Y$ are independent under the null hypothesis $p_1 := P(X \in I_1 | Y \in J_1) = P(X \in I_1 | Y \in J_2) =: p_2$. Then the conclusion in Theorem 4.1.3 about the conditional hypergeometric distribution of the frequency count $N_{11} = \sum_{i=1}^{n} 1_{I_1}(X_i) 1_{J_1}(Y_i)$, given the marginal frequencies $\sum_{i=1}^{n} 1_{I_1}(X_i) = n_1.$ and $\sum_{i=1}^{n} 1_{J_1}(Y_i) = n_{.1}$, remains valid.*

The hypothesis $H_0 : p_1 = p_2$ of homogeneity will, consequently, be rejected if the realization $n_{11}$ is significantly large or, equivalently, if the corresponding right–tail $p$–value derived from the hypergeometric distribution is too small.

Like the Welch test in Section 2.3, Fisher's exact test is a *conditional* test, whose critical region for the rejection of $H_0$ depends on the data and is, therefore, a random variable itself. For a survey of exact inference for contingency tables we refer to Agresti (1992).

## General Contingency Tables and Homogeneity

In the sequel we will consider contingency tables of arbitrary dimensions.

**4.1.6 Lemma.** *We consider a random experiment which yields exactly one out of $I$ different possible outcomes $A_1, \ldots, A_I$ where the corresponding probabilities are $p_1, \ldots, p_I$ with $\sum_{i=1}^{I} p_i = 1$. Suppose this experiment is carried out independently $n$ times. The probability that simultaneously $A_1$ will occur exactly $n_1$ times, $A_2$ exactly $n_2$ times, \ldots, and $A_I$ exactly $n_I$ times with $\sum_{i=1}^{I} n_i = n$ is then given by*

$$
B(n, p_1, \ldots, p_I)(\{(n_1, \ldots, n_I)\}) := \binom{n}{n_1, \ldots, n_I} p_1^{n_1} \cdots p_I^{n_I}
$$

$$
:= \frac{n!}{n_1! \cdots n_I!} p_1^{n_1} \cdots p_I^{n_I}.
$$

**Proof:** Exercise 7. □

The distribution $B(n, p_1, \ldots, p_I)$ is the *multinomial* or *polynomial distribution* with $I$ cells and cell probabilities $p_1, \ldots, p_I$. It obviously generalizes the binomial distribution $B(n, p) = B(n, p, 1 - p)$, which is the case $I = 2$ of only two cells. The factors

$$\binom{n}{n_1, \ldots, n_I} = \frac{n!}{n_1! \cdots n_I!}$$

with $\sum_{i=1}^{I} n_i = n$ are called *polynomial coefficients*. Note that a histogram as defined in Section 1.1 follows a multinomial distribution if the underlying observations are independent and identically distributed.

Let now $\boldsymbol{X}_1, \ldots, \boldsymbol{X}_J$ be independent and each multinomial distributed random vectors with identical number $I$ of cells but possibly different number of experiments $n_{.j}$, $j = 1, \ldots, J$, i.e.,

$$P\{\boldsymbol{X}_j = (n_1, \ldots, n_I)^T\} = B\left(n_{.j}, p_{1j}, \ldots, p_{Ij}\right)\left(\{(n_1, \ldots, n_I)\}\right), \quad j = 1, \ldots, J.$$

By means of the realization of $(\boldsymbol{X}_1, \ldots, \boldsymbol{X}_J)$, displayed in the table below, we want to test the hypothesis that the cell probabilities of the $J$ multinomial distributions coincide:

$$H_0 : p_{i1} = p_{i2} = \ldots = p_{iJ} =: p_i, \qquad i = 1, \ldots, I. \tag{4.2}$$

Assuming that this hypothesis $H_0$ is valid, a plausible estimator of the probability $p_i$ of the $i$-th cell or group in the table

| | | | vectors | | | | |
|---|---|---|---|---|---|---|---|
| | | 1 | 2 | $\cdots$ | $j$ | $\cdots$ | $J$ | |
| groups | 1 | | | | | | | |
| | 2 | | | | | | | |
| | $\vdots$ | | | | | | | |
| | $i$ | | | | $n_{ij}$ | | | $\Sigma = n_{i.}$ |
| | $\vdots$ | | | | | | | |
| | $I$ | | | | | | | |
| | | | | | $\Sigma = n_{.j}$ | | | $\Sigma = n$ |

is the relative frequency of the $i$-th group

$$\hat{p}_i := \frac{\sum_{j=1}^{J} n_{ij}}{\sum_{i=1}^{I} \sum_{j=1}^{J} n_{ij}} =: \frac{n_{i.}}{n}.$$

The $j$-th column $(n_{1j}, \ldots, n_{Ij})^T$ in the above table displays the realization of the random vector $\boldsymbol{X}_j, j = 1, \ldots, J$. The estimator $(\hat{p}_1, \ldots, \hat{p}_I)$ is under (4.2) the *maximum–likelihood* estimator of the underlying vector $(p_1, \ldots, p_I)$: The probability of the given realization of $(\boldsymbol{X}_1, \ldots, \boldsymbol{X}_J)$ as a function of $(p_1, \ldots, p_I)$ is maximized for the argument $(\hat{p}_1, \ldots, \hat{p}_I)$. This is the content of the following result.

**4.1.7 Theorem.** *Suppose* $\boldsymbol{X}_1, \ldots, \boldsymbol{X}_J$ *are independent. Then we have under the hypothesis (4.2)*

$$\max_{0 \leq p_1, \ldots, p_I \leq 1, \sum_{i=1}^I p_i = 1} P\{\boldsymbol{X}_j = (n_{1j}, \ldots, n_{Ij})^T, \ j = 1, \ldots, J\}$$

$$= \max_{0 \leq p_1, \ldots, p_I \leq 1, \sum_{i=1}^I p_i = 1} \prod_{j=1}^J B(n_{.j}, \ p_1, \ldots, p_I)(\{(n_{1j}, \ldots, n_{Ij})\})$$

$$= \prod_{j=1}^J B(n_{.j}, \ \hat{p}_1, \ldots, \hat{p}_I)(\{(n_{1j}, \ldots, n_{Ij})\}).$$

**Proof:** By the strict monotonicity of the logarithm, the maximization problem above is equivalent to the maximization problem

$$\max_{0 \leq p_i \leq 1, \, 1 \leq i \leq I} \sum_{j=1}^J \log\left\{\binom{n_{.j}}{n_{1j}, \ldots, n_{Ij}} p_1^{n_{1j}} \ldots p_I^{n_{Ij}}\right\}$$

under the constraint $\sum_{i=1}^I p_i = 1$. We assume in the following that $n_{i.} = \sum_{j=1}^J n_{ij} > 0$, $i = 1, \ldots, I$. Since $p_1 + \cdots + p_I = 1$ or $p_I = 1 - (p_1 + \cdots + p_{I-1})$, respectively, the maximization problem above is equivalent to the maximization of the function

$$l(p_1, \ldots, p_{I-1})$$

$$:= \sum_{j=1}^J \log\left\{\binom{n_{.j}}{n_{1j}, \ldots, n_{Ij}} p_1^{n_{1j}} \cdots p_{I-1}^{n_{I-1j}} (1 - (p_1 + \cdots + p_{I-1}))^{n_{Ij}}\right\}$$

$$= \sum_{j=1}^J \log\binom{n_{.j}}{n_{1j}, \ldots, n_{Ij}} + \sum_{i=1}^{I-1} n_{i.} \log(p_i) + n_{I.} \log(1 - (p_1 + \cdots + p_{I-1}))$$

on the set of arguments $M := \{(p_1, \ldots, p_{I-1})^T \in [0, 1]^{I-1} : p_1 + \cdots + p_{I-1} \leq 1\}$. Since $n_{i.} > 0$, $i = 1, \ldots, I$, the function $l$ is $-\infty$ if $p_i = 0$ for some $i$. It, consequently, attains its maximum on the interior of the set $M$, i.e., on $\{(p_1, \ldots, p_{I-1})^T \in (0, 1)^{I-1} : p_1 + \cdots + p_{I-1} < 1\}$.

Taking partial derivatives of $l$ and equating them to zero, we obtain the following necessary condition for the maximizer $(p_1^0, \ldots, p_{I-1}^0)^T$:

$$0 = \frac{\partial l}{\partial p_i}(p_1^0, \ldots, p_{I-1}^0) = \frac{n_{i\cdot}}{p_i^0} - \frac{n_{I\cdot}}{1 - (p_1^0 + \cdots + p_{I-1}^0)}, \qquad i = 1, \ldots, I-1,$$

i.e.,

$$\frac{n_{i\cdot}}{n_{I\cdot}}(1 - (p_1^0 + \cdots + p_{I-1}^0)) = p_i^0, \qquad i = 1, \ldots, I-1,$$

or

$$p_i^0 = c\,\frac{n_{i\cdot}}{n_{I\cdot}}, \qquad i = 1, \ldots, I-1,$$

with $c = 1 - (p_1^0 + \cdots + p_{I-1}^0) \in (0,1)$. Substituting $p_i^0$, in the above equation we obtain

$$1 - c\,\frac{n_{1\cdot} + \cdots + n_{I-1\cdot}}{n_{I\cdot}} = c$$

or $c = n_{I\cdot}/(n_{1\cdot} + \cdots + n_{I\cdot}) = n_{I\cdot}/n$. This yields

$$p_i^0 = \frac{n_{i\cdot}}{n} = \hat{p}_i, \qquad i = 1, \ldots, I-1,$$

and, thus, the assertion.                                                       □

## The Chi–Square Goodness–of–Fit Test

The following results on the relationship between multinomial distributed random vectors and the chi–square distribution will be quite useful. A suitable version of the multivariate central limit theorem implies the next theorem, cf Section 2.7 in Serfling (1980).

**4.1.8 Theorem.** *Let $X := (n_1, \ldots, n_I)^T$ be a multinomial distributed random vector with the parameters $n$ and $p_1, \ldots, p_I$, where $p_i > 0$ for $i = 1, \ldots, I$. Then we have, as $n \to \infty$,*

$$X_n := n^{1/2}\left(\frac{n_1}{n} - p_1, \ldots, \frac{n_I}{n} - p_I\right)^T \xrightarrow{\mathcal{D}} N(0, \Sigma),$$

*where $\Sigma = (\sigma_{ij})$ is the $I \times I$–matrix with*

$$\sigma_{ij} = \begin{cases} p_i(1 - p_i) & \text{for } i = j \\ -p_i p_j & \text{for } i \neq j. \end{cases}$$

A common statistic for testing for particular values of cell probabilities $p_i$ is the *chi–square statistic*

$$X_n^2 := \sum_{i=1}^{I} \frac{(n_i - np_i)^2}{np_i} = n\sum_{i=1}^{I} \frac{1}{p_i}\left(\frac{n_i}{n} - p_i\right)^2,$$

where $np_i = E(n_i)$ is the expected frequency of the $i$-th cell, $i = 1, \ldots, I$, see Exercise 8. The statistic $X_n^2$ can obviously be written as a quadratic form in $\boldsymbol{X}_n$

$$X_n^2 = \boldsymbol{X}_n^T \boldsymbol{C} \boldsymbol{X}_n,$$

where

$$\boldsymbol{C} := \begin{pmatrix} 1/p_1 & & 0 \\ & \ddots & \\ 0 & & 1/p_I \end{pmatrix}.$$

Use this representation together with the asymptotic normality of $\boldsymbol{X}_n$ and the equation $\boldsymbol{C}\boldsymbol{\Sigma}\boldsymbol{C}\boldsymbol{\Sigma} = \boldsymbol{C}\boldsymbol{\Sigma}$, and show that

$$X_n^2 \xrightarrow[D]{} \chi_{I-1}^2, \tag{4.3}$$

see Exercises 23 and 27, Chapter 3 and Exercise 18 in Chapter 6. This convergence is the reason for calling $X_n^2$ as the chi–square statistic. As a rule of thumb, $X_n^2$ is sufficiently close to a chi–square distribution if $np_i > 5$ for each cell probability $p_i$, note the warning by SAS in Figure 4.1.4. Observe that the degrees of freedom $I - 1$ of the chi–square distribution coincide with the numbers of *free* parameters among $p_1, \ldots, p_I$, since $p_1 + \cdots + p_I = 1$ implies that $I - 1$ of the parameters $p_1, \ldots, p_I$ determine the left hand side.

Suppose that we want to test the hypothesis that a sample $X_1, \ldots, X_n$ of independent and identically distributed random variables comes from a particular theoretical distribution function $F$. To this end we divide the real line into disjoint intervals $A_1, \ldots, A_I$ and compare their empirical frequencies $n_i/n = \sum_{j=1}^n 1_{A_i}(X_j)/n$ with the theoretical probabilities $p_i := F(A_i)$, $i = 1, \ldots, I$. If $n$ is sufficiently large and if actually $p_i = P\{X_1 \in A_i\}$, then the corresponding test statistic $X_n^2$ is approximately $\chi_{I-1}^2$ distributed. If, however, $p_i \neq P\{X_1 \in A_i\}$ for at least one index $i$, then $X_n^2$ will have the tendency to attain large values. The approximate $p$–value $p = 1 - \chi_{I-1}^2(X_n^2)$ will, consequently, have the tendency to attain small values and, thus, will tend to reject the hypothesis that $F$ is the underlying distribution function. This is a version of the *chi–square goodness–of–fit test*.

**4.1.9 Example** (Claim Size Data). The following bar chart shows the grouped claim sizes exceeding 30,000 German marks of a liability insurance company in 1983. The data are taken from Falk et al. (1994), Section 6.2.

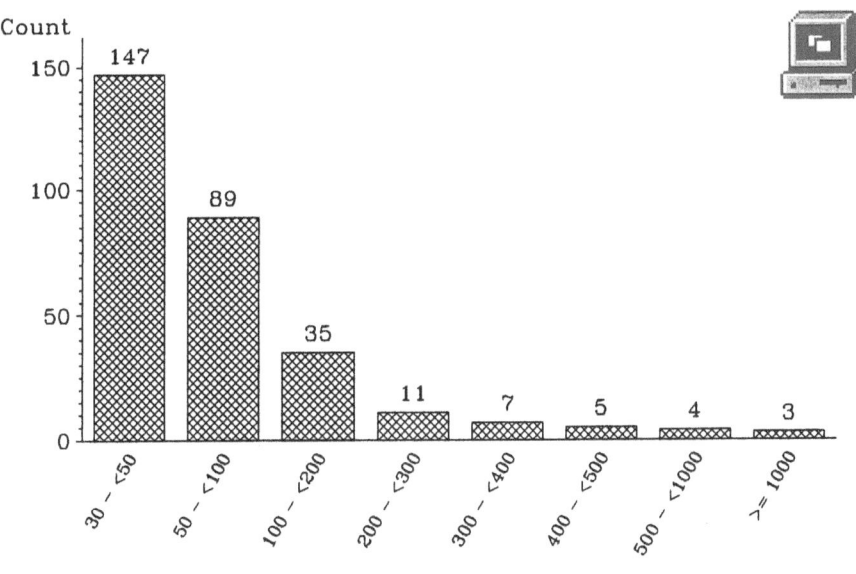

Figure 4.1.5. Bar chart of the claim sizes.

```
***    Program 4_1_5    ***;
TITLE1 'Bar Chart';
TITLE2 'Claim Size Data';
LIBNAME datalib 'c:\data';

AXIS1 LABEL=('Count') MAJOR=(N=4);
AXIS2 LABEL=('Interval') VALUE=(H=1.6 A=60);
PROC GCHART DATA=datalib.claim;
    VBAR interval / FREQ=count FREQ DESCENDING
                    RAXIS=AXIS1 MAXIS=AXIS2;
RUN; QUIT;
```

This output produced by PROC GCHART shows the following special features:

The text strings of the horizontal axis are rotated by an angle of 60 degrees. This is performed by the value 'A=60' in the AXIS2 statement.

The option 'FREQ' in the VBAR statement puts the counts of the frequency variable on top of the bars.

The frequencies are displayed in descending order by the option DESCENDING.

Since the input data set is already a frequency table, the variable containing the frequencies has to be specified in the 'FREQ=count' option.

Note that for easier reception the widths of the bars in the displayed bar chart are smaller than the widths of the pertaining intervals.

We model the claim sizes as realizations of $n = 301$ independent and identically *Pareto distributed* random variables $X_1, \ldots, X_n$, i.e.,

$$P\{X_i \le x\} = 1 - \left(\frac{x}{\sigma}\right)^{-\alpha}, \qquad x \ge \sigma,$$

with the scale parameter $\sigma = 30$ and some shape parameter $\alpha > 0$. We apply the chi–square goodness–of–fit test to the above bar chart with $\alpha = 1.5$. The theoretical probabilities of the above intervals $A_i = [s_i, s_{i+1})$ are

$$p_i = \left(\frac{s_i}{\sigma}\right)^{-\alpha} - \left(\frac{s_{i+1}}{\sigma}\right)^{-\alpha} = \sigma^{\alpha}(s_i^{-\alpha} - s_{i+1}^{-\alpha}), \qquad i = 1, \ldots, 8.$$

The chi–square statistic is $X_{301}^2 = 14.4697$ for the above parameters $\alpha = 1.5$ and $\sigma = 30$. We have 7 degrees of freedom and, hence, the approximate $p$–value is $1 - \chi_7^2(14.4697) = 0.0434$. The hypothesis $H_0$ that the claim sizes were independently generated by the Pareto distribution with parameters $\sigma = 30$ and $\alpha = 1.5$ would, consequently, be rejected at level 5 %.

| INT | COUNT | expected | diff | p_i | x2_n | p |
|-----|-------|----------|------|-----|------|---|
| 1 | 147 | 161.108 | -14.1078 | 0.53524 | 1.2354 | . |
| 2 | 89 | 90.433 | -1.4328 | 0.30044 | 1.2581 | . |
| 3 | 35 | 31.973 | 3.0272 | 0.10622 | 1.5447 | . |
| 4 | 11 | 7.968 | 3.0319 | 0.02647 | 2.6984 | . |
| 5 | 7 | 3.336 | 3.6640 | 0.01108 | 6.7225 | . |
| 6 | 5 | 1.759 | 3.2414 | 0.00584 | 12.6967 | . |
| 7 | 4 | 2.860 | 1.1403 | 0.00950 | 13.1513 | . |
| 8 | 3 | 1.564 | 1.4360 | 0.00520 | 14.4697 | 0.043431 |

**Figure 4.1.6.** Chi–square goodness–of–fit test of the claim size data; hypothesis: Pareto distributed with $\sigma = 30$ and $\alpha = 1.5$.

```
***    Program 4_1_6    ***;
TITLE1 'Chi-Square Goodness-of-Fit Test';
TITLE2 'Claim Size Data';
LIBNAME datalib 'c:\data';
                        ↓
```

```
                                        ↑
 DATA data1;
    sigma=30; alpha=1.5;
    SET datalib.claim;
    p_i=(lowbound/sigma)**(-alpha) - (uprbound/sigma)**(-alpha);
    expected=p_i*301;
    diff=count-expected;
    x2_n+(diff**2/(301*p_i));               * recursive summation;
    IF _N_=8 THEN p=1-PROBCHI(x2_n,7);      * p-value;

 PROC PRINT DATA=data1 NOOBS;
    VAR no count expected diff p_i x2_n p;
 RUN; QUIT;
```

The computation of the interval probabilities of the Pareto distribution requires the specification of the parameters $\sigma$ (here sigma=30) and $\alpha$ (here alpha=1.5) as well as of the lower and upper interval bounds (lowbound, up-

rbound).

Note that the DATA step is automatically repeated until each of the eight values of the variables lowbound and uprbound of the file 'claim' is used.

## The Chi–Square Test for Homogeneity

The convergence (4.3) suggests the statistic

$$X^2 := \sum_{j=1}^{J} \sum_{i=1}^{I} \frac{(n_{ij} - n_{\cdot j}\hat{p}_i)^2}{n_{\cdot j}\hat{p}_i},$$

for testing the hypothesis $H_0 : p_{i1} = \ldots = p_{iJ}, i = 1, \ldots, I$, in (4.2). We assume $J$ independent and multinomial distributed random vectors, and we denote by $(\hat{p}_1, \ldots, \hat{p}_I)$ with $\hat{p}_i = n_{i\cdot}/n, i = 1, \ldots, I$, the maximum–likelihood estimator of $(p_1, \ldots, p_I)$. The test statistic $X^2$ is *Pearson's chi–square statistic* for testing homogeneity. By the convergence (4.3) and the convolution theorem of the chi–square distribution in Remark 2.1.8 one expects that $X^2$ is approximately chi–square distributed under the hypothesis $H_0$ in (4.2) if the sample sizes $n_{\cdot j}, j = 1, \ldots, J$, are large enough. The degrees of freedom $df$ should be $J(I-1)$ minus the number of free estimated parameters, i.e.,

$$df = J(I - 1) - (I - 1) = (I - 1)(J - 1).$$

The next result shows that this is actually true.

**4.1.10 Theorem.** *If $p_i > 0, i = 1, \ldots, I$, then we have under the hypothesis (4.2)*

$$X^2 \xrightarrow[D]{} \chi^2_{(I-1)(J-1)},$$

*as $n_{\cdot j} \to \infty, j = 1, \ldots, J$.*

**Proof:** See, for instance, Section 6c in Rao (1973). □

**4.1.11 Example** (Suicide Data). Heuer (1979) investigated suicide rates in West Germany. The following *block chart* displays frequencies of suicides by sex by $I = 6$ different methods. The problem is to decide whether sex has an influence on the method of suicide, i.e., whether the hypothesis (4.2) of identical preferences $p_{i1} = p_{i2}$ for the $i$–th method, $i = 1, \ldots, 6$, of males and females is to be rejected. The data are analyzed in Friendly (1994).

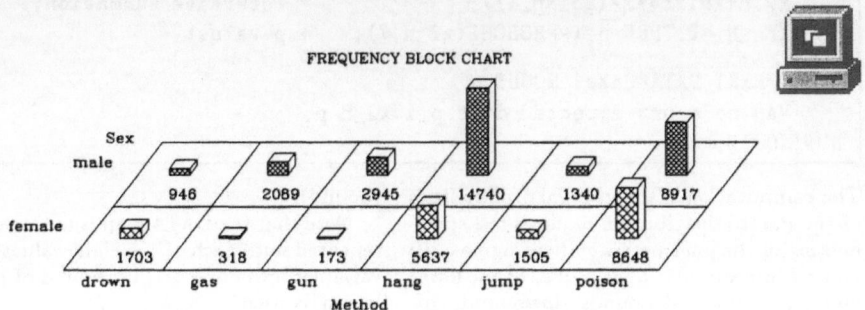

**Figure 4.1.7.** Block chart of suicide data.

```
***    Program 4_1_7    ***;
TITLE1 'Block Chart';
TITLE2 'Suicide Data';
LIBNAME datalib 'c:\data';

DATA datalib.suicide;
    LABEL sex='Sex' method='Method';
    sex='  male'  ;  INPUT method $ count @@;   OUTPUT;
    sex='  female';  INPUT count @@;  OUTPUT;
    CARDS;
 poison    8917    8648
 gas       2089     318
 hang     14740    5637
 drown      946    1703
 gun       2945     173
 jump      1340    1505
 ;

GOPTION HTEXT=2 HPOS=120 VPOS=75;
PROC GCHART DATA=datalib.suicide;
    BLOCK method / GROUP=sex FREQ=count FREQ PATTERNID=GROUP;
RUN; QUIT;
```

The variable 'sex' is first set to 'male' in the DATA step and the frequencies in the first column are read from the CARDS section. The specification @@ suppresses the automatic line feed at the end of each INPUT statement.

These values of the variables 'sex', 'method' and 'count' are written to the file 'datalib.suicide' by the option OUTPUT. Then the variable sex is set to 'female' and the frequencies in the second column are read from the CARDS section. Again, these values of sex, method and count are written to datalib.suicide.

The statements VBAR, HBAR and BLOCK of the procedure GCHART entail the drawing of multidimensional contingency tables by using the options GROUP and SUBGROUP.

The option PATTERNID=GROUP changes patterns every time the value of the group variable (here sex) changes. All blocks in each group use the same pattern, but a different pattern is used for each group. The values of the character variables appear in alphabetical order.

Pearson's chi-square test for the hypothesis $H_0$ that the preferences for a particular method of suicide coincide for males and females yields the following result:

```
                    The FREQ Procedure

              Statistics for Table of METHOD by SEX


     Statistic                     DF      Value      Prob
     ------------------------------------------------------
     Chi-Square                     5    4965.7301    <.0001
     Likelihood Ratio Chi-Square    5    5400.5748    <.0001
     Mantel-Haenszel Chi-Square     1     980.4999    <.0001
     Phi Coefficient                       0.3185
     Contingency Coefficient               0.3035
     Cramer's V                            0.3185


                    Sample Size = 48961
```

**Figure 4.1.8.** Pearson's chi–square test for homogeneity of the suicide data.

Program 4_1_8, which produces the above output, uses the data 'datalib.suicide' from Program 4_1_7 and is analogous to Program 4_1_2. The options in the TABLES statement are NOPRINT and CHISQ.

Pearson's chi–square statistic is in this case $X^2 = 4\,965.73$ with $(I-1)(J-1) = 5 \cdot 1 = 5$ degrees of freedom. The corresponding $p$–value is almost 0. Therefore, the hypothesis of identical preferences $p_{i1} = p_{i2}$, $i = 1, \ldots, 6$, for a particular method of suicide of males and females is rejected.

## The Chi–Square Test for Independence

In the sequel we will derive the chi–square test for independence. Its application is quite similar to that of the above chi–square test for homogeneity. It aims, however, at a different problem.

We consider now just *one* multinomial distributed random variable with $I \cdot J$ cells. We can represent this random variable as a $I \times J$–dimensional random matrix $(n_{ij})$, with a fixed total sum $n$ of entries $\sum_{i=1}^{I} \sum_{j=1}^{J} n_{ij} = n$. By $p_{ij}$ we denote the probability of occurrence of the cell with row index $i$ and column index $j$. Further,

$$p_{i.} := \sum_{j=1}^{J} p_{ij}, \quad p_{.j} := \sum_{i=1}^{I} p_{ij}$$

are the probabilities of occurrence of the $i$–th row and the $j$–th column, respectively. If these two events are independent, then we have

$$p_{ij} = p_{i.} \, p_{.j} \, .$$

We want to test the hypothesis that this independence is actually true. Our null hypothesis is, therefore,

$$H_0 : p_{ij} = p_{i.}p_{.j} \, , \qquad i = 1, \dots, I, \quad j = 1, \dots, J. \tag{4.4}$$

The maximum–likelihood estimator of $p_{ij}$ is under $H_0$

$$\hat{p}_{ij} = \frac{n_{i.}}{n} \frac{n_{.j}}{n} =: \hat{p}_{i.}\hat{p}_{.j} \, ,$$

cf Exercise 12. In the general case, i.e., without the assumption of independence, the maximum–likelihood estimator of $p_{ij}$ is

$$\tilde{p}_{ij} := \frac{n_{ij}}{n},$$

cf Exercise 12. With these two estimators of the cell probability matrix $(p_{ij})$, we can check the hypothesis $H_0$ by means of the *likelihood ratio*

$$L_n := \prod_{i \leq I, j \leq J} \left( \frac{\tilde{p}_{ij}}{\hat{p}_{ij}} \right)^{n_{ij}} .$$

The denominator of this statistic uses the maximum–likelihood estimator $\hat{p}_{ij}$ under $H_0$ and the numerator uses the maximum–likelihood estimator $\tilde{p}_{ij}$ in the general case. The likelihood ratio $L_n$ will, therefore, be greater than one in general and the hypothesis $H_0$ will be rejected if it is significantly larger than 1. This is a version of a *likelihood ratio test*. As $n \to \infty$, the test statistic

$$2 \log(L_n) = 2 \sum_{i=1}^{I} \sum_{j=1}^{J} n_{ij} \log \left( \frac{\tilde{p}_{ij}}{\hat{p}_{ij}} \right)$$

is chi–square distributed with $(I-1)(J-1)$ degrees of freedom if $p_{ij} > 0$, as shown in Theorem 4.1.12 below.

Since $\log(1+\varepsilon) \approx \varepsilon - \varepsilon^2/2$ as $\varepsilon \to 0$, under $H_0$, we get the approximation

$$2\log(L_n) = -2\sum_{i=1}^{I}\sum_{j=1}^{J} n_{ij} \log\left(1 + \left(\frac{\hat{p}_{ij}}{\tilde{p}_{ij}} - 1\right)\right)$$

$$\approx -2\sum_{i=1}^{I}\sum_{j=1}^{J} n_{ij} \left(\frac{\hat{p}_{ij}}{\tilde{p}_{ij}} - 1 - \frac{1}{2}\left(\frac{\hat{p}_{ij}}{\tilde{p}_{ij}} - 1\right)^2\right)$$

$$= \sum_{i=1}^{I}\sum_{j=1}^{J} n_{ij} \left(\frac{\hat{p}_{ij}}{\tilde{p}_{ij}} - 1\right)^2$$

$$\approx \sum_{i=1}^{I}\sum_{j=1}^{J} \frac{(n_{ij} - n\hat{p}_{ij})^2}{n\hat{p}_{ij}} = \sum_{i=1}^{I}\sum_{j=1}^{J} \frac{(n_{ij} - n_{\cdot j}n_{i\cdot}/n)^2}{n_{\cdot j}n_{i\cdot}/n} = X^2.$$

One will expect, therefore, that under the independence (4.4) the distribution of $2\log(L_n)$ is close to the chi–square distribution with $(I-1)(J-1)$ degrees of freedom if $n$ is large. This is actually true, i.e., Theorem 4.1.10 carries over as $n \to \infty$, see e.g. Section 6d in Rao (1973).

**4.1.12 Theorem.** *If $p_{ij} > 0$ for $i = 1, \ldots, I$ and $j = 1, \ldots, J$, then we have, under the hypothesis (4.4), as $n \to \infty$*

$$2\log(L_n) \xrightarrow{D} \chi^2_{(I-1)(J-1)}.$$

**4.1.13 Example** (Hair Color Data). To understand the nature of the association between hair and eye color, Snee (1974) collected these data among 592 students in a statistics course. The following table of eyes by hair color summarizes the data. For a discussion of this table we refer to Friendly (1994).

| Frequency | black | blond | brown | red | Total |
|-----------|-------|-------|-------|-----|-------|
| blue      | 20    | 94    | 84    | 17  | 215   |
| brown     | 68    | 7     | 119   | 26  | 220   |
| green     | 5     | 16    | 29    | 14  | 64    |
| hazel     | 15    | 10    | 54    | 14  | 93    |
| Total     | 108   | 127   | 286   | 71  | 592   |

Statistics for Table of eyes by hair

| Statistic | DF | Value | Prob |
|---|---|---|---|
| Chi-Square | 9 | 138.2898 | <.0001 |
| Likelihood Ratio Chi-Square | 9 | 146.4436 | <.0001 |
| Mantel-Haenszel Chi-Square | 1 | 8.6058 | 0.0034 |
| Phi Coefficient | | 0.4833 | |
| Contingency Coefficient | | 0.4352 | |
| Cramer's V | | 0.2790 | |

Sample Size = 592

**Figure 4.1.9.** Table and Pearson's chi–square test for independence of eye color and hair color.

```
***    Program 4_1_9    ***;
TITLE1 'Chi-Square Test for Independence';
TITLE2 'Hair Color Data';

DATA data1;
    INPUT eyes $ @@;
    DO hair='black', 'brown', 'red', 'blond';
       INPUT count @@; OUTPUT;
    END;
    CARDS;
  brown      68   119  26    7
  blue       20    84  17   94
  hazel      15    54  14   10
  green       5    29  14   16
  ;

PROC FREQ DATA=data1;
    TABLES eyes*hair / NOROW NOCOL NOPERCENT CHISQ;
    WEIGHT count;
RUN; QUIT;
```

The DATA step in this program is of particular interest, as it extends the way of reading data in Program 4_1_7.

The eye color is read in the first INPUT statement from the first column of the CARDS section. The subsequent DO loop does not use numbers for counting the loops such as DO i=1 TO 4, but character variables. First, 'hair' is set to 'black' and the pertaining frequencies are read by the INPUT statement from the second column of the CARDS section. Then hair is set to 'brown' etc.

In this case we have $J = 4$ and $I = 4$. Pearson's chi–square statistic is 138.29 with $(I - 1)(J - 1) = 9$ degrees of freedom. The likelihood ratio test statistic is $2 \log L_n = 146.444$ with 9 degrees of freedom as well. The corresponding $p$–values are almost 0, the hypothesis of the independence of hair color and eye color has to be rejected, consequently.

## 4.2 Categorical Regression

In the following we introduce the simplest example of a *categorical regression*. Consider a random variable $Z$ that can take only two values; a popular example is the creditworthiness of a client, who is financially sound enough to justify the extension of a credit or not. Such a random variable is called *dichotomous* or *binary* and we can code its two possible outcomes by 0 and 1, i.e., $Z \in \{0, 1\}$. We provide in the sequel various models, where the probability that $Z$ takes the value 1 is explained by a list of variables.

**4.2.1 Example** (Divorce Data). The following table is taken from a study of divorce patterns by Thornes and Collard (1979). A sample of about 500 people who had requested for divorce and a similar number of married people were asked the following two questions about their premarital and extramarital sexual experience: (1) "Did you have sexual intercourse with anyone other than your later husband/wife before your marriage?" (2) "Did you have any affairs or brief sexual encounters with another man/woman during your (former) marriage?'

Can the current marital status in this example, i.e., the probability of petitioning for divorce, be explained by gender, reported premarital and extramarital sex? Do the data follow such a pattern? For an analysis of these data we refer to Friendly (1994).

| Marital Status | Gender | Premarital Sex | Extramarital Sex | COUNT |
|---|---|---|---|---|
| divorced | female | no | no | 214 |
| divorced | female | no | yes | 36 |
| divorced | female | yes | no | 54 |
| divorced | female | yes | yes | 17 |
| divorced | male | no | no | 68 |
| divorced | male | no | yes | 17 |
| divorced | male | yes | no | 60 |
| divorced | male | yes | yes | 28 |
| married | female | no | no | 322 |
| married | female | no | yes | 4 |

| married | female | yes | no  | 25  |
|---------|--------|-----|-----|-----|
| married | female | yes | yes | 4   |
| married | male   | no  | no  | 130 |
| married | male   | no  | yes | 4   |
| married | male   | yes | no  | 42  |
| married | male   | yes | yes | 11  |

**Figure 4.2.1.** Divorce data.

```
***    Program 4_2_1    ***;
TITLE 'Divorce Data';
LIBNAME datalib 'c:\data';

DATA datalib.divorce;
    DO status='divorced', 'married';
        DO gender='female', 'male';
            DO presex='no', 'yes';
                DO extrasex='no', 'yes';
                    INPUT count @@; OUTPUT;
            END; END; END; END;
        LABEL status='Marital*Status' gender='Gender'
              presex='Premarital*Sex'
              extrasex='Extramarital*Sex';
        CARDS;
  214   36   54   17   68   17   60   28
  322    4   25    4  130    4   42   11
  ;

PROC PRINT DATA=datalib.divorce NOOBS SPLIT='*';
RUN; QUIT;
```

The nonstandard way of reading external data, which is used in the above DATA step, is again an extension of the Programs 4\_1\_7 and 4\_1\_9. This technique, which is quite common in agricultural experiments (split–plot design, cf Program 5\_2\_5), can consid-erably accelerate the processing of data.

The option SPLIT='\*' in the PROC PRINT statement determines the single char-acter in the text of the label, where SAS splits its display.

## Dummy and Effect Coding

We assume the special case that the probability for the outcome $Z = 1$ is a linear response function of some explanatory variables $x_1, \ldots, x_{p-1}$:

$$P(Z = 1|\boldsymbol{x}) = \beta_0 + x_1\beta_1 + \cdots + x_{p-1}\beta_{p-1} = (1, \boldsymbol{x}^T)\boldsymbol{\beta},$$

where $x = (x_1, \ldots, x_{p-1})^T$, $\beta = (\beta_0, \beta_1, \ldots, \beta_{p-1})^T$. We replace categorical data such as gender, extra– or premarital sex in Example 4.2.1 by *dummy variables*: Suppose a class variable $A$ such as gender has $p$ distinct *levels* or *categories*. We can code the actual level of $A$ by the vector $x^A$, whose $j$-th component is

$$x_j^A = \begin{cases} 1 & \text{if } A \text{ has level } j \\ 0 & \text{elsewhere,} \end{cases}$$

$j = 1, \ldots, p-1$. The $p$-th level is then coded by the vector $(0, \ldots, 0)^T$ with $p-1$ zeros. The vector $x^A = (x_1^A, \ldots, x_{p-1}^A)^T$ represents, therefore, all possible levels of the variable $A$. A simple model, which represents only the influence of $A$ on the probability for the outcome $Z = 1$, is

$$P(Z = 1|x^A) = \beta_0 + x_1^A\beta_1 + \cdots + x_{p-1}^A\beta_{p-1}.$$

If $A$ has level $j \in \{1, \ldots, p-1\}$, we have

$$P(Z = 1|x^A) = \beta_0 + \beta_j$$

and for the level $p$

$$P(Z = 1|x^A) = \beta_0.$$

This is the *dummy coding*. The coding

$$x_j^A = \begin{cases} 1 & \text{if } A \text{ has level } j, \\ -1 & \text{if } A \text{ has level } p, \\ 0 & \text{elsewhere,} \end{cases}$$

$j = 1, \ldots, p-1$, is called *effect coding* (see, for example, the bottom of page 165 and Section 8.1 in Jobson (1992) for further discussion). In this case we obtain

$$P(Z = 1|x^A) = \beta_0 + \beta_j$$

for the levels $j = 1, \ldots, p-1$, and for $j = p$

$$P(Z = 1|x^A) = \beta_0 - \beta_1 - \cdots - \beta_{p-1}.$$

## Linear Main Effects

Suppose an analysis contains *several* class variables $A, B, C$, say, with different levels $p_A, p_B$ and $p_C$. With the corresponding dummy variables $x_i^A$, $x_j^B$, $x_k^C$, $i = 1, \ldots, p_A - 1$, $j = 1, \ldots, p_B - 1$, $k = 1, \ldots, p_C - 1$ we obtain the coding vector

$$x = \left(x_1^A, \ldots, x_{p_A-1}^A, x_1^B, \ldots, x_{p_B-1}^B, x_1^C, \ldots, x_{p_C-1}^C\right)^T.$$

Denoting by

$$\beta = \left(\beta_0, \beta_1^A, \ldots, \beta_{p_A-1}^A, \beta_1^B, \ldots, \beta_{p_B-1}^B, \beta_1^C, \ldots, \beta_{p_C-1}^C\right)^T$$

the vector of *main effects*, we now obtain the *linear model*

$$P(Z = 1|\boldsymbol{x}) = (1, \boldsymbol{x}^T)\beta.$$

*Crossed effects* or *interactions* of two or more class variables, such as synergistic pairs, can be included into the above linear regression model by considering for instance products $x_i^A \cdot x_j^B \cdot x_k^C$ of dummy variables and dummy coding.

## The Linear Model

We denote all possible different outcomes of the coding vector $\boldsymbol{x}$ by $\boldsymbol{x}_1, \ldots, \boldsymbol{x}_I$. The vector of the dependent variable $P(Z = 1|\cdot)$ is

$$\alpha := \Big(P(Z = 1|\boldsymbol{x}_1), \ldots, P(Z = 1|\boldsymbol{x}_I)\Big)^T,$$

and

$$\boldsymbol{X} := \begin{pmatrix} 1 & \boldsymbol{x}_1^T \\ \vdots & \vdots \\ 1 & \boldsymbol{x}_I^T \end{pmatrix}$$

denotes the design matrix. The *linear categorical regression model* is now

$$\alpha = \boldsymbol{X}\beta. \tag{4.5}$$

We suppose in the following that the columns of the design matrix $\boldsymbol{X}$ are linearly independent vectors. If our model (4.5) includes, for example, an intercept $\beta_0$ together with $p-1$ binary explanatory variables, we require, therefore, that

$$Rank\,(\boldsymbol{X}) = p.$$

## Generalized Linear Model (GLIM)

A popular generalization of the linear model

$$\alpha_i = (1, \boldsymbol{x}_i^T)\beta, \qquad i = 1, \ldots, I,$$

is the model

$$g(\alpha_i) = (1, \boldsymbol{x}_i^T)\beta$$

or

$$\alpha_i = g^{-1}((1, \boldsymbol{x}_i^T)\beta),$$

where $g : (0,1) \longrightarrow \mathbb{R}$ is a one–to–one function, called *link function*. With $g(\alpha) := (g(\alpha_1), \ldots, g(\alpha_I))^T$ we obtain a *generalized linear model (GLIM)*

$$g(\alpha) = X\beta. \tag{4.6}$$

Note that even if the link function $g$ is known, the individual probabilities $\alpha_i$ are typically unknown and must be estimated from the data.

## Estimation of $\alpha_i$

Suppose we have $n_i > 0$ independent realizations $Z_1^{(i)}, \ldots, Z_{n_i}^{(i)}$ of $Z$ for the coding vector $x_i$. A natural estimator of $\alpha_i = P(Z = 1|x_i)$ is the relative frequency

$$\hat{\alpha}_i := \frac{1}{n_i} \sum_{j=1}^{n_i} Z_j^{(i)} = \frac{\text{number of } Z_1^{(i)}, \ldots, Z_{n_i}^{(i)} \text{ with value } 1}{n_i}.$$

If $n_i$ is fixed, the random frequency $n_i \hat{\alpha}_i$ is binomial distributed with parameters $n_i$ and $\alpha_i$. Equation (4.6) now suggests the model

$$g(\hat{\alpha}_i) = (1, x_i^T)\beta + \varepsilon_i, \qquad i = 1, \ldots, I,$$

or, respectively,

$$g(\hat{\alpha}) = X\beta + \varepsilon \tag{4.7}$$

with $\hat{\alpha} = (\hat{\alpha}_1, \ldots, \hat{\alpha}_I)^T$ and the vector of errors $\varepsilon = (\varepsilon_1, \ldots, \varepsilon_I)^T$.

As an example we fit to the estimated probabilities of divorce $\hat{\alpha}_i$ in Example 4.2.1 a function of the three binary class variables *gender* $(A)$ with the levels female and male $(f \,|\, m)$, *premarital sex* $(B)$ with the levels no and yes $(n \,|\, y)$ and *extramarital sex* $(C)$ with the levels no and yes $(n \,|\, y)$. Using effect coding and including the interaction $BC$ of premarital and extramarital sex we obtain the model

$$g \begin{pmatrix} \hat{\alpha}_1 \\ \hat{\alpha}_2 \\ \hat{\alpha}_3 \\ \hat{\alpha}_4 \\ \hat{\alpha}_5 \\ \hat{\alpha}_6 \\ \hat{\alpha}_7 \\ \hat{\alpha}_8 \end{pmatrix} = \begin{pmatrix} 1 & 1 & 1 & 1 & 1 \\ 1 & 1 & 1 & -1 & -1 \\ 1 & 1 & -1 & 1 & -1 \\ 1 & 1 & -1 & -1 & 1 \\ 1 & -1 & 1 & 1 & 1 \\ 1 & -1 & 1 & -1 & -1 \\ 1 & -1 & -1 & 1 & -1 \\ 1 & -1 & -1 & -1 & 1 \end{pmatrix} \begin{pmatrix} \beta_0 \\ \beta_A \\ \beta_B \\ \beta_C \\ \beta_{BC} \end{pmatrix} + \begin{pmatrix} \varepsilon_1 \\ \varepsilon_2 \\ \varepsilon_3 \\ \varepsilon_4 \\ \varepsilon_5 \\ \varepsilon_6 \\ \varepsilon_7 \\ \varepsilon_8 \end{pmatrix}.$$

From the table in Figure 4.2.1 we obtain the vector of estimated probabilities of divorce

$$
\begin{pmatrix} \hat{\alpha}_1 \\ \hat{\alpha}_2 \\ \hat{\alpha}_3 \\ \hat{\alpha}_4 \\ \hat{\alpha}_5 \\ \hat{\alpha}_6 \\ \hat{\alpha}_7 \\ \hat{\alpha}_8 \end{pmatrix}
=
\begin{pmatrix} 214/536 \\ 36/40 \\ 54/79 \\ 17/21 \\ 68/198 \\ 17/21 \\ 60/102 \\ 28/39 \end{pmatrix}
\approx
\begin{pmatrix} 0.4 \\ 0.9 \\ 0.68 \\ 0.81 \\ 0.34 \\ 0.81 \\ 0.59 \\ 0.72 \end{pmatrix}.
$$

The highest probability of divorce is $\hat{\alpha}_2 = 0.9$, which corresponds to women, who had no premarital sex with a partner different from their later husbands, but who had extramarital sex during their marriage.

For the estimation of the unknown vector $\beta$ in the general model (4.7) we will use a weighted–least–squares estimator. The covariance matrix $\Sigma_{\hat{\alpha}}$ of $\hat{\alpha}$ is

$$
\Sigma_{\hat{\alpha}} =
\begin{pmatrix} \dfrac{\alpha_1(1-\alpha_1)}{n_1} & & 0 \\ & \ddots & \\ 0 & & \dfrac{\alpha_I(1-\alpha_I)}{n_I} \end{pmatrix}
$$

if we assume that we have $n_i > 0$ independent realizations of $Z$ which are coded by $x_i$, and that all realizations are independent as well.

**4.2.2 Theorem.** *Suppose that the link function $g : (0,1) \to \mathbb{R}$ is continuously differentiable. Put $n := \sum_{i=1}^{I} n_i$, and assume that $n_i \to \infty$ and $n/n_i \to \lambda_i \in [1,\infty)$, $i = 1,\ldots,I$. Then we have*

$$
n^{1/2}\left(g(\hat{\alpha}) - g(\alpha)\right) \xrightarrow{\mathcal{D}} N(0,\Sigma),
$$

*where*

$$
\Sigma =
\begin{pmatrix} \lambda_1 g'(\alpha_1)^2 \alpha_1(1-\alpha_1) & & 0 \\ & \ddots & \\ 0 & & \lambda_I g'(\alpha_I)^2 \alpha_I(1-\alpha_I) \end{pmatrix}.
$$

**Proof:** Taylor's formula implies for $i = 1,\ldots,I$

$$
\begin{aligned}
g(\hat{\alpha}_i) - g(\alpha_i) &= g'(\alpha_i)(\hat{\alpha}_i - \alpha_i) + (g'(\xi_i) - g'(\alpha_i))(\hat{\alpha}_i - \alpha_i) \\
&= g'(\alpha_i)(\hat{\alpha}_i - \alpha_i) + r_{n_i},
\end{aligned}
$$

with $\xi_i$ between $\hat{\alpha}_i$ and $\alpha_i$ and $n_i^{1/2} r_{n_i}$ converging to 0 in probability. The assertion is now immediate from the independence of the $\hat{\alpha}_i$, their asymptotic normality $n_i^{1/2}(\hat{\alpha}_i - \alpha_i) \xrightarrow[\mathcal{D}]{} N(0, \alpha_i(1 - \alpha_i))$ and Slutzky's Lemma (Exercise 15). □

## Weighted–Least–Squares Estimation

By the preceding result, a plausible estimator $\hat{\beta}$ of $\beta$ is that vector, which minimizes the *weighted residual sum of squares*:

$$R_g(\beta) := \sum_{i=1}^{I} \frac{(g(\hat{\alpha}_i) - (1, x_i^T)\beta)^2}{\hat{\sigma}_{ii}^2}$$

$$= (g(\hat{\alpha}) - X\beta)^T \hat{\Sigma}^{-1} (g(\hat{\alpha}) - X\beta). \tag{4.8}$$

The $I \times I$–matrix

$$\hat{\Sigma} := (\hat{\sigma}_{ij}^2) := \begin{pmatrix} \frac{n}{n_1} g'(\hat{\alpha}_1)^2 \hat{\alpha}_1(1 - \hat{\alpha}_1) & & 0 \\ & \ddots & \\ 0 & & \frac{n}{n_I} g'(\hat{\alpha}_I)^2 \hat{\alpha}_I(1 - \hat{\alpha}_I) \end{pmatrix}$$

is a reasonable estimator of the matrix $\Sigma$ defined in Theorem 4.2.2. We assume in the following $\hat{\sigma}_{ii}^2 > 0$ for $i = 1, \ldots, I$. Using the weights $1/\hat{\sigma}_{ii}^2$ in $R_g(\beta)$ aims at standardizing each summand. The weighted–least–squares problem (4.8) leads to the normal equations

$$X^T \hat{\Sigma}^{-1} X \hat{\beta} = X^T \hat{\Sigma}^{-1} g(\hat{\alpha}). \tag{4.9}$$

As the $p$ columns of matrix $X$ are independent, we can solve this equation and obtain the *weighted–least–squares estimator*

$$\hat{\beta} := \left(X^T \hat{\Sigma}^{-1} X\right)^{-1} X^T \hat{\Sigma}^{-1} g(\hat{\alpha}), \tag{4.10}$$

by repeating the arguments of Section 3.3, just replace $X$ by $\hat{\Sigma}^{-1/2} X$, where $\hat{\Sigma}^{-1/2} \hat{\Sigma}^{-1/2} = \hat{\Sigma}^{-1}$. We now have the following result.

**4.2.3 Theorem.** *Suppose we are in the model $g(\alpha) = X\beta$ with continuously differentiable response function $g : (0, 1) \to \mathbb{R}$ and $g'(\alpha_i) \neq 0$, $\alpha_i(1 - \alpha_i) \neq 0$, $i = 1, \ldots, I$. If $n = \sum_{i=1}^{I} n_i \to \infty$, $n/n_i \to \lambda_i \in [1, \infty)$, $i = 1, \ldots, I$, the estimator $\beta$ defined in (4.10) satisfies*

$$n^{1/2}(\hat{\beta} - \beta) \xrightarrow[\mathcal{D}]{} N\left(0, (X^T \Sigma^{-1} X)^{-1}\right),$$

*with $\Sigma$ as in Theorem 4.2.2.*

**Proof:** By Theorem 4.2.2 we have $n^{1/2}(g(\hat{\alpha}) - g(\alpha)) \xrightarrow{D} N(0, \Sigma)$. The equation $g(\alpha) = X\beta$ implies, therefore,

$$
\begin{aligned}
n^{1/2}(X^T\Sigma^{-1}X)^{-1}&X^T\Sigma^{-1}(g(\hat{\alpha}) - g(\alpha)) \\
&= n^{1/2}(X^T\Sigma^{-1}X)^{-1}X^T\Sigma^{-1}(g(\hat{\alpha}) - X\beta) \\
&= n^{1/2}\left\{(X^T\Sigma^{-1}X)^{-1}X^T\Sigma^{-1}g(\hat{\alpha}) - \beta\right\} \\
&\xrightarrow{D} N\left(0, (X^T\Sigma^{-1}X)^{-1}X^T\Sigma^{-1}\Sigma\Sigma^{-1}X(X^T\Sigma^{-1}X)^{-1}\right) \\
&= N(0, (X^T\Sigma^{-1}X)^{-1})
\end{aligned}
$$

(see Theorem 3.3.7 and Definition 2.1.2). As the entries $\hat{\sigma}_{ij}$ of the matrix $\hat{\Sigma}$ obviously converge in probability to the entries $\sigma_{ij}$ of the matrix $\Sigma$, by Slutzky's Lemma, we can replace in the preceding result $(X^T\Sigma^{-1}X)^{-1}X^T$ $\Sigma^{-1}$ by $(X^T\hat{\Sigma}^{-1}X)^{-1}X^T\hat{\Sigma}^{-1}$ (Exercise 15). The assertion is now immediate from the definition (4.10) of $\hat{\beta}$.                                             □

The practical significance of Theorem 4.2.3 is the fact that we can assume the estimate $\hat{\beta}$ to be approximately normally distributed for large sample sizes $n_i$ with covariance matrix $n^{-1}(X^T\hat{\Sigma}^{-1}X)^{-1}$. The elements $s_i^2$ in the main diagonal of this matrix are, in particular, estimates of the variances of the components $\hat{\beta}_i$ of $\hat{\beta}$. From this we can immediately derive tests for the null hypothesis $H_0 : \beta_i = 0$; see the end of Section 3.3 for details.

A crucial problem is the specification of the design matrix $X$. On the one hand, its dimension and, therefore, the number of crossed effects should be small as the number of available data is limited. The model has to be complex enough, on the other hand, to reflect the characteristic patterns in the data set. The following result can be derived from Theorem 4.2.3. It provides a goodness–of–fit measure for the design matrix $X$ which helps to decide in this trade–off situation.

**4.2.4 Theorem.** *Assume the model $g(\alpha) = X\beta$. Then, under the conditions of Theorem 4.2.3, we have*

$$
n\, R_g(\hat{\beta}) = n(g(\hat{\alpha}) - X\hat{\beta})^T\hat{\Sigma}^{-1}(g(\hat{\alpha}) - X\hat{\beta}) \xrightarrow{D} \chi^2_{I-p}.
$$

**Proof:** Exercise 17.                                                                              □

## Logit, Probit and Log–Linear Models

We will now compile several examples of popular link functions $g : (0,1) \to \mathbb{R}$, with

$$
g(\alpha) = \beta_0 + x_1\beta_1 + \cdots + x_{p-1}\beta_{p-1} = (1, x^T)\beta, \qquad x^T = (x_1, \ldots, x_{p-1}).
$$

The basic model is

$$\alpha = F((1, \boldsymbol{x}^T)\boldsymbol{\beta}), \tag{4.11}$$

where $F$ is some distribution function. Under suitable regularity conditions on $F$, the corresponding link function is its inverse function $g(z) = F^{-1}(z)$, $z \in (0, 1)$; see Section 1.6 for details.

The distribution function

$$F(t) = \frac{1}{1 + \exp(-t)}, \qquad t \in \mathbb{R},$$

is that of the *logistic distribution*, which is typically used in models of growth processes. This yields the link function $g(z) = F^{-1}(z) = \log(z/(1 - z))$, $z \in (0, 1)$. The corresponding model $g(\alpha) = (1, \boldsymbol{x}^T)\boldsymbol{\beta}$ is for $\alpha \in (0, 1)$

$$\log\left(\frac{\alpha}{1 - \alpha}\right) = (1, \boldsymbol{x}^T)\boldsymbol{\beta} \tag{4.12}$$

or

$$\alpha = \frac{1}{1 + \exp(-(1, \boldsymbol{x}^T)\boldsymbol{\beta})} = \frac{\exp((1, \boldsymbol{x}^T)\boldsymbol{\beta})}{1 + \exp((1, \boldsymbol{x}^T)\boldsymbol{\beta})}$$

which is called the *logit model*.

If $F$ is the distribution function of the standard normal distribution $F(t) = \Phi(t)$, $t \in \mathbb{R}$, we obtain the link function $g(z) = \Phi^{-1}(z)$ and, thus, the model

$$\Phi^{-1}(\alpha) = (1, \boldsymbol{x}^T)\boldsymbol{\beta} \tag{4.13}$$

or

$$\alpha = \Phi((1, \boldsymbol{x}^T)\boldsymbol{\beta}) = (2\pi)^{-\frac{1}{2}} \int_{-\infty}^{(1, \boldsymbol{x}^T)\boldsymbol{\beta}} \exp(-t^2/2)\, dt.$$

This is the *probit model*, which is, for example, commonly used in pharmacology. The levels $x_i$ are typically different quantities of a drug given to guinea pigs in order to evaluate the corresponding reaction. The probability $\alpha$ could, for instance, be the corresponding survival probability.

The distribution function $F(t) = \exp(t)$, $t \leq 0$, of the *reverse exponential distribution* yields the link function $g(z) = \log(z)$, $z \in (0, 1)$. This gives the *log–linear model*

$$\log(\alpha) = (1, \boldsymbol{x}^T)\boldsymbol{\beta}, \tag{4.14}$$

or

$$\alpha = \exp((1, \boldsymbol{x}^T)\boldsymbol{\beta}),$$

which is quite popular in various fields. Let, for example, $\alpha = \alpha(i)$ be the probability that an organism, which is alive at time $i$, will die in the time interval $(i, i+1]$. A log–linear model is

$$\alpha(i) = \lambda(i)\exp((1, \boldsymbol{x}^T)\boldsymbol{\beta}), \qquad i = 0, 1, 2, \ldots$$

This model consists of a factor $\lambda(\cdot) > 0$ depending on the time $i$ and an independent variable $x$. It is a discrete version of the *Cox model*, which is quite popular in *survival analysis* (Cox (1972)).

**4.2.5 Example.** We fit the log–linear model (4.14) with effects coding and included crossed effects of premarital and extramarital sex to the divorce data of Example 4.2.1. The weighted–least–squares estimator $\hat{\beta}$ from (4.10) is then given by

$$\hat{\beta} = \begin{pmatrix} \hat{\beta}_0 \\ \hat{\beta}_A \\ \hat{\beta}_B \\ \hat{\beta}_C \\ \hat{\beta}_{BC} \end{pmatrix} = \begin{pmatrix} \hat{\beta}_0 \\ \hat{\beta}_1 \\ \hat{\beta}_2 \\ \hat{\beta}_3 \\ \hat{\beta}_4 \end{pmatrix} = \begin{pmatrix} -0.4704 \\ 0.0669 \\ -0.1074 \\ -0.2515 \\ -0.1595 \end{pmatrix}.$$

The goodness–of–fit measure for this model is $nR_g(\hat{\beta}) = 0.11$ with $8 - 5 = 3$ degrees of freedom. This is quite close to 0, since the corresponding $p$-value is $1 - \chi_3^2(0.11) = 0.9909$; compare with Theorem 4.2.4. Therefore, the model

$$\log(\hat{\alpha}) = X\hat{\beta} + \varepsilon$$

fits the data quite well. All chi–square test statistics $\beta_i^2/s_i^2$ for the null hypothesis $\beta_i = 0$ have $p$-values approximately equal to 0, so that the intercept and all the effects are included in the model.

We can interpret the parameter estimates as follows: The parameter $\hat{\beta}_3 = -0.2515$, which corresponds to no extramarital sex, has among $\hat{\beta}_1, \ldots, \hat{\beta}_4$ the largest influence on the divorce probability. The combination extramarital sex | no premarital sex increases the probability of divorce, as the interaction effect is $-\hat{\beta}_4 = 0.1595$, which is larger than the main effect of no premarital sex $\hat{\beta}_2 = -0.1074$. The effect of extramarital sex $-\hat{\beta}_3 = 0.2515$ is, however, reduced for the combination extramarital sex | premarital sex by the crossed effect $\hat{\beta}_4 = -0.1595$. An affair seems to be easier forgiven in this combination. The gender seems to have no dominating influence on the probability of divorce, as the effect $\hat{\beta}_1 = 0.0669$ is close to 0. The combination no extramarital sex | no premarital sex has the smallest probability of divorce, cf the values for the response function of samples 1 and 4 in the SAS ouput below.

The CATMOD Procedure

Data Summary

| Response | STATUS | Response Levels | 2 |
|---|---|---|---|
| Weight Variable | COUNT | Populations | 8 |

```
        Data Set            DIVORCE    Total Frequency  1036
        Frequency Missing   0          Observations       16
```

### Population Profiles

| Sample | GENDER | PRESEX | EXTRASEX | Sample Size |
|--------|--------|--------|----------|-------------|
| 1 | female | no  | no  | 536 |
| 2 | female | no  | yes | 40  |
| 3 | female | yes | no  | 79  |
| 4 | female | yes | yes | 21  |
| 5 | male   | no  | no  | 198 |
| 6 | male   | no  | yes | 21  |
| 7 | male   | yes | no  | 102 |
| 8 | male   | yes | yes | 39  |

### Response Profiles

| Response | STATUS |
|----------|--------|
| 1 | divorced |
| 2 | married  |

### Analysis of Variance

| Source | DF | Chi-Square | Pr > ChiSq |
|--------|----|------------|------------|
| Intercept       | 1 | 250.75 | <.0001 |
| GENDER          | 1 | 4.96   | 0.0259 |
| PRESEX          | 1 | 13.18  | 0.0003 |
| EXTRASEX        | 1 | 78.63  | <.0001 |
| PRESEX*EXTRASEX | 1 | 31.58  | <.0001 |
| Residual        | 3 | 0.11   | 0.9909 |

### Analysis of Weighted Least Squares Estimates

| Effect | Parameter | Estimate | Standard Error | Chi-Square | Pr > ChiSq |
|--------|-----------|----------|----------------|------------|------------|
| Intercept | 1 | -0.4704 | 0.0297 | 250.75 | <.0001 |
| GENDER    | 2 |  0.0669 | 0.0300 | 4.96   | 0.0259 |

```
PRESEX              3    -0.1074    0.0296    13.18    0.0003
EXTRASEX            4    -0.2515    0.0284    78.63    <.0001
PRESEX*EXTRASEX     5    -0.1595    0.0284    31.58    <.0001
```

### Predicted Values for Response Functions

| | | | | -----Observed---- | | ----Predicted---- | |
|---|---|---|---|---|---|---|---|
| | | | Function | | Standard | | Standard |
| GENDER | PRESEX | EXTRASEX | Number | Function | Error | Function | Error |
| female | no | no | 1 | -0.91816 | 0.052983 | -0.92195 | 0.048557 |
| female | no | yes | 1 | -0.10536 | 0.052705 | -0.09983 | 0.048664 |
| female | yes | no | 1 | -0.38046 | 0.076553 | -0.38801 | 0.062657 |
| female | yes | yes | 1 | -0.21131 | 0.105851 | -0.20407 | 0.079399 |
| male | no | no | 1 | -1.06876 | 0.098262 | -1.05573 | 0.065871 |
| male | no | yes | 1 | -0.21131 | 0.105851 | -0.23362 | 0.067388 |
| male | yes | no | 1 | -0.53063 | 0.082842 | -0.52179 | 0.064885 |
| male | yes | yes | 1 | -0.33136 | 0.100366 | -0.33786 | 0.078182 |

### Predicted Values for Response Functions

| | | | Function | |
|---|---|---|---|---|
| GENDER | PRESEX | EXTRASEX | Number | Residual |
| female | no | no | 1 | 0.003788 |
| female | no | yes | 1 | -0.00553 |
| female | yes | no | 1 | 0.007544 |
| female | yes | yes | 1 | -0.00723 |
| male | no | no | 1 | -0.01303 |
| male | no | yes | 1 | 0.022307 |
| male | yes | no | 1 | -0.00883 |
| male | yes | yes | 1 | 0.006504 |

**Figure 4.2.2.** Weighted–least–squares estimation for the divorce data in Figure 4.2.1.

```
***    Program 4_2_2    ***;
TITLE1 'Weighted-Least-Squares Estimation';
TITLE2 'Divorce Data';
LIBNAME datalib 'c:\data';

PROC CATMOD DATA=datalib.divorce;
    WEIGHT count;
    RESPONSE 1 0 LOG;
    MODEL status=gender presex|extrasex / WLS NODESIGN PRED=PROB;
RUN; QUIT;
```

The CATMOD procedure provides a wide variety of categorical data analyses. It analyzes data that can be represented by a contingency table.

The MODEL statement determines the dependent variable, here it is 'status'. By 'gender presex|extrasex' all main effects and the interaction of presex and extrasex are computed (cf Program 5_2_4). Effect coding is the default. The option 'WLS' computes weighted–least–squares estimates.

PRED=PROB prints the observed and predicted values of the response function for each population.

If a WEIGHT statement is used, then CATMOD uses the values of the WEIGHT variable as the frequency counts. They need not be integers.

The statement RESPONSE 1 0 LOG codes the link function $(1, 0)(\log(\alpha), \log(1-\alpha))^T = \log(\alpha)$, cf the comments on Program 4_2_4.

# Maximum–Likelihood Estimation in Logit Models

**4.2.6 Example** (O–Ring Data). On January 28, 1986 the space shuttle *Challenger* exploded. This accident was caused by a combustion gas leak through a field–joint in one of the booster rockets, which was sealed by a device called an O–ring. Each of the two solid rocket motors was equipped with three such O–rings.

The night before the accident there was a three–hour teleconference between people at Morton Thiokol, the manufacturer of the rocket motor, NASA's Marshall Space Flight Center and Kennedy Space Center. The discussion focussed on the forecast of 31° F temperature for launch time the following morning, and the effect of low temperature on O–ring performance. The following data played an important role in the discussion. These are those shuttle flights that experienced thermal distress at least on one of the field–joint O–rings together with the temperature at launch in Fahrenheit:

| Flight number | Date | Temperature at launch |
|---|---|---|
| STS-2 | 11/12/81 | 70° |
| 41-B | 02/03/84 | 57° |
| 41-C | 04/06/84 | 63° |
| 41-D | 08/30/84 | 70° |
| 51-C | 01/24/85 | 53° |
| 61-A | 10/30/85 | 75° |
| 61-C | 01/12/86 | 58° |

Though some participants recommended that the launch be postponed until the temperature rose above 53° F — the lowest temperature experienced in previous launches — the final recommendation of Morton Thiokol was to launch the Challenger on schedule. The recommendation transmitted to NASA stated that "Temperature data [are] not conclusive on predicting primary O–ring blow-by".

After the accident a commission was appointed by President Ronald Reagan to find the cause. This commission, headed by William Rogers, determined the thermal distress of O–rings to be the cause. It noted that a mistake in the analysis of the above data was that the flights with no incidents were left out; it was felt that these flights did not contribute any information about the temperature effect. The following figure visualizes all flights with the temperature at launch plotted against the binary variable $Z \in \{0, 1\}$, where $Z = 0$ represents no thermal distress and $Z = 1$ means thermal distress of at least one O–ring. The plot reveals that this was a fatal error. For an exhaustive analysis of these data we refer to Dalal et al. (1989).

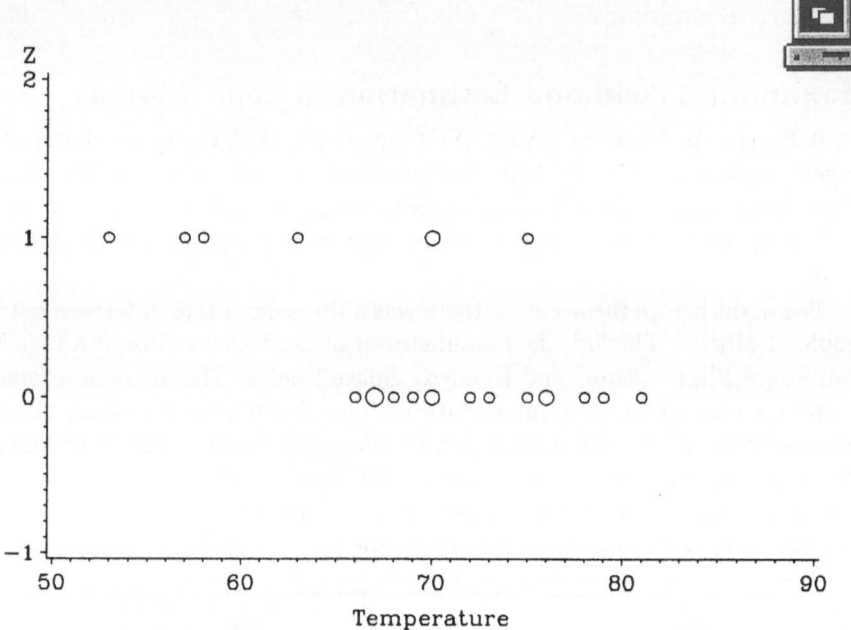

**Figure 4.2.3.** Scatterplot of O–ring data; larger bubbles represent two or more identical observations.

```
***    Program 4_2_3    ***;
TITLE1 'Scatter Plot';
TITLE2 'O-Ring Data';
LIBNAME datalib 'c:\data';

AXIS1 ORDER=(-1 TO 2) LABEL=('Z');
AXIS2 LABEL=('Temperature');
SYMBOL1 V=DOT C=G H=1 I=NONE;
PROC GPLOT DATA=datalib.oring(OBS=23);
    BUBBLE erosion*temp=f / VAXIS=AXIS1 HAXIS=AXIS2;
RUN; QUIT;
```

The ORDER option in the AXIS statement of the above program shifts the displayed bubbles vertically closer to the center of the graph. The option '(OBS=23)' excludes the 24th observation (Challenger flight). The

BUBBLE statement determines the size of the displayed bubbles depending on the value of the variable 'f' (frequency of identical data).

We choose a logit model for the probability $\alpha = \alpha(x)$ that at least one O–ring fails at launch temperature $x$, i.e.,

$$\log\left(\frac{\alpha}{1-\alpha}\right) = \beta_0 + \beta_1 x \quad \text{or} \quad \alpha = \frac{\exp(\beta_0 + \beta_1 x)}{1 + \exp(\beta_0 + \beta_1 x)}.$$

The temperatures vary, however, from launch to launch within a wide range, so that a number of temperatures $x_i$ occur only once in the sample of $I = 23$ flights. In this case we cannot compute a weighted–least–squares estimator $\hat{\beta}$ as in (4.10) or (4.8), since these single values yield $\hat{\sigma}_{ii}^2 = 0$.

We use, therefore, the maximum–likelihood principle for the derivation of an estimator of $\beta$. To this end we assume that we observe independent copies $Z_1, \ldots, Z_I$ of a binary random variable $Z$, where the conditional probability $\alpha = \alpha(x) = P(Z = 1|x)$, given $x \in \mathbb{R}^{p-1}$, is the logit model

$$\log\left(\frac{\alpha}{1-\alpha}\right) = (1, x^T)\beta \quad \text{or} \quad \alpha = \frac{\exp((1, x^T)\beta)}{1 + \exp((1, x^T)\beta)}.$$

The $I \times p$–design matrix is in this case

$$X = \begin{pmatrix} 1 & x_1^T \\ \vdots & \vdots \\ 1 & x_I^T \end{pmatrix},$$

and the unknown $p$-dimensional vector

$$\beta = \begin{pmatrix} \beta_0 \\ \vdots \\ \beta_{p-1} \end{pmatrix}.$$

The following representation will be quite useful:

$$P(Z = z|x) = \exp\left(\log\left(\frac{\alpha}{1-\alpha}\right)z + \log(1 - \alpha)\right)$$

$$= \exp\left((1, x^T)\beta z - \log(1 + \exp((1, x^T)\beta))\right), \quad z \in \{0, 1\}.$$

Let now $z_1, \ldots, z_I$ be realizations of the independent copies $Z_1, \ldots, Z_I$ of the random variable $Z$ with design points $x_1, \ldots, x_I$. The corresponding log–likelihood function is then given by

$$
\begin{aligned}
l(\beta) &:= \log \Big( \prod_{i=1}^{I} P(Z_i = z_i | x_i) \Big) \\
&= \sum_{i=1}^{I} \log(P(Z_i = z_i | x_i)) \\
&= \sum_{i=1}^{I} \Big( (1, x_i^T)\beta z_i - \log(1 + \exp((1, x_i^T)\beta)) \Big).
\end{aligned}
$$

**4.2.7 Lemma.** *The gradient of the log–likelihood function* $l(\beta), \beta \in \mathbb{R}^p$, *is given by*

$$
\operatorname{grad} l(\beta) = \begin{pmatrix} \dfrac{\partial l}{\partial \beta_0}(\beta) \\ \vdots \\ \dfrac{\partial l}{\partial \beta_{p-1}}(\beta) \end{pmatrix} = X^T(z - \alpha), \tag{4.15}
$$

*where* $z = (z_1, \ldots, z_I)^T$, *and* $\alpha = (\alpha_1, \ldots, \alpha_I)^T$ *is the vector of logit probabilities at the design points* $x_1, \ldots, x_I$.

**Proof:** (Exercise 18).                                                                          $\square$

A solution $\hat{\beta}$ of the equation

$$
\operatorname{grad} l(\beta) = 0 \quad \text{or} \quad X^T(z - \alpha) = 0
$$

is a maximum–likelihood estimator of $\beta$. Note that (4.15) is not a linear equation in $\beta$ since $\alpha$ and $\beta$ interact via the nonlinear logit model. An explicit solution cannot be computed in general. An iteration process typically provides an appropriate solution instead. If $I$ is large, under suitable regularity conditions, the maximum–likelihood estimator $\hat{\beta}$ is approximately normally distributed with mean vector $\beta$ and covariance matrix

$$
\Sigma_{\hat{\beta}} = \left( X^T \begin{pmatrix} \alpha_1(1 - \alpha_1) & & 0 \\ & \ddots & \\ 0 & & \alpha_I(1 - \alpha_I) \end{pmatrix} X \right)^{-1}.
$$

If we replace the probability $\alpha_i$ by $\hat{\alpha}_i = \exp((1, x_i^T)\hat{\beta})/(1 + \exp((1, x_i^T)\hat{\beta}))$ in the above matrix, we obtain an estimator $\hat{\Sigma}_{\hat{\beta}}$ which converges as $I \to \infty$

in probability to $\Sigma_{\hat{\beta}}$ (cf Theorem 2.2 in Chapter 7 in Fahrmeir and Hamerle (1984) and Section 2.2 in Fahrmeir and Tutz (1994)):

$$\hat{\Sigma}_{\hat{\beta}} = \left( \boldsymbol{X}^T \begin{pmatrix} \hat{\alpha}_1(1-\hat{\alpha}_1) & & 0 \\ & \ddots & \\ 0 & & \hat{\alpha}_I(1-\hat{\alpha}_I) \end{pmatrix} \boldsymbol{X} \right)^{-1}.$$

**4.2.8 Example.** In the case of the O–ring data in Example 4.2.6 we take a logit model for the probability $\alpha = \alpha(x)$ of the event that at least one O–ring fails at the launch temperature, i.e.,

$$\log \left( \frac{\alpha}{1-\alpha} \right) = \beta_0 + x\beta_1 \quad \text{or} \quad \alpha = \frac{\exp(\beta_0 + x\beta_1)}{1 + \exp(\beta_0 + x\beta_1)}.$$

We obtain the maximum–likelihood estimate

$$\hat{\beta}^T = (\hat{\beta}_0, \hat{\beta}_1) = (15.0429, -0.2322)$$

with estimated standard deviations 7.3784 and 0.1082. The estimated probability for failure of at least one O–ring at the temperature $x = 31°$ F is then

$$\hat{\alpha}(31) = \frac{\exp(\hat{\beta}_0 + 31\hat{\beta}_1)}{1 + \exp(\hat{\beta}_0 + 31\hat{\beta}_1)} = 0.9996.$$

A launch temperature of at least 70° F would reduce this (estimated) probability to 0.23. The estimated standard deviations of $\hat{\beta}_0$ and $\hat{\beta}_1$ indicate, however, that the deviation of $\hat{\alpha}$ from the true probability can be quite large (see Exercise 19).

Analysis of maximum likelihood estimates

| Effect | Parameter | Estimate | Standard Error | Chi-Square | Prob |
|--------|-----------|----------|----------------|------------|------|
| Intercept | 1 | 15.0429 | 7.3784 | 4.16 | 0.0415 |
| TEMP | 2 | -0.2322 | 0.1082 | 4.60 | 0.0320 |

**Figure 4.2.4.** Maximum–likelihood estimates in the logit model for O–ring data.

```
***    Program 4_2_4    ***;
TITLE1 'ML Estimates in a Logit Model';
TITLE2 'O-Ring Data';
LIBNAME datalib 'c:\data';

PROC CATMOD DATA=datalib.oring(OBS=23);
   RESPONSE CLOGITS;
   DIRECT temp;
   MODEL erosion=temp /
      ML NOITER NODESIGN NOPROFILE NORESPONSE;
RUN; QUIT;
```

If the dependent variable $Z$ attains the two values $a, b$ with $a < b$, then SAS puts $\alpha(x) = P(Z = a|x)$ by default. In this example we have $a = 0$, $b = 1$. The option RESPONSE CLOGITS yields here $\alpha(x) = P(Z = b|x)$, which is in accordance with the example. The DIRECT statement declares the independent variable 'temp' to be treated as a continuous numeric variable.

# Neural Networks

The rapidly increasing capacity of modern computers enables the calculation of nonlinear regression models, which are considerably more complex than the GLIM in (4.6), within a reasonable time. A quite popular and computer-intensive example is the following *feedforward neural network*. We consider the case, where the probability that the binary random variable $Z$ attains the value 1 satisfies, conditional on the vector $x = (x_1, \ldots, x_{p-1})^T$ of observations, the equation

$$P(Z = 1|x) = f\left(\omega_0 + \sum_{k=1}^{m} \omega_k\, g(\beta_{k0} + \beta_{k1}\, x_1 + \ldots + \beta_{k\,p-1}\, x_{p-1})\right).$$

Here, $f, g$ are functions and $\omega_k, \beta_{kj}$ are weights. This iterated GLIM is obviously an extension of the model (4.6). The functions $f$ and $g$ are usually prescribed, the number $m$ and the weights are, however, unknown.

Neural networks originated in Biophysics, aiming at a mathematical modeling of nervous activities (McCulloch and Pitts (1943)). As a consequence, they have their own terminology, which we partially explain for the above model: Each of the $p - 1$ observations $x_1, \ldots, x_{p-1}$ enters an *input neuron*. These $p - 1$ input neurons feed their input jointly as the linear combination $\beta_{k0} + \beta_{k1}\, x_1 + \ldots + \beta_{k\,p-1}\, x_{p-1}$ forward to the $k$-th neuron in the *hidden layer*,

$k = 1, \ldots, m$. As this process is invisible, the number $m$ as well as the *connection weights* $\beta_{kj}$ are unknown. The neurons in this hidden layer are, therefore, called *hidden neurons*. The output of the $k$-th hidden neuron is $g(\beta_{k0} + \beta_{k1} x_1 + \ldots + \beta_{k\,p-1} x_{p-1})$, where $g$ is an *activation function*. This again is a typical distribution function or, more generally, a *sigmoidal function*, in which case it satisfies $\lim_{z \to \infty} g(z) = 1, \lim_{z \to -\infty} g(z) = 0$, it is Borel measurable and bounded. The $m$ hidden neurons jointly feed their outputs as the linear combination $\omega_0 + \sum_{k=1}^{m} \omega_k\, g(\beta_{k0} + \beta_{k1} x_1 + \ldots + \beta_{k\,p-1} x_{p-1})$ to the *ouput neuron*, which finally provides the output $f(\omega_0 + \sum_{k=1}^{m} \omega_k\, g(\beta_{k0} + \beta_{k1} x_1 + \ldots + \beta_{k\,p-1} x_{p-1}))$ using the activation function $f$. Typical examples of sigmoidal functions are the logistic function $h(z) = (1 + \exp(-z))^{-1}$, the hyperbolic tangent based function $h(z) = (\tanh(z) + 1)/2 = (1 + \exp(-2z))^{-1}$, or just $h(z) = 1$ if $z$ exceeds some threshold $c$, and $h(z) = 0$ elsewhere, $z \in \mathbb{R}$.

The following picture illustrates the *architecture* of the feedforward neural network with one hidden layer:

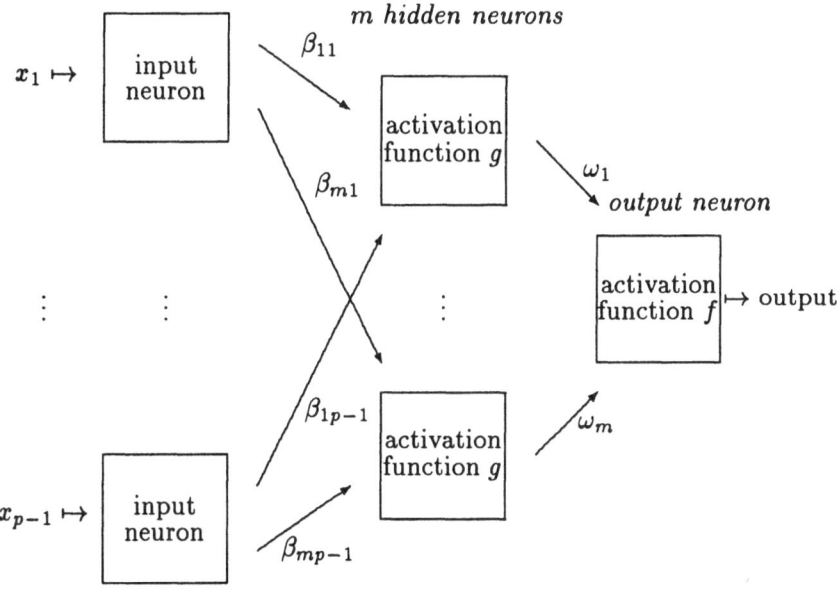

Suppose we have a training sample with $n_i$ realizations $Z_1^{(i)}, \ldots, Z_{n_i}^{(i)}$ of $Z$ for the coding vector $x_i = (x_{i1}, \ldots, x_{i\,p-1})^T$. A measure of performance for the goodness–of–fit of the neural net is the sum of squares

$$E := \sum_{i=1}^{I} \sum_{j=1}^{n_i} \left( Z_j^{(i)} - f\left( \omega_0 + \sum_{k=1}^{m} \omega_k\, g\left( \beta_{k0} + \beta_{k1} x_{i1} + \ldots + \beta_{k\,p-1} x_{i\,p-1} \right) \right) \right)^2$$

of the errors between the output of the net and its *targets*.

The choice of $m$ yields a typical trade–off situation. By choosing $m$ too large, the data will be overfitted and the model has reduced its ability to predict future values by sticking to the training sample. If $m$ is chosen too small, the error will be too large. After the determination of $m$, the weights $\omega_k$ and $\beta_{kj}$ are chosen such that $E$ is minimized, i.e., the least squares estimators are computed.

For a review of neural networks from a statistical point of view we refer to Cheng and Titterton (1994) and the monograph by Fine (1999).

### Exercises

**1.** Show by elementary computations that the function $g(k) := H(K, W, n)(\{k\})$ first increases and then decreases.

**2.** (i) Show that an $H(K, W, n)$ distributed random variable $X$ satisfies

$$E(X) = n\frac{W}{K}, \quad Var(X) = n\frac{W}{K}\left(1 - \frac{W}{K}\right)\frac{K - n}{K - 1}.$$

Compare it with the expectation and variance of a $B(n, W/K)$ distributed random variable. What do you notice?

(ii) If $W = W(K)$ satisfies $\lim_{K \to \infty} W(K)/K = p \in [0, 1]$, prove that

$$\lim_{K \to \infty} H(K, W, n)\{k\} = B(n, p)(\{k\}), \quad k \in \{0, \ldots, n\}.$$

Give a heuristic explanation of this result.

**3.** The parameter $W$ of the hypergeometric distribution $H(K, W, n)$ is commonly unknown in applications and has to be estimated from the data. Take, for instance, the quality check of a lot consisting of $K$ units of an article. An unknown number $W$ of them is faulty. This number $W$ has to be estimated by means of $k$ faulty units in a sample of size $n$. Compute the maximum–likelihood estimator $\hat{W}$ of the corresponding hypergeometric distribution, i.e.,

$$\max_{W} H(K, W, n)(\{k\}) = H(K, \hat{W}, n)(\{k\}).$$

**4.** Consider an arbitrary probability space $(\Omega, \mathcal{A}, P)$ and let $(B_i)_{i \in I}$, $I = \{1, \ldots, n\}$ or $I = \mathbb{N}$, be a partition of $\Omega$ with $P(B_i) > 0$, $i \in I$. Establish the expansion

$$P(A) = \sum_{i \in I} P(B_i)P(A|B_i), \quad A \in \mathcal{A},$$

where we denote by $P(A|B) = P(A \cap B)/P(B)$ the conditional probability of $A$ given $B$.

**5.** Complete the proof of Lemma 4.1.4.

**6.** Let $X_{11},\ldots,X_{1n_1}$, $X_{21},\ldots,X_{2n_2}$ be independent and identically $B(1,p)$ distributed random variables. Show that with $S := \sum_{i=1}^{n_1} X_{1i} + \sum_{j=1}^{n_2} X_{2j}$ and $S \in \{0,\ldots,n_1+n_2\}$

$$P(X_{11} + \cdots + X_{1n_1} \in \cdot | S = s) = H(n_1 + n_2, n_1, s)(\cdot).$$

**7.** (i) Show that there are

$$\binom{n}{n_1,\ldots,n_I} := \frac{n!}{n_1!\ldots n_I!}$$

different combinations for the distribution of $n$ different balls over $I$ urns in such a way that the $i$-th urn contains $n_i$ balls, $i = 1,\ldots,I$, $n_1 + \cdots + n_I = n$. (ii) Prove Lemma 4.1.6.

**8.** Suppose the random vector $N := (N_1,\ldots,N_I)^T$ is $B(n,p_1,\ldots,p_I)$ distributed. Compute (i) the distribution of $N_i$ and of $(N_i, N_j)^T$, $i \neq j$, (ii) the expected frequency $E(N_i)$, (iii) the covariance matrix $Cov(N)$ and its rank.
Hint to the rank: By Theorem 3.3.7 it suffices to compute the rank of the matrix $\Sigma := n^{-1} Cov(DN)$, where $D$ is the diagonal matrix with the entries $\sqrt{p_i}, i = 1,\ldots,I$. Check that $x^T \Sigma x = 0$ if and only if $x$ is a multiple of $(\sqrt{p_1},\ldots,\sqrt{p_I})^T$ and note that $x^T \Sigma x = 0$ if and only if $\Sigma x = 0$ ($\Sigma$ is positive semidefinite by Exercise 20, Chapter 3; cf Exercise 18, Chapter 6).

**9.** (Relationship between the Poisson and the polynomial distribution) Let $X_1,\ldots,X_n$ be independent and Poisson distributed random variables with the parameters $\lambda_i > 0$. Put $S_n := \sum_{i=1}^{n} X_i$, $\lambda := \sum_{i=1}^{n} \lambda_i$ and show that

$$P((X_1,\ldots,X_n) \in \cdot | S_n = k) = B(k, \lambda_1/\lambda, \ldots, \lambda_n/\lambda)(\cdot),$$

where $P(B|A)$ denotes the conditional probability of the event $B$ given the event $A$ (see Exercise 4).

**10.** (Convolution theorem for the polynomial distribution) Let $N_1$, $N_2$ be independent and $B(n_k, p_1,\ldots,p_I)$ distributed random variables, $k = 1,2$. Show that $N_1 + N_2$ follows the $B(n_1 + n_2, p_1,\ldots,p_I)$ distribution.

**11.** (Economy data) Plot a $2 \times 2$ table of the variable *unempld* with the two classes 'low', if *unempld* $< 8$ and 'high' elsewhere by the variable *nukes* with the two classes 'no' if *nukes* $= 0$, and 'yes' otherwise. Are these two variables independent? Plot the frequencies.

**12.** Show that the maximum–likelihood estimator of $p_{ij}$ is under the null hypothesis (4.4)

$$\hat{p}_{ij} = \frac{n_{i \cdot}}{n} \, \frac{n_{\cdot j}}{n}.$$

In the general case, the maximum–likelihood estimator of $p_{ij}$ is $\tilde{p}_{ij} = n_{ij}/n$.

**13.** (Ethno data) Dowdall (1974) studied the effect of ethnic background on role attitude of women of ages 15 to 64 in Rhode Island. Respondents were asked whether they thought it was alright for a woman to have a job instead of taking care of the home and the children while the husband worked. The following data, taken from Rice (1995), page 500, display the responses by ethnic origin of the respondents; the data are also discussed in Haberman (1978).

| Ethnic Origin | Yes | No |
|---|---|---|
| Italian | 78 | 47 |
| Northern European | 56 | 29 |
| Other European | 43 | 29 |
| English | 53 | 32 |
| Irish | 43 | 30 |
| French Canadian | 36 | 22 |
| French | 42 | 23 |
| Portugese | 29 | 7 |

Compute first the expected frequency of each cell under the assumption of independence. Is there any association between ethnic background and response? If so, describe it.

**14.** Derive, from the table in Example 4.1.13, a prediction of the eye color, given the hair color.

**15.** (Slutzky's Lemma) Let $(X_n)_{n \geq 1}$ be a sequence of random variables, which converges in distribution to the random variable $X$, denoted by $X_n \to_D X$. Hence, we have $\lim_{n \to \infty} F_{X_n}(t) = F_X(t)$ for each point of continuity $t$ of $F_X$, where $F_{X_n}, F_X$ are the distribution functions of $X_n$ and $X$, $n \in \mathbb{N}$. Let $(Y_n)_{n \geq 1}$ be another sequence of random variables that converge in probability to some constant $c \in \mathbb{R}$, i.e., $\lim_{n \to \infty} P\{|Y_n - c| > \varepsilon\} = 0$, $\varepsilon > 0$. Show that

(i) $X_n + Y_n \to_D X + c$,

(ii) $X_n Y_n \to_D cX$,

(iii) $X_n/Y_n \to_D X/c$ if $c \neq 0$.

As a consequence one obtains that convergence in probability of $(Z_n)_{n \geq 1}$ to $Z$ implies convergence in distribution of $(Z_n)_{n \geq 1}$ to $Z$. The reverse implication is not true in general. Give an example.

**16.** Suppose the sequence $X_n, n \in \mathbb{N}$ converges in probability to $c \in \mathbb{R}$. Let $f : \mathbb{R} \to \mathbb{R}$ be an arbitrary continuous function, which is differentiable in $c$. Prove the following assertions:

(i) We have the stochastic expansion

$$f(X_n) = f(c) + f'(c)(X_n - c) + (X_n - c)r_n,$$

where $r_n$ converges in probability to zero.

(ii) Suppose $\sqrt{n}(X_n - c) \to_D X$, where $X$ is $N(0, \sigma^2)$ distributed. This implies $\sqrt{n}(f(X_n) - f(c)) \to_D f'(c)X$.

Hint to (i): Check first that the continuity of $f$ implies that $f(Z_n) \to f(Z)$ in probability whenever $Z_n \to Z$ in probability.

**17.** Prove Theorem 4.2.4. Hint: Plug (4.10) in the quadratic form (4.8) and replace $\hat{\Sigma}^{-1}$ by $\Sigma^{-1}$, cf the proof of Theorem 4.2.3. This quadratic form has then a representation which makes Exercise 23 in Chapter 3 applicable.

**18.** Prove Lemma 4.2.7.

**19.** Plot in the framework of a risk analysis the function $\hat{\alpha}(x)$ for $x \in [20, 100]$ in the example of the O–ring data as well as "confidence" curves for $\alpha(x)$, i.e., plot intervals for $\alpha(x)$ for each $x$. Hint: Denote by $F(y) = 1/(1 + \exp(-y)), y \in \mathbb{R}$, the logistic distribution function. We have approximately

$$\begin{aligned}
\hat{\alpha}(x) - \alpha(x) &= F(\hat{\beta}_0 + \hat{\beta}_1 x) - F(\beta_0 + \beta_1 x) \\
&\approx F'(\hat{\beta}_0 + \hat{\beta}_1 x)(\hat{\beta}_0 - \beta_0 + (\hat{\beta}_1 - \beta_1)x).
\end{aligned}$$

Choose $\Delta = \Delta_t(x) = 2t \max\{\sigma_0, \sigma_1 |x|\}$, $t > 0$, where $\sigma_i$ denotes the standard deviation of $\hat{\beta}_i$, $i = 0, 1$. We have

$$\begin{aligned}
&P\{|\hat{\beta}_0 - \beta_0 + (\hat{\beta}_1 - \beta_1)x| > \Delta\} \\
&\le P\{|\hat{\beta}_0 - \beta_0| > \Delta/2\} + P\{|(\hat{\beta}_1 - \beta_1)x)| > \Delta/2\} \\
&\le P\{|\hat{\beta}_0 - \beta_0| > t\sigma_0\} + P\{|\hat{\beta}_1 - \beta_1| > t\sigma_1\} \\
&\approx 2(1 - \Phi(t)) + 2(1 - \Phi(t)) = 4(1 - \Phi(t)).
\end{aligned}$$

For example for $t = 2$ we obtain $4(1 - \Phi(2)) \approx 0.1820$. The upper and lower curves are then defined by

$$\alpha^+(x) := \hat{\alpha}(x) + F'(\hat{\beta}_0 + \hat{\beta}_1 x)\Delta_t(x)$$

and

$$\alpha^-(x) := \hat{\alpha}(x) - F'(\hat{\beta}_0 + \hat{\beta}_1 x)\Delta_t(x).$$

# Chapter 5

# Analysis of Variance

The analysis of variance is a statistical toolbox for the comparison of two or more independent samples. It can, therefore, be viewed as a generalization of the comparison of two independent samples as in Section 2.3.

## 5.1 The One–Way Analysis of Variance

The one–way analysis of variance investigates independent measurements from several *treatments* or *levels* of one *factor*.

**5.1.1 Example** (Course Data; Fahrmeir und Hamerle (1984), Example 1.1, Chapter 5). Four different ways of teaching a course are to be evaluated in order to find out their effect on performance levels. To this end, a group of 32 participants is randomly split into four subgroups of eight members. Each subgroup is taught the course in one of four different ways. When the course is finished, each participant undergoes the same final examination. The following table lists the scores of each participant in the four subgroups I – IV, representing the different teaching methods, in the final examination.

| I | II | III | IV |
|---|---|---|---|
| 16 | 16 | 2 | 5 |
| 18 | 12 | 10 | 8 |
| 20 | 10 | 9 | 8 |
| 15 | 14 | 10 | 11 |
| 20 | 18 | 11 | 1 |
| 15 | 15 | 9 | 9 |
| 23 | 12 | 10 | 5 |
| 19 | 13 | 9 | 9 |

**Figure 5.1.1.** Printout of course data; scores per participant in subgroups I – IV.

```
***    Program 5_1_1    ***;
TITLE1 'Printout';
TITLE2 'Course Data';
LIBNAME datalib 'c:\data';

PROC TRANSPOSE DATA=datalib.course OUT=printout;
    BY method;
PROC TRANSPOSE DATA=printout OUT=printout(KEEP= i ii iii iv);
    ID method;
PROC PRINT DATA=printout NOOBS;
RUN; QUIT;
```

A trick is needed for the above printout of the data, since the titles of the columns I – IV are themselves entries of the variable 'method' in the SAS data file. This data file contains the two variables 'method' and 'scores', and the total number of $(4 \times 8=)$ 32 observations.

The trick is the double application of PROC TRANSPOSE, which leads to the above listing using PROC PRINT.

The question, whether the four ways of teaching the course have different effects on performance levels, has to be answered on the basis of these data. In this example we have one factor, namely the way of teaching the course, with four different treatments or levels. A visual comparison using boxplots is already quite informative.

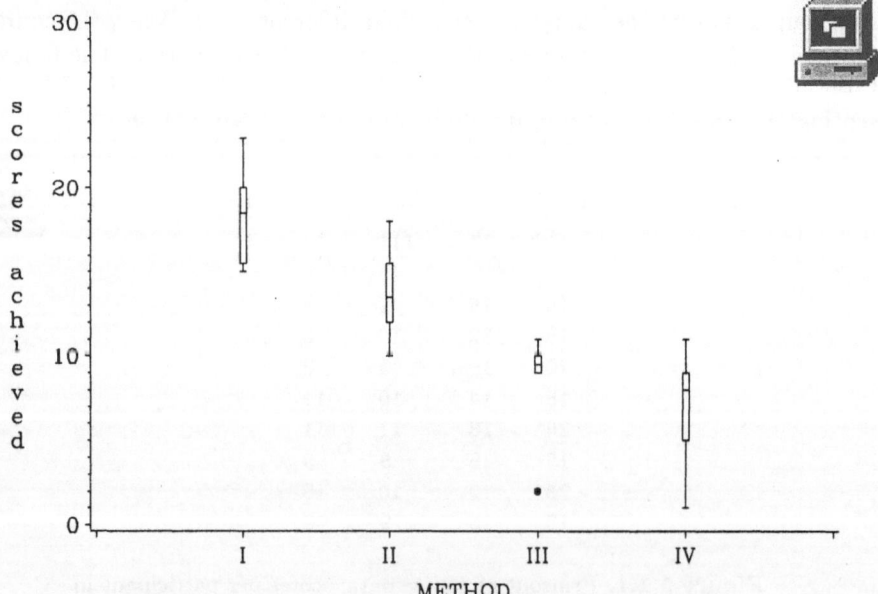

Figure 5.1.2. Boxplots of course data in Figure 5.1.1.

```
***    Program 5_1_2    ***;
TITLE1 'Boxplots';
TITLE2 'Course Data';
LIBNAME datalib 'c:\data';

AXIS1 ORDER=(' ' 'I' 'II' 'III' 'IV' ' ');
SYMBOL1 V=DOT C=GREEN I=BOXT;
PROC GPLOT DATA=datalib.course;
    PLOT scores*method=1 / HAXIS=AXIS1;
RUN; QUIT;
```

The ORDER option in the AXIS statement causes the four boxplots to be shifted to the center of the graph.

## The Model

We assume that we have $J \geq 2$ measurements $Y_{ij}$ from each of $I \geq 2$ different levels of a factor $A$ according to the standard model of the *one–way analysis of variance*

$$Y_{ij} = \mu_i + \varepsilon_{ij}, \qquad i = 1, \ldots, I, \quad j = 1, \ldots, J, \qquad (5.1)$$

where $\mu_i \in \mathbb{R}$ are the means of the $I$ levels and the *measurement errors* $\varepsilon_{ij}$ are independent and identically $N(0, \sigma^2)$ distributed random variables. We want to test the null hypothesis that all means coincide, i.e., $H_0 : \mu_1 = \cdots = \mu_I$.

The following version of model (5.1) is quite common. It splits each mean $\mu_i$ into an overall mean and an individual effect. Define the *overall mean effect* as

$$\mu := E\left(\frac{1}{IJ}\sum_{i=1}^{I}\sum_{j=1}^{J} Y_{ij}\right) = \frac{1}{I}\sum_{i=1}^{I}\left(\frac{1}{J}\sum_{j=1}^{J} E(Y_{ij})\right) = \frac{1}{I}\sum_{i=1}^{I}\mu_i$$

and the *effect* of level $i$ as

$$\alpha_i := \mu_i - \mu, \qquad i = 1, \ldots, I.$$

Then we obviously have $\sum_{i=1}^{I} \alpha_i = 0$ and model (5.1) becomes

$$Y_{ij} = \mu + \alpha_i + \varepsilon_{ij} \qquad (5.2)$$

with the overall mean effect $\mu$, the effect $\alpha_i$ of level $i$ and independent and $N(0, \sigma^2)$ distributed measurement errors $\varepsilon_{ij}$. The effects $\alpha_i$ satisfy the constraint $\sum_{i=1}^{I} \alpha_i = 0$.

## Null Hypothesis and Test Statistic

The null hypothesis for the one–way analysis of variance is now

$$H_0 : \alpha_1 = \alpha_2 = \cdots = \alpha_I = 0.$$

The overall sample mean

$$\hat{\mu} := \frac{1}{IJ} \sum_{i=1}^{I} \sum_{j=1}^{J} Y_{ij} =: \bar{Y}_{..}$$

is obviously an unbiased estimator of $\mu$ in model (5.2). Denote by

$$\hat{\mu}_i := \frac{1}{J} \sum_{j=1}^{J} Y_{ij} =: \bar{Y}_{i.}, \qquad i = 1, \ldots, I,$$

the sample mean of level $i$. Then

$$\hat{\alpha}_i := \hat{\mu}_i - \hat{\mu} = \bar{Y}_{i.} - \bar{Y}_{..}, \qquad i = 1, \ldots, I,$$

is an unbiased estimator of the effect $\alpha_i$. The basic idea of the analysis of variance is now the comparison of the sum of squares *among* the groups

$$SS_A := J \sum_{i=1}^{I} \hat{\alpha}_i^2 = J \sum_{i=1}^{I} (\bar{Y}_{i.} - \bar{Y}_{..})^2$$

with the sum of squares *within* the groups

$$SS_R := \sum_{i=1}^{I} \sum_{j=1}^{J} (Y_{ij} - \hat{\mu}_i)^2 = \sum_{i=1}^{I} \sum_{j=1}^{J} (Y_{ij} - \bar{Y}_{i.})^2.$$

The following result shows in particular that $SS_A/(I-1)$ is an unbiased estimator of $\sigma^2$ under the null hypothesis, whereas $SS_R/(I(J-1))$ is an unbiased estimator of $\sigma^2$ in the general model (5.2); see also the discussion of the mean values of $SS_A$ and $SS_R$ after the proof of Theorem 5.1.2. This leads to the statistic

$$F := \frac{SS_A/(I-1)}{SS_R/(I(J-1))}, \tag{5.3}$$

for testing the null hypothesis $H_0$ of identical means in each group. It will be rejected if $F$ is significantly larger than 1.

**5.1.2 Theorem.** *Assume the model (5.2). Then we have:*

*(i) $SS_R/\sigma^2$ is $\chi^2_{I(J-1)}$ distributed.*

(ii) $SS_A/\sigma^2$ is $\chi^2_{I-1}$ distributed under $H_0 : \alpha_1 = \alpha_2 = \cdots = \alpha_I = 0$.

(iii) $SS_R$ and $SS_A$ are independent.

**Proof:** (i) The random variables

$$Z_i := \frac{1}{\sigma^2} \sum_{j=1}^{J} (Y_{ij} - \bar{Y}_{i.})^2, \qquad i = 1, \ldots, I,$$

are obviously independent and by Theorem 2.2.1 (ii), chi–square distributed with $J - 1$ degrees of freedom. The assertion is now a consequence of the representation

$$\frac{SS_R}{\sigma^2} = \sum_{i=1}^{I} Z_i$$

and the convolution theorem for the chi–square distribution, see Remark 2.1.8.
(ii) The random variables

$$W_i := J^{1/2} \bar{Y}_{i.}/\sigma = J^{-1/2} \sum_{j=1}^{J} Y_{ij}/\sigma, \qquad i = 1, \ldots, I,$$

are independent and under the null hypothesis $N(J^{1/2}\mu/\sigma, 1)$ distributed. With $\bar{W} := \sum_{i=1}^{I} W_i/I$ we obtain from Theorem 2.2.1 (ii) that

$$\frac{SS_A}{\sigma^2} = \frac{1}{\sigma^2} \sum_{i=1}^{I} J(\bar{Y}_{i.} - \bar{Y}_{..})^2 = \sum_{i=1}^{I} (W_i - \bar{W})^2$$

is chi–square distributed with $I - 1$ degrees of freedom.
(iii) The independence of $SS_R$ and $SS_A$ can be seen as follows: $SS_R$ is the sum of the random variables $\sigma^2 Z_i$, $i = 1, \ldots, I$, and $SS_A$ is a function of the variables $\bar{Y}_{i.} = \sigma W_i/J^{1/2}$, $i = 1, \ldots, I$. The random vectors $(\sigma^2 Z_i, \bar{Y}_{i.})$, $i = 1, \ldots, I$, are obviously independent. Since $\sigma^2 Z_i$ and $\bar{Y}_{i.}$ are by Theorem 2.2.1 (i) independent as well, we obtain that $SS_R$ and $SS_A$ are functions of the independent vectors $(Z_1, \ldots, Z_I)$ and $(W_1, \ldots, W_I)$, respectively, and thus they are independent, too. $\qquad \square$

For *arbitrary* effects $\alpha_i \in \mathbb{R}$, the mean values of $SS_R$ and $SS_A$ are $E(SS_R) = I(J-1)\sigma^2$ and $E(SS_A) = J \sum_{i=1}^{I} \alpha_i^2 + (I-1)\sigma^2$ (see Exercise 2). This implies the inequality

$$E(SS_A/(I-1)) \geq E(SS_R/(I(J-1))) = \sigma^2,$$

and we have equality if and only if $\alpha_i = 0$ for each $i = 1, \ldots, I$, i.e., under the
null hypothesis $H_0$. It is, therefore, reasonable to test $H_0$ by means of the $F$
statistic

$$F = \frac{SS_A/(I-1)}{SS_R/(I(J-1))}. \tag{5.4}$$

This statistic is by Definition 2.1.9 of the $F$ distribution, $F_{I-1,I(J-1)}$ distributed
if $H_0$ is true. The null hypothesis is, consequently, rejected if $F$ is significantly
large or, equivalently, if the corresponding $p$-value

$$p = 1 - F_{I-1,I(J-1)}(F)$$

is significantly small, for instance if $p \leq 0.05$. $\qquad\qquad\qquad\qquad\qquad$ □

## The ANOVA Table

Those numbers, which are relevant for the one–way analysis of variance, are
typically presented in a table called *ANOVA table*:

|  | degrees of freedom (df) | sum of squares | $F$ statistic |
|---|---|---|---|
| variation among the groups ($SS_A$) | $I - 1$ | $J \sum_{i=1}^{I}(\bar{Y}_{i.} - \bar{Y}_{..})^2$ | $\frac{SS_A/(I-1)}{SS_R/(I(J-1))}$ |
| residual sum of squares ($SS_R$) | $I(J-1)$ | $\sum_{i=1}^{I}\sum_{j=1}^{J}(Y_{ij} - \bar{Y}_{i.})^2$ | |
| total variation ($SS_T$) | $IJ - 1$ | $\sum_{i=1}^{I}\sum_{j=1}^{J}(Y_{ij} - \bar{Y}_{..})^2$ | |

The *variance decomposition* of the one–way analysis of variance

$$SS_T = SS_A + SS_R$$

*(corrected total sum of squares = model sum of squares + error sum of squares)*

is immediate from the equation $\sum_{j=1}^{J}(Y_{ij} - \bar{Y}_{i.}) = 0$, $i = 1, \ldots, I$:

$$SS_T = \sum_{i=1}^{I}\sum_{j=1}^{J}(Y_{ij} - \bar{Y}_{..})^2 = \sum_{i=1}^{I}\sum_{j=1}^{J}((Y_{ij} - \bar{Y}_{i.}) + (\bar{Y}_{i.} - \bar{Y}_{..}))^2$$

$$= \sum_{i=1}^{I}\sum_{j=1}^{J}\left((Y_{ij} - \bar{Y}_{i.})^2 + (\bar{Y}_{i.} - \bar{Y}_{..})^2 + 2(Y_{ij} - \bar{Y}_{i.})(\bar{Y}_{i.} - \bar{Y}_{..})\right)$$

$$= SS_R + SS_A + 2\sum_{i=1}^{I}(\bar{Y}_{i.} - \bar{Y}_{..})\sum_{j=1}^{J}(Y_{ij} - \bar{Y}_{i.})$$

$$= SS_A + SS_R.$$

Theorem 5.1.2 and the convolution theorem for the chi–square distribution imply now that $SS_T/\sigma^2$ is $\chi^2_{IJ-1}$ distributed under $H_0 : \alpha_1 = \cdots = \alpha_I = 0$. Example 5.1.1 yields the following ANOVA table:

The ANOVA Procedure

Class Level Information

| Class | Levels | Values |
|-------|--------|--------|
| METHOD | 4 | I II III IV |

Number of observations    32

Dependent Variable: SCORES    scores achieved

| Source | DF | Sum of Squares | Mean Square | F Value | Pr > F |
|--------|----|----------------|-------------|---------|--------|
| Model | 3 | 621.3750000 | 207.1250000 | 25.60 | <.0001 |
| Error | 28 | 226.5000000 | 8.0892857 | | |
| Corrected Total | 31 | 847.8750000 | | | |

| R-Square | Coeff Var | Root MSE | SCORES Mean |
|----------|-----------|----------|-------------|
| 0.732862 | 23.82548 | 2.844167 | 11.93750 |

| Source | DF | Anova SS | Mean Square | F Value | Pr > F |
|--------|----|----------|-------------|---------|--------|
| METHOD | 3 | 621.3750000 | 207.1250000 | 25.60 | <.0001 |

**Figure 5.1.3.** ANOVA table for the course data in Example 5.1.1.

```
***    Program 5_1_3    ***;
TITLE1 'ANOVA Table';
TITLE2 'Course Data';
LIBNAME datalib 'c:\data';

PROC ANOVA DATA=datalib.course;
    CLASS method;
    MODEL scores=method;
RUN; QUIT;
```

The crucial statements in PROC ANOVA are the CLASS and the MODEL statement. Any classification variable must be declared first in the CLASS statement. It is required and must appear before the MODEL statement. The MODEL statement names the dependent variable and the independent level variable by the syntax *dependent variable = independent level*. PROC ANOVA requires balanced data, i.e., equal number of observations for every level of the classification variable.

Since $p = 1 - F_{3,28}(25.60) = 0.0001$, the null hypothesis $H_0$ of equality of effects of the four types of teaching on performance levels is rejected.

## Unbalanced Data

We assumed hitherto the same number $J$ of observations for each of the $I$ treatments, which is called the case of *balanced* data. In the general case of possibly *unbalanced* data with unequal number of observations for the levels of the classification variable we define the following obvious generalizations of $SS_R$ and $SS_A$.

Suppose we have $J_i \geq 2$ observations for the level $i$. Then we put

$$SS_R = \sum_{i=1}^{I} \sum_{j=1}^{J_i} (Y_{ij} - \bar{Y}_{i.})^2$$

and

$$SS_A = \sum_{i=1}^{I} J_i (\bar{Y}_{i.} - \bar{Y}_{..})^2,$$

where again

$$\bar{Y}_{i.} = \frac{1}{J_i} \sum_{j=1}^{J_i} Y_{ij}$$

is the sample mean of the $i$–th group and

$$\bar{Y}_{..} = \frac{1}{\sum_{i=1}^{I} J_i} \sum_{i=1}^{I} \sum_{j=1}^{J_i} Y_{ij}$$

is the overall sample mean. The overall mean effect is $\mu = \sum_{i=1}^{I} J_i \mu_i / \sum_{i=1}^{I} J_i$, and the effects $\alpha_i = \mu_i - \mu$ of level $i$ satisfy $\sum_{i=1}^{I} J_i \alpha_i = 0$. Theorem 5.1.2 remains valid, but the degrees of freedom of the $\chi^2$ distribution of $SS_R/\sigma^2$ are now $\sum_{i=1}^{I} J_i - I$ (Exercise 3). For arbitrary effects $\alpha_i \in \mathbb{R}$ we have (Exercise 2)

$$E(SS_R) = \left( \sum_{i=1}^{I} J_i - I \right) \sigma^2, \qquad E(SS_A) = (I - 1)\sigma^2 + \sum_{i=1}^{I} J_i \alpha_i^2.$$

The test statistic

$$F = \frac{SS_A/(I-1)}{SS_R/(\sum_{i=1}^{I} J_i - I)}$$

is $F_{I-1, \sum_{i=1}^{I} J_i - I}$ distributed under the null hypothesis $\alpha_i = 0, i = 1, \ldots, I$. This null hypothesis is rejected if $F$ is significantly large, i.e., if the corresponding $p$–value $1 - F_{I-1, \sum_{i=1}^{I} J_i - I}(F)$ is significantly small.

## Tukey's Studentized Range Test

Tukey's studentized range test is an alternative to the $F$ statistic in (5.3) for testing the null hypothesis $H_0 : \alpha_1 = \cdots = \alpha_I = 0$ in model (5.2) with balanced data. It tests the null hypothesis

$$H_0 : \alpha_i = \alpha_j, \qquad i, j = 1, \ldots, I,$$

of equal effects in each group. This hypothesis coincides with the null hypothesis $\alpha_i = 0, \ i = 1, \ldots, I$, with the constraint $\sum_{i=1}^{I} \alpha_i = 0$ in model (5.2). Tukey's test reveals in addition those pairs of effects $\alpha_{i_1}, \alpha_{i_2}$ which are significantly different.

**5.1.3 Definition.** Let $X_1, \ldots, X_m, Z$ be independent random variables, $m \geq 2$, where the $X_i$ are $N(0, 1)$ distributed and $Z$ is chi–square distributed with $n$ degrees of freedom. The distribution of the random variable

$$T = \frac{\max_{1 \leq i, j \leq m} |X_i - X_j|}{\sqrt{Z/n}} = \frac{X_{m:m} - X_{1:m}}{\sqrt{Z/n}}$$

is called the *studentized range distribution* with parameters $m, n$. We denote its distribution function by $T_{m,n}$.

A table of the distribution function $T_{m,n}$ and of its quantiles $T_{m,n}^{-1}(q)$ is given in Miller (1981), pages 37 ff and 234 ff. Note that for $m = 2$ the statistic $T$ is quite closely related to the two sample $t$ test statistic in Theorem 2.3.5 if $X_1, X_2$ are taken to be the standardized sample means of the two samples and $Z/n$ is the pooled sample variance. Tukey's test can, therefore, be viewed as a generalization of the $t$ test for testing the equality of the underlying means in more than two independent samples. We test the hypothesis $H_0 : \alpha_1 = \cdots = \alpha_I$ in model (5.2) via Tukey's test as follows. The sample means

$$\bar{Y}_{i\cdot} = \frac{1}{J} \sum_{j=1}^{J} Y_{ij}, \qquad i = 1, \ldots, I,$$

in each group are independent and $N(\alpha_i + \mu, \sigma^2/J)$ distributed random variables. By Theorem 2.2.1 (i) they are independent of the variance estimator

$$S^2 := \frac{1}{I(J-1)} \sum_{i=1}^{I} \sum_{j=1}^{J} (Y_{ij} - \bar{Y}_{i.})^2.$$

Theorem 5.1.2 yields that $I(J-1)S^2/\sigma^2$ is chi–square distributed with $I(J-1)$ degrees of freedom in the model (5.2). The studentized range

$$T = \frac{\sqrt{J} \max_{1 \le i_1, i_2 \le I} |\bar{Y}_{i_1.} - \bar{Y}_{i_2.}|}{S}$$

is, therefore, $T_{I,I(J-1)}$ distributed under the hypothesis of equal effects. Tukey's test now rejects this hypothesis if the corresponding $p$–value

$$p = 1 - T_{I,I(J-1)}(T)$$

is significantly small. This means that there is at least one pair of effects $\alpha_{i_1}, \alpha_{i_2}$ such that the estimate $|\bar{Y}_{i_1.} - \bar{Y}_{i_2.}|$ for the distance $|\alpha_{i_1} - \alpha_{i_2}|$ is *significantly* large. The estimate is called significantly large at error level $\alpha$ if

$$|\bar{Y}_{i_1.} - \bar{Y}_{i_2.}| > \frac{S}{\sqrt{J}} T_{I,I(J-1)}^{-1}(1-\alpha).$$

This is equivalent to the condition that 0 does *not* belong to the interval with endpoints

$$\bar{Y}_{i_1.} - \bar{Y}_{i_2.} \mp \frac{S}{\sqrt{J}} T_{I,I(J-1)}^{-1}(1-\alpha).$$

**5.1.4 Example** (SO$_2$ Data). In each of the seven Bavarian counties the concentration of sulfur dioxide (SO$_2$) in the air was measured in milligram per cubic meter, mg/m$^3$. A total of 12 measurements in each county was taken at different points in July 1993 and in April 1994. Does the factor *county* have an influence on the SO$_2$ concentration? And if so, in which counties are the SO$_2$ concentrations significantly high? The following chorographic map visualizes the means of the twelve measurements in each county, where the radius of each ball represents the corresponding mean. As the square of a ball is a quadratic function of its radius, the balls grow overproportionately with the radius and, thus, the map is deceptive.

Figure 5.1.4. Chorographic map of $SO_2$ means. The means are represented by the radii of the balls.

```
***    Program 5_1_4    ***;
TITLE1 'Chorographic Map';
TITLE2 'SO2 Data';
LIBNAME datalib 'c:\data';

DATA bavanno;
    LENGTH TEXT $40 FUNCTION $8 mtext $6;
    SET datalib.so2means;
    IF cnty=9100 THEN Y=Y-100;
    IF cnty=9100 THEN X=X+100;
    IF cnty=9300 THEN Y=Y+100;
    IF cnty=9300 THEN X=X+100;
    XSYS='2'; YSYS='2'; POSITION='5'; WHEN='A';
    FUNCTION='PIE'; ANGLE=0; ROTATE=360; STYLE='P5X45';
        SIZE=so2mean*100; OUTPUT;
    Y=Y+200; FUNCTION='TEXT'; TEXT=county;
        SIZE=1; STYLE='CENTB'; OUTPUT;
    Y=Y-350; FUNCTION='TEXT'; mtext=LEFT(so2mean);
        TEXT=mtext; OUTPUT;
RUN;
                            ↓
```

```
                                            ↑
PATTERN1 V=MEMPTY C=G REPEAT=7;
PROC GMAP DATA=datalib.so2means MAP=datalib.bavaria;
    CHORO cnty / ANNOTATE=bavanno LEVELS=7 NOLEGEND;
    ID cnty;
RUN; QUIT;
```

The GMAP procedure with the statement CHORO produces two–dimensional color maps that show variations of a variable with respect to an area. Three–dimensional maps are produced using the statements BLOCK, SURFACE or PRISM. The option LEVELS specifies the number of response levels to be graphed. The ID statement specifies the variable in the MAP data set (here datalib.bavaria) and in the DATA set (here datalib.so2means) that defines unit areas on the map. The PATTERN statement defines the color and fill pattern used to fill unit areas. V = MEMPTY requests an empty pattern. Annotation of the graphs produced by the GMAP procedure is specified by the ANNOTATE option and the pertaining DATA step (here bavanno).

### The GLM Procedure

### Class Level Information

| Class  | Levels | Values                                                                                         |
|--------|--------|------------------------------------------------------------------------------------------------|
| COUNTY | 7      | Central Franconi Lower Bavaria Lower Franconia Swabia Upper Bavaria Upper Franconia Upper Palatinate |

Number of observations      84

Dependent Variable: SO2     SO2

| Source          | DF | Sum of Squares | Mean Square | F Value | Pr > F |
|-----------------|----|----------------|-------------|---------|--------|
| Model           | 6  | 0.00028514     | 0.00004752  | 2.80    | 0.0161 |
| Error           | 77 | 0.00130600     | 0.00001696  |         |        |
| Corrected Total | 83 | 0.00159114     |             |         |        |

| R-Square | Coeff Var | Root MSE | SO2 Mean |
|----------|-----------|----------|----------|
| 0.179206 | 53.38638  | 0.004118 | 0.007714 |

| Source | DF | Type I SS | Mean Square | F Value | Pr > F |
|--------|----|-----------|-------------|---------|--------|
| COUNTY | 6 | 0.00028514 | 0.00004752 | 2.80 | 0.0161 |

| Source | DF | Type III SS | Mean Square | F Value | Pr > F |
|--------|----|-------------|-------------|---------|--------|
| COUNTY | 6 | 0.00028514 | 0.00004752 | 2.80 | 0.0161 |

The GLM Procedure

Tukey's Studentized Range (HSD) Test for SO2

NOTE: This test controls the Type I experimentwise error rate, but it
      generally has a higher Type II error rate than REGWQ.

| | |
|---|---|
| Alpha | 0.1 |
| Error Degrees of Freedom | 77 |
| Error Mean Square | 0.000017 |
| Critical Value of Studentized Range | 3.88807 |
| Minimum Significant Difference | 0.0046 |

Means with the same letter are not significantly different.

| Tukey Grouping | | Mean | N | COUNTY |
|---|---|---|---|---|
|   | A | 0.010167 | 12 | Upper Franconia |
|   | A | | | |
| B | A | 0.009417 | 12 | Central Franconi |
| B | A | | | |
| B | A | 0.009250 | 12 | Upper Palatinate |
| B | A | | | |
| B | A | 0.007917 | 12 | Lower Franconia |
| B | A | | | |
| B | A | 0.006500 | 12 | Swabia |
| B | | | | |
| B | | 0.005500 | 12 | Lower Bavaria |
| B | | | | |
| B | | 0.005250 | 12 | Upper Bavaria |

Figure 5.1.5. Tukey's test for $SO_2$ means.

```
***    Program 5_1_5    ***;
TITLE1 'Tukey Test';
TITLE2 'SO2 Data';
LIBNAME datalib 'c:\data';

PROC GLM DATA=datalib.so2;
   CLASS county;
   MODEL so2=county;
   MEANS county / TUKEY ALPHA=0.1 LINES;
RUN; QUIT;
```

The SAS procedure GLM is used here as an alternative to ANOVA. Note that you get the same result with the procedure ANOVA. The GLM procedure is only needed if the data do not fit into a balanced design. The CLASS statement specifies the factor and has to precede the MODEL statement. The option TUKEY performs Tukey's studentized range test on the main effect means in the MEANS statement, ALPHA determines the level of significance. The option LINES presents the means in descending order and indicates non-significant subsets by line segments beside the corresponding means. An alternative option is CLDIF, which presents the results as confidence intervals for all pairwise differences between means.

If we assume that the $SO_2$ data were generated according to model (5.2), we have $J = 12$ independent measurements for each of the $I = 7$ levels of the factor *county*. The sample means $\bar{Y}_{i\cdot}$ have the expectations $\mu + \alpha_i$, $i = 1,\ldots,7$, and each sample mean is the average of $J = 12$ independent observations. We test the null hypothesis of equal effects $\alpha_{i_1} = \alpha_{i_2}$, $1 \leq i_1, i_2 \leq 7$, via the studentized range

$$T = \frac{\sqrt{12}\, \max_{1 \leq i_1,i_2 \leq 7} |\bar{Y}_{i_1\cdot} - \bar{Y}_{i_2\cdot}|}{S},$$

which has the distribution $T_{7,77}$ under the null hypothesis. We obtain $S^2 = 0.000017$ and the critical value $T_{7,77}^{-1}(1 - \alpha) = 3.888$ with error level $\alpha = 0.1$ for the statistic $T$. This value is attained if the distance between two means $\bar{Y}_{i_1\cdot}, \bar{Y}_{i_2\cdot}$ is at least $T_{7,77}^{-1}(1 - 0.1)\, S/\sqrt{12} = 0.0046$. The above output lists the seven means in decreasing order and coupled in groups A, B. Only those means, which belong to different groups, have a significant distance larger than 0.0046. This grouping visualizes the fact that only the distances between the means of Upper Franconia and Lower Bavaria or Upper Bavaria differ significantly.

The global $F$ statistic in (5.4) has for these data the $p$–value 0.0161 and, therefore, rejects the null hypothesis of equal means in the seven counties. Tukey's test provides us with the additional information about pairs of means which differ significantly. But recall that you have to fix the error level before.

## The Kruskal–Wallis Test

The Wilcoxon test in Section 2.4 was introduced as a nonparametric alternative to the $t$ test for the comparison of the means of two independent samples. The Kruskal–Wallis test is a generalization of the Wilcoxon test. It compares in a nonparametric way the means of $I \geq 2$ independent samples. By this test we can in particular test the null hypothesis $H_0 : \alpha_1 = \ldots = \alpha_I = 0$ of identical effects in the one–way analysis of variance model (5.2) with unbalanced data and *without* the special assumption of *normal* errors $\varepsilon_{ij}$. The Kruskal–Wallis approach starts by assembling all data $Y_{i1}, \ldots, Y_{iJ_i}$, $i = 1, \ldots, I$, in one sample, where $R_{ij}$ denotes now the rank of $Y_{ij}$ in the joint sample. Ties among the data can be broken by computing average ranks, as we did in Section 2.4. The sample mean $\bar{Y}_{i\cdot}$ of the $i$–th group is now replaced by the mean of the ranks in this group

$$\bar{R}_{i\cdot} := \frac{1}{J_i} \sum_{j=1}^{J_i} R_{ij}, \qquad i = 1, \ldots, I.$$

The counterpart of the sample overall mean $\bar{Y}_{\cdot\cdot}$ is

$$\bar{R}_{\cdot\cdot} := \frac{1}{N} \sum_{i=1}^{I} \sum_{j=1}^{J_i} R_{ij} = \frac{1 + 2 + \ldots + N}{N} = \frac{N+1}{2},$$

where $N := \sum_{i=1}^{I} J_i$ and $SS_A$ is replaced by

$$SRS_A := \sum_{i=1}^{I} J_i (\bar{R}_{i\cdot} - \bar{R}_{\cdot\cdot})^2.$$

**5.1.5 Theorem.** *Suppose that the random variables* $Y_{ij}$, $j = 1, \ldots, J_i$, $i = 1, \ldots, I$, *are independent and identically distributed with a common continuous distribution function. Then we have:*

$$\frac{12}{N(N+1)} SRS_A \xrightarrow[\mathcal{D}]{} \chi^2_{I-1}$$

*as* $J_i \to \infty$, $i = 1, \ldots, I$, *where* $N = \sum_{i=1}^{I} J_i$.

**Proof:** See Lehmann (1975), Section 5.2.                                    □

If the distribution function $F$ of the $Y_{ij}$ is continuous, the distribution of $SRS_A$ is independent of $F$, see the arguments in Section 2.4. It can then be computed exactly, at least in theory. The above chi–square approximation is, however, sufficiently close for practical purposes if $I = 3$ and $J_i \geq 5$ or if $I > 3$ and $J_i \geq 4$, see page 207 of Lehmann (1975). Suppose that in model (5.2)

$$Y_{ij} = \mu + \alpha_i + \varepsilon_{ij}, \qquad j = 1, \ldots, J_i, \quad i = 1, \ldots, I,$$

the random variables $\varepsilon_{ij}$ are independent and have a common continuous distribution function. The hypothesis $H_0 : \alpha_1 = \ldots = \alpha_I = 0$ is then rejected if the value of the statistic $(12/(N(N+1)))SRS_A$ is significantly large. By utilizing the chi–square approximation of Theorem 5.1.5 we reject $H_0$ if the $p$–value

$$p = 1 - \chi^2_{I-1}\left(\frac{12}{N(N+1)} SRS_A\right)$$

is significantly small. This is the *Kruskal–Wallis test*.

The statistic $(12/(N(N+1)))SRS_A$ is in the case $I = 2$ the square of the Wilcoxon test statistic $Z$ from Section 2.4 with $m = J_1$ and $n = J_2$: The identity $\sum_{j=1}^{J_2} R_{2j} = N(N+1)/2 - \sum_{j=1}^{J_1} R_{1j}$ implies

$$SRS_A = J_1\left(\frac{1}{J_1}\sum_{j=1}^{J_1} R_{1j} - \frac{N+1}{2}\right)^2 + J_2\left(\frac{1}{J_2}\sum_{j=1}^{J_2} R_{2j} - \frac{N+1}{2}\right)^2$$

$$= J_1\left(\frac{1}{J_1}\sum_{j=1}^{J_1} R_{1j} - \frac{N+1}{2}\right)^2$$

$$+ J_2\left(\frac{1}{J_2}\left(\frac{N(N+1)}{2} - \sum_{j=1}^{J_1} R_{1j}\right) - \frac{N+1}{2}\right)^2$$

$$= \frac{1}{J_1}\left(\sum_{j=1}^{J_1} R_{1j} - \frac{J_1(N+1)}{2}\right)^2 + \frac{1}{J_2}\left(\frac{J_1(N+1)}{2} - \sum_{j=1}^{J_1} R_{1j}\right)^2$$

$$= \left(\sum_{j=1}^{J_1} R_{1j} - \frac{J_1(N+1)}{2}\right)^2\left(\frac{1}{J_1} + \frac{1}{J_2}\right)$$

$$= \frac{N(N+1)}{12} Z^2.$$

We apply the Kruskal–Wallis test to the $SO_2$ data in Example 5.1.4 for testing, whether the factor *county* has any influence on the $SO_2$ concentration in the air. The test statistic $(12/(84 \cdot 85))SRS_A$ has the value 15.283 with the approximate $p$–value $1 - \chi^2_6(15.283) = 0.0182$ and, thus, the hypothesis $H_0$ of identical means will be rejected. This coincides with the result of Tukey's test and the $F$ test in Figure 5.1.5.

The NPAR1WAY Procedure

Wilcoxon Scores (Rank Sums) for Variable SO2
Classified by Variable COUNTY

| COUNTY | N | Sum of Scores | Expected Under HO | Std Dev Under HO |
|---|---|---|---|---|
| Upper Bavaria | 12 | 323.00 | 510.0 | 77.682792 |
| Lower Bavaria | 12 | 365.50 | 510.0 | 77.682792 |
| Upper Palatinate | 12 | 589.00 | 510.0 | 77.682792 |
| Upper Franconia | 12 | 641.50 | 510.0 | 77.682792 |
| Central Franconia | 12 | 621.00 | 510.0 | 77.682792 |
| Lower Franconia | 12 | 604.00 | 510.0 | 77.682792 |
| Swabia | 12 | 426.00 | 510.0 | 77.682792 |

Average scores were used for ties.

Wilcoxon Scores (Rank
Sums) for Variable SO2
Classified by Variable COUNTY

| COUNTY | Mean Score |
|---|---|
| Upper Bavaria | 26.916667 |
| Lower Bavaria | 30.458333 |
| Upper Palatinate | 49.083333 |
| Upper Franconia | 53.458333 |
| Central Franconia | 51.750000 |
| Lower Franconia | 50.333333 |
| Swabia | 35.500000 |

Average scores were
used for ties.

Kruskal-Wallis Test

| Chi-Square | 15.2826 |
|---|---|
| DF | 6 |
| Pr > Chi-Square | 0.0182 |

Figure 5.1.6. Kruskal–Wallis test for the $SO_2$ data from Example 5.1.4.

```
***    Program 5_1_6    ***;
TITLE1 'Kruskal-Wallis Test';
TITLE2 'SO2 Data';
LIBNAME datalib 'c:\data';

PROC NPAR1WAY DATA=datalib.so2 WILCOXON;
    CLASS county;
    VAR so2;
RUN; QUIT;
```

This program is completely analogous to Program 2_4_1. Using the option WILCOXON, SAS performs a Kruskal–Wallis test by default if the CLASS variable has more than two levels.

## 5.2   The Two–Way Analysis of Variance

We consider in the following the analysis of variance with *two* factors, where factor $A$ has $I \geq 2$ different levels and factor $B$ has $K \geq 2$ different levels. We assume the same number $J \geq 1$ of independent observations for each of all $I \times K$ possible factor combinations. This is the case of a balanced design.

**5.2.1 Example** (pH Data; Exercise 1 in Chapter 1). A study by the faculty of forestry of the University of Munich investigated among others the effect of watering and liming on the hydrogen–ion concentration in forest floor. The measurements were taken in pH, used in expressing both acidity and alkalinity on a scale with values from 0 to 14, where 7 represents neutrality, numbers less than 7 increasing acidity, and numbers greater than 7 increasing alkalinity. The factor *watering* has three different levels, *no* (no additional watering), *ac* (additional acid watering), and *w* (additional neutral watering) and the *liming* has the two levels *n* (no additional liming) and *y* (additional liming). To compare the effects, 6 relatively homogeneous plots of land were selected and $J = 16$ observations were taken from each of the 6 possible cross effects. The data are listed in Exercise 1 of Chapter 1. Do the two factors *watering* and *liming* have an effect on the *pH* values?

The following table lists the means of the 16 *pH* observations in each group. They vary considerably, but is this variation significantly large?

```
--------------------------------------
|                 |      LIMING       |
|                 |-------------------|
|                 |   n    |    y     |
|                 |--------+----------|
|                 |  MEAN  |   MEAN   |
|-----------------+--------+----------|
|WATERING         |        |          |
|ac               |   3.95|    6.95|
|no               |   4.02|    6.46|
|w                |   4.12|    7.26|
--------------------------------------
```

**Figure 5.2.1.** Table of the means *watering/liming*; pH data.

```
***    Program 5_2_1    ***;
TITLE1 'Table of Means';
TITLE2 'pH Data';
LIBNAME datalib 'c:\data';

PROC TABULATE DATA=datalib.ph FORMAT=8.2 NOSEPS;
    CLASS watering liming;
    VAR  ph;
    TABLE watering, liming*(MEAN*ph=' ');
RUN; QUIT;
```

The file datalib.ph contains the three variables 'ph', 'watering' (with the levels a, n, w) and 'liming' (n, y). The procedure TAB-ULATE has quite a complex syntax. For its description we refer to the SAS Procedures-Guide (SAS OnlineDoc).

## The Model

Our model for the balanced two–factorial design is completely analogous to that of the one–way analysis of variance in (5.1). We assume that the observations $Y_{ikj}$ from the combination of the $i$–th level of factor $A$ and the $k$–th level of $B$ are linear combinations of the group mean $\mu_{ik}$ and of homoscedastic deviations $\varepsilon_{ikj}$ from the means. We assume, therefore, the model

$$Y_{ikj} = \mu_{ik} + \varepsilon_{ikj}, \qquad i = 1, \ldots, I, \quad k = 1, \ldots, K, \quad j = 1, \ldots, J, \qquad (5.5)$$

where $\mu_{ik}$ are arbitrary real numbers and the $\varepsilon_{ikj}$ are independent and identically $N(0, \sigma^2)$ distributed random variables.

In analogy to model (5.2) we again split the mean $\mu_{ik}$ into an overall mean effect $\mu$ and individual effects $\alpha_i, \beta_k, \gamma_{ik}$, which are due to the levels $i$ and $k$ and the crossing of both. To this end we put

$$\mu := \frac{1}{IK} \sum_{i=1}^{I} \sum_{k=1}^{K} \mu_{ik}, \qquad \alpha_i := \frac{1}{K} \sum_{k=1}^{K} \mu_{ik} - \mu,$$

$$\beta_k := \frac{1}{I} \sum_{i=1}^{I} \mu_{ik} - \mu, \qquad \gamma_{ik} := \mu_{ik} - \alpha_i - \beta_k - \mu.$$

Then we have

$$\sum_{i=1}^{I} \alpha_i = 0 = \sum_{k=1}^{K} \beta_k$$

$$\sum_{i=1}^{I} \gamma_{ik} = 0 = \sum_{k=1}^{K} \gamma_{ik}, \qquad k = 1, \ldots, K, \quad i = 1, \ldots, I. \tag{5.6}$$

Our model for the two–factorial design is now

$$Y_{ikj} = \mu + \alpha_i + \beta_k + \gamma_{ik} + \varepsilon_{ikj}, \tag{5.7}$$

where the effects $\alpha_i, \beta_k$ and $\gamma_{ik}$ satisfy the above constraints (5.6).

The effect $\alpha_i$ is determined only by the $i$-th level of factor $A$ and is, therefore, the *main effect* of this level. The effect $\beta_k$ is, consequently, the main effect of the $k$-th level of factor $B$, which is completely analogous to the one–way analysis of variance model. But here we have to take into account the additional effect $\gamma_{ik}$, generated by the interaction of the $i$-th and $k$-th levels of $A$ and $B$. This *cross effect* is quite commonly known as the *synergy* effect.

## The Hypotheses

We test at first the hypothesis of no cross effects

$$H_{0,\gamma} : \gamma_{ik} = 0, \qquad i = 1, \ldots, I, \quad k = 1, \ldots, K.$$

Then we investigate the main effects of both factors $A$ and $B$ separately. The two null hypotheses are in this case

$$H_{0,\alpha} : \alpha_i = 0, \qquad i = 1, \ldots, I, \quad \text{(factor } A \text{ has no main effect)},$$

and

$$H_{0,\beta} : \beta_k = 0, \qquad k = 1, \ldots, K, \quad \text{(factor } B \text{ has no main effect)}.$$

Denote by

$$\bar{Y}_{\cdots} := \frac{1}{IKJ} \sum_{i=1}^{I} \sum_{k=1}^{K} \sum_{j=1}^{J} Y_{ikj}$$

the overall sample mean and by

$$\bar{Y}_{i\cdots} := \frac{1}{KJ} \sum_{k=1}^{K} \sum_{j=1}^{J} Y_{ikj}, \quad \bar{Y}_{\cdot k\cdot} := \frac{1}{IJ} \sum_{i=1}^{I} \sum_{j=1}^{J} Y_{ikj}, \quad \bar{Y}_{ik\cdot} := \frac{1}{J} \sum_{j=1}^{J} Y_{ikj}$$

the sample means of the $i$–th level of $A$, the $k$–th level of $B$ and of the crossed levels.

## Parameter Estimates

By these sample means we define unbiased estimators of $\mu, \alpha_i, \beta_k$ and $\gamma_{ik}$. It is obvious that

$$\hat{\mu}_{ik} := \frac{1}{J} \sum_{j=1}^{J} Y_{ikj} = \bar{Y}_{ik\cdot}.$$

is an unbiased estimator of $\mu_{ik}$ in model (5.5). Consequently,

$$\hat{\mu} := \frac{1}{IK} \sum_{i=1}^{I} \sum_{k=1}^{K} \hat{\mu}_{ik} = \bar{Y}_{\cdots}$$

is an unbiased estimator of $\mu$ as well, and so is

$$\hat{\alpha}_i := \frac{1}{K} \sum_{k=1}^{K} \hat{\mu}_{ik} - \hat{\mu} = \bar{Y}_{i\cdots} - \bar{Y}_{\cdots}$$

for $\alpha_i$,

$$\hat{\beta}_k := \frac{1}{I} \sum_{i=1}^{I} \hat{\mu}_{ik} - \hat{\mu} = \bar{Y}_{\cdot k\cdot} - \bar{Y}_{\cdots}$$

for $\beta_k$ and

$$\hat{\gamma}_{ik} := \hat{\mu}_{ik} - \hat{\alpha}_i - \hat{\beta}_k - \hat{\mu} = \bar{Y}_{ik\cdot} - \bar{Y}_{i\cdots} - \bar{Y}_{\cdot k\cdot} + \bar{Y}_{\cdots}$$

for $\gamma_{ik}$. The estimators $\hat{\mu}, \hat{\alpha}_i, \hat{\beta}_k, \hat{\gamma}_{ik}$ are actually maximum likelihood estimators for $\mu, \alpha_i, \beta_k, \gamma_{ik}$ in model (5.7) respectively (Exercise 10).

## A Graphic Test for $H_{0,\gamma}$

A graphic examination of the hypothesis $H_{0,\gamma}$ that all cross effects $\gamma_{ik}$ vanish, can be based on the $I$ polygons $s_1, \ldots, s_I$ which are generated by connecting the subsequent points

$$(k, \hat{\mu}_{ik}), \qquad k = 1, \ldots, K,$$

by a straight line for the $i$-th polygon $s_i$. If actually all $\gamma_{ik}$ vanish, we obtain from (5.7) the representation

$$\mu_{ik} = \mu + \alpha_i + \beta_k.$$

The polygons $s_i$ will approximately be parallels in this case, since the distances $s_{i_1}(k) - s_{i_2}(k)$, $k = 1, \ldots, K$, between the two polygons $s_{i_1}, s_{i_2}$ at the connecting points tend to be constant:

$$s_{i_1}(k) - s_{i_2}(k) = \hat{\mu}_{i_1 k} - \hat{\mu}_{i_2 k} \approx \mu_{i_1 k} - \mu_{i_2 k} = \alpha_{i_1} - \alpha_{i_2}.$$

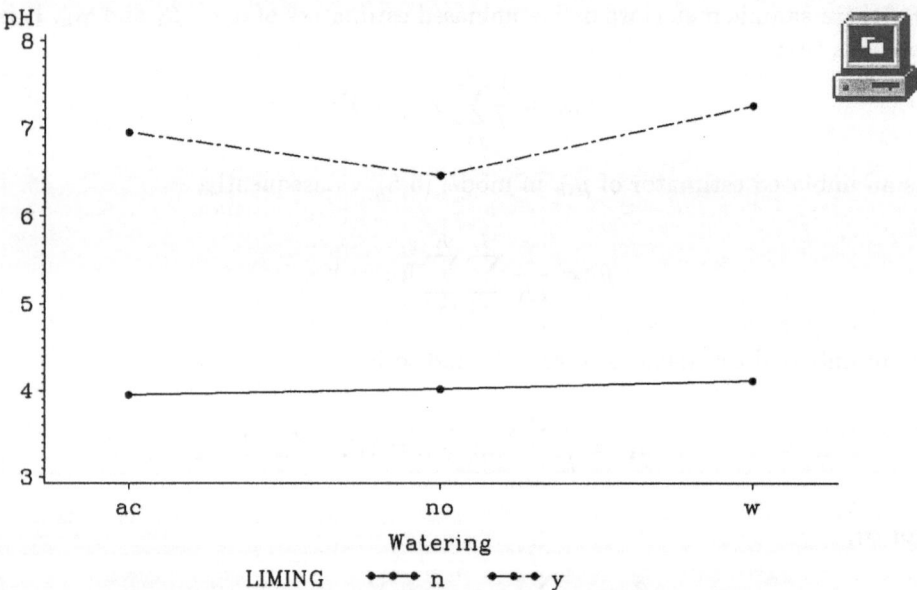

**Figure 5.2.2.** Are the polygons corresponding to the pH data in Example 5.2.1 approximately parallel? Factors *watering/liming*.

```
***    Program 5_2_2    ***;
TITLE1 'Graphic Test'; TITLE2 'pH Data';
LIBNAME datalib 'c:\data';

PROC MEANS DATA=datalib.ph NOPRINT;
    VAR    ph;
    CLASS liming watering;
    OUTPUT OUT=phmean MEAN=ph;
                                       ↓
```

```
                                        ↑
AXIS1 OFFSET=(10 PCT)    LABEL=('Watering');
AXIS2 LABEL=('pH');
SYMBOL1 V=DOT C=G I=JOIN L=1;
SYMBOL2 V=DOT C=G I=JOIN L=8;
PROC GPLOT DATA=phmean(WHERE=(_TYPE_=3));
    PLOT ph*watering=liming / HAXIS=AXIS1 VAXIS=AXIS2;
RUN; QUIT;
```

The syntax of the PLOT statement is analogous to that in the Programs 2_1_1-2_1_3. It generates a polygon for each of the two levels of the variable 'liming' (n, y). The OFF-SET option in the AXIS statement specifies the distance from the first and last tick mark to the end of the axisline (here 10 % of the axis line).

If we assume a two–factorial design for the $SO_2$ data in Example 5.1.4 with the two factors *county* having 7 levels and *date* having 2 levels, we obtain the following plot of the two polygons based on date. This plot suggests to reject the hypothesis $H_{0,\gamma}$ of no cross effects.

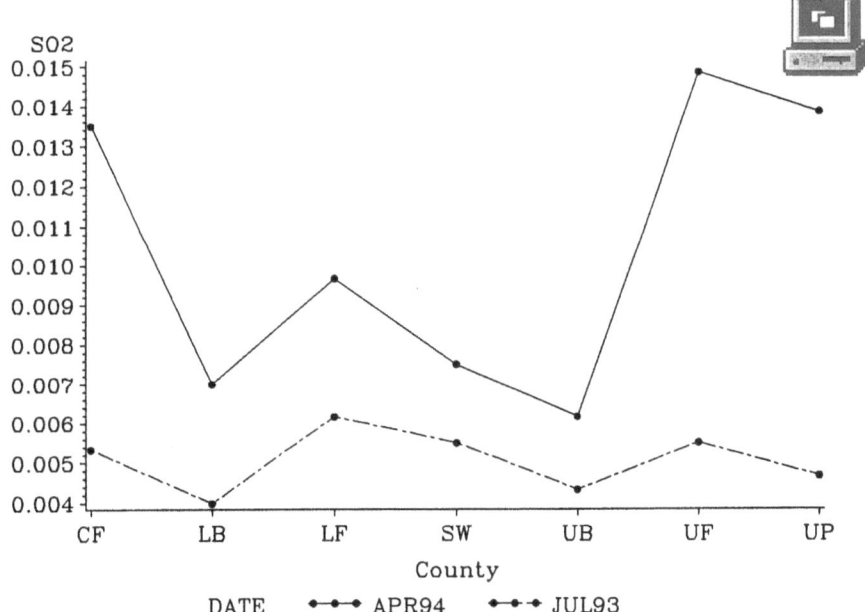

**Figure 5.2.3.** Nonparallel polygons based on different months for the $SO_2$ data in Example 5.1.4; factors *date/county*.

The same considerations apply to the $K$ polygons with the connecting points

$$(i, \hat{\mu}_{ik}), \qquad i = 1, \ldots, I.$$

Note that we do not require a specific assumption on the distribution of $Y_{ikj}$ for the above considerations, but only the representation $E(Y_{ikj}) = \mu_{ik} = \mu + \alpha_i + \beta_k + \gamma_{ik}$ with the constraint (5.6).

## The Sums of Squares

Denote by

$$SS_A := KJ \sum_{i=1}^{I} \hat{\alpha}_i^2 = KJ \sum_{i=1}^{I} (\bar{Y}_{i..} - \bar{Y}_{...})^2,$$

$$SS_B := IJ \sum_{k=1}^{K} \hat{\beta}_k^2 = IJ \sum_{k=1}^{K} (\bar{Y}_{.k.} - \bar{Y}_{...})^2$$

and

$$SS_{AB} := J \sum_{i=1}^{I} \sum_{k=1}^{K} \hat{\gamma}_{ik}^2 = J \sum_{i=1}^{I} \sum_{k=1}^{K} (\bar{Y}_{ik.} - \bar{Y}_{i..} - \bar{Y}_{.k.} + \bar{Y}_{...})^2$$

the variations among the groups, by

$$SS_T := \sum_{i=1}^{I} \sum_{k=1}^{K} \sum_{j=1}^{J} (Y_{ikj} - \hat{\mu})^2$$

the total variation and by

$$SS_R = \sum_{i=1}^{I} \sum_{k=1}^{K} \sum_{j=1}^{J} (Y_{ikj} - \hat{\mu}_{ik})^2$$

the residual sum of squares.

The analysis of the two–factorial design is also based on the skillful comparison of sums of squares. The equation

$$\hat{\mu}_{ik} = \hat{\mu} + \hat{\alpha}_i + \hat{\beta}_k + \hat{\gamma}_{ik},$$

implies the representation

$$SS_R = \sum_{i=1}^{I} \sum_{k=1}^{K} \sum_{j=1}^{J} (Y_{ikj} - (\hat{\mu} + \hat{\alpha}_i + \hat{\beta}_k + \hat{\gamma}_{ik}))^2,$$

and, thus, $SS_R$ is the sum of the squares of the residuals, i.e. the distances between $Y_{ikj}$ and its *ex post facto* prediction $\hat{\mu} + \hat{\alpha}_i + \hat{\beta}_k + \hat{\gamma}_{ik}$.

Straightforward computations yield the decomposition

$$SS_T = SS_A + SS_B + SS_{AB} + SS_R$$

(Exercise 11). The expected values of the above sums are given in the following result.

**5.2.2 Theorem.** *Assuming model (5.7) for the two-factorial design, the above sums of squares have the expected values*

$$E(SS_A) = (I-1)\sigma^2 + KJ\sum_{i=1}^{I}\alpha_i^2,$$

$$E(SS_B) = (K-1)\sigma^2 + IJ\sum_{k=1}^{K}\beta_k^2,$$

$$E(SS_{AB}) = (I-1)(K-1)\sigma^2 + J\sum_{i=1}^{I}\sum_{k=1}^{K}\gamma_{ik}^2,$$

$$E(SS_R) = IK(J-1)\sigma^2.$$

**Proof:** Exercise 12. $\qquad\qquad\qquad\qquad\qquad\qquad\qquad\qquad\qquad\qquad\square$

## The Test Statistics

The above result shows that $SS_A/(I-1)$, $SS_B/(K-1)$ and $SS_{AB}/((I-1)(K-1))$ are unbiased estimators of $\sigma^2$ only under the null hypotheses $H_{0,\alpha} : \alpha_i = 0$, $H_{0,\beta} : \beta_k = 0$ and $H_{0,\gamma} : \gamma_{ik} = 0$, respectively. The sum $SS_R/(IK(J-1))$ is, however, an unbiased estimator of $\sigma^2$ in the general model (5.7) if $J \geq 2$. Therefore, the natural choice is to test these hypotheses by means of the ratios of the corresponding sums of squares, i.e.,

$$F_\alpha := \frac{SS_A/(I-1)}{SS_R/(IK(J-1))} \qquad \text{for testing } H_{0,\alpha}$$

$$F_\beta := \frac{SS_B/(K-1)}{SS_R/(IK(J-1))} \qquad \text{for testing } H_{0,\beta}$$

$$F_\gamma := \frac{SS_{AB}/((I-1)(K-1))}{SS_R/(IK(J-1))} \qquad \text{for testing } H_{0,\gamma}.$$

A null hypothesis is rejected if the corresponding $F$ statistic attains a significantly large value, i.e., if its $p$-value is significantly small. The following result provides us with the distributions of the above statistics. This knowledge is necessary for the computation of the $p$-value. Note that the condition $J \geq 2$ is required only for part (i) of Theorem 5.2.3.

**5.2.3 Theorem.** *If we assume model (5.7) for the two-factorial design, then we have:*

(i) $SS_R/\sigma^2$ is $\chi^2_{IK(J-1)}$ distributed if $J \geq 2$.

(ii) $SS_A/\sigma^2$ is $\chi^2_{I-1}$ distributed under the null hypothesis $H_{0,\alpha} : \alpha_i = 0$, $i = 1, \ldots, I$.

(iii) $SS_B/\sigma^2$ is $\chi^2_{K-1}$ distributed under the null hypothesis $H_{0,\beta} : \beta_k = 0$, $k = 1, \ldots, K$.

(iv) $SS_A, SS_B$ and $SS_A + SS_B$ are independent of $SS_{AB}$. The standardized sum $SS_{AB}/\sigma^2$ is $\chi^2_{(I-1)(K-1)}$ distributed under the null hypothesis $H_{0,\gamma} : \gamma_{ik} = 0$, $i = 1, \ldots, I$, $k = 1, \ldots, K$. $(SS_A + SS_B)/\sigma^2$ is $\chi^2_{I+K-2}$ distributed if $H_{0,\alpha}, H_{0,\beta}$ and $H_{0,\gamma}$ are simultaneously true.

(v) $SS_A, SS_B$ and $SS_{AB}$ are independent of $SS_R$.

(vi) $SS_T/\sigma^2$ is $\chi^2_{IKJ-1}$ distributed if $H_{0,\alpha}, H_{0,\beta}$ and $H_{0,\gamma}$ are simultaneously true.

The above result together with the definition of the $F$ distribution in 2.1.9 yield the distributions of the test statistics $F_\alpha, F_\beta, F_\gamma$ for the model (5.7):

$$F_\alpha \text{ is under } H_{0,\alpha}, \quad F_{I-1,IK(J-1)} \text{ distributed,}$$
$$F_\beta \text{ is under } H_{0,\beta}, \quad F_{K-1,IK(J-1)} \text{ distributed,}$$
$$F_\gamma \text{ is under } H_{0,\gamma}, \quad F_{(I-1)(K-1),IK(J-1)} \text{ distributed.}$$

**Proof of Theorem 5.2.3:** Part (vi) is an immediate consequence of Theorem 2.2.1 (ii). The assertions (i) – (iii) and (v) can be shown along the lines of the proof of Theorem 5.1.2; observe that $SS_A, SS_B$ and $SS_{AB}$ are functions of $\bar{Y}_{ik\cdot}$, so that (v) is a consequence of Theorem 2.2.1 (i). The proof of (iv), however, requires different arguments. We will apply Lemma 3.3.24 on independence and use the distributions of the projections of normally distributed random vectors. Define now the $IK \times I$-matrix

$$A^T := \begin{pmatrix} \mathbf{1}_K & & 0 \\ & \ddots & \\ 0 & & \mathbf{1}_K \end{pmatrix}, \quad \mathbf{1}_K = (1, \ldots, 1)^T \in \mathbb{R}^K,$$

the $n \times n$-matrix

$$E_n := \begin{pmatrix} 1 & \cdots & 1 \\ \vdots & \ddots & \vdots \\ 1 & \cdots & 1 \end{pmatrix},$$

the $IK \times K$–matrix

$$B^T := \begin{pmatrix} I_K \\ \vdots \\ I_K \end{pmatrix}$$

and the symmetric $IK \times IK$–matrix

$$C := I_{IK} - \frac{1}{K}A^T A - \frac{1}{I}B^T B + \frac{1}{IK}E_{IK}.$$

Note that

$$A^T A = \begin{pmatrix} E_K & & 0 \\ & \ddots & \\ 0 & & E_K \end{pmatrix}$$

and

$$B^T B = \begin{pmatrix} I_K & \cdots & I_K \\ \vdots & & \vdots \\ I_K & \cdots & I_K \end{pmatrix}.$$

Next we show that $C$ is an idempotent matrix with $\mathrm{rank}(C) = (I-1)(K-1)$. Using the equations

$$\begin{aligned}
(A^T A)(A^T A) &= K A^T A, \ (B^T B)(B^T B) = I B^T B, \\
(A^T A)(B^T B) &= E_{IK} = (B^T B)(A^T A), \\
(A^T A)E_{IK} &= K E_{IK} = E_{IK}(A^T A), \\
(B^T B)E_{IK} &= I E_{IK} = E_{IK}(B^T B),
\end{aligned}$$

straightforward computations show that it is idempotent. As the main diagonal of $C$ has the constant entry $1 - K^{-1} - I^{-1} + (IK)^{-1}$, we obtain from Exercise 27 in Chapter 3,

$$\mathrm{rank}(C) = tr(C) = IK\left(1 - \frac{1}{K} - \frac{1}{I} + \frac{1}{IK}\right) = (I-1)(K-1). \qquad \square$$

Consequently the matrix $C$ is symmetric and, thus, it is a projection matrix with rank equal to $(I-1)(K-1)$ (see Exercises 27 and 28 in Chapter 3). The $IK$–dimensional random vector

$$Z := \sqrt{J}(\bar{Y}_{11.}, \bar{Y}_{12.}, \ldots, \bar{Y}_{1K.}, \bar{Y}_{21.}, \ldots, \bar{Y}_{IK.})^T = \sqrt{J}(\bar{Y}_{ik.})$$

satisfies

$$CZ = \sqrt{J}(\bar{Y}_{ik.} - \bar{Y}_{i..} - \bar{Y}_{.k.} + \bar{Y}_{...})$$

and, thus, we obtain

$$SS_{AB} = J \sum_{i=1}^{I} \sum_{k=1}^{K} (\bar{Y}_{ik.} - \bar{Y}_{i..} - \bar{Y}_{.k.} + \bar{Y}_{...})^2 = \|CZ\|^2 = (CZ)^T CZ.$$

Denote by $\mu := (\mu_{ik}) = (\mu + \alpha_i + \beta_k + \gamma_{ik})$ the vector of expected values of $(\bar{Y}_{ik.}) = Z/\sqrt{J}$. The convolution theorem 2.1.5 of the normal distribution then implies that the random vector $\tilde{Z} := (Z - \sqrt{J}\mu)/\sigma$ of dimension $IK$ is standard normally distributed. Lemma 3.3.24 yields, therefore, that

$$C\tilde{Z} = \sqrt{J}(\bar{Y}_{ik.} - \bar{Y}_{i..} - \bar{Y}_{.k.} + \bar{Y}_{...})/\sigma - \sqrt{J}C\mu/\sigma$$

and

$$\tilde{Z} - C\tilde{Z} = \sqrt{J}(\bar{Y}_{i..} + \bar{Y}_{.k.} - \bar{Y}_{...})/\sigma + \sqrt{J}(C\mu - \mu)/\sigma$$

are independent and, thus, $SS_{AB}$ is independent of

$$\sum_{i=1}^{I} (\bar{Y}_{i..} + \bar{Y}_{.1.} - \bar{Y}_{...})^2 - \frac{1}{I} \Big( \sum_{i=1}^{I} (\bar{Y}_{i..} + \bar{Y}_{.1.} - \bar{Y}_{...}) \Big)^2$$

$$= \sum_{i=1}^{I} (\bar{Y}_{i..} - \bar{Y}_{...})^2 + 2\bar{Y}_{.1.} \sum_{i=1}^{I} (\bar{Y}_{i..} - \bar{Y}_{...}) + I\bar{Y}_{.1.}^2 - I\bar{Y}_{.1.}^2.$$

$$= \sum_{i=1}^{I} (\bar{Y}_{i..} - \bar{Y}_{...})^2 = SS_A/(KJ)$$

and of

$$\sum_{k=1}^{K} (\bar{Y}_{1..} + \bar{Y}_{.k.} - \bar{Y}_{...})^2 - \frac{1}{K} \Big( \sum_{k=1}^{K} (\bar{Y}_{1..} + \bar{Y}_{.k.} - \bar{Y}_{...}) \Big)^2 = SS_B/(IJ)$$

as well as of $SS_A + SS_B$. Hence we have shown that $SS_{AB}$ is independent of $SS_A, SS_B$ and $SS_A + SS_B$.

Under the null hypothesis $H_{0,\gamma} : \gamma_{ik} = 0$, the constraints (5.6) entail $C\mu = 0$ and, hence,

$$SS_{AB}/\sigma^2 = (CZ)^T CZ/\sigma^2 = (C\tilde{Z})^T C\tilde{Z}.$$

Since $\text{rank}(C) = (I-1)(K-1)$, from Lemma 3.3.24, we finally obtain that $SS_{AB}/\sigma^2$ is $\chi^2_{(I-1)(K-1)}$ distributed under $H_{0,\gamma}$.

The $IK \times IK$–matrices

$$C_1 := \frac{1}{K} A^T A - \frac{1}{IK} E_{IK} \qquad C_2 := \frac{1}{I} B^T B - \frac{1}{IK} E_{IK}$$

are projection matrices with the particular property $C_1C_2 = C_2C_1 = 0$. The sum $C_1+C_2$ is, therefore, a projection matrix as well and has $\text{rank}(C_1+C_2) = \text{tr}\,(C_1 + C_2) = I + K - 2$. If $H_{0,\alpha}$, $H_{0,\beta}$ and $H_{0,\gamma}$ are simultaneously true, the vector $\mu$ of expected values has constant entry $\mu$, i.e., we have $\mu = (\mu)$ and $(C_1 + C_2)\mu = 0$. Lemma 3.3.24 now implies that $||(C_1 + C_2)\tilde{Z}||^2$ is $\chi^2_{I+K-2}$ distributed, which means that

$$||(C_1 + C_2)\tilde{Z}||^2 = ||(C_1 + C_2)Z||^2/\sigma^2 = ||C_1 Z||^2/\sigma^2 + ||C_2 Z||^2/\sigma^2$$
$$= J||C_1(\bar{Y}_{ik.})||^2/\sigma^2 + J||C_2(\bar{Y}_{ik.})||^2/\sigma^2 = SS_A/\sigma^2 + SS_B/\sigma^2$$

is $\chi^2_{I+K-2}$ distributed.                                                                 □

The following ANOVA table for the two–factorial design lists the relevant variables for testing the hypotheses $H_{0,\alpha}$, $H_{0,\beta}$ and $H_{0,\gamma}$:

| | degrees of freedom (df) | sum of squares | test statistic |
|---|---|---|---|
| main effect of factor $A$ | $I - 1$ | $SS_A$ | $F_\alpha = \dfrac{SS_A/(I - 1)}{SS_R/(IK(J - 1))}$ |
| main effect of factor $B$ | $K - 1$ | $SS_B$ | $F_\beta = \dfrac{SS_B/(K - 1)}{SS_R/(IK(J - 1))}$ |
| cross effect | $(I - 1)$ $\times(K - 1)$ | $SS_{AB}$ | $F_\gamma = \dfrac{SS_{AB}/((I - 1)(K - 1))}{SS_R/(IK(J - 1))}$ |
| residual sum of squares | $IK(J - 1)$ | $SS_R$ | |

The pH data in Example 5.2.1 yield the following ANOVA table:

```
                    The ANOVA Procedure


                 Class Level Information


          Class            Levels    Values


          LIMING              2      n y


          WATERING            3      ac no w



       Number of observations      96
```

Dependent Variable: PH

| Source | DF | Sum of Squares | Mean Square | F Value | Pr > F |
|---|---|---|---|---|---|
| Model | 5 | 201.6411969 | 40.3282394 | 241.08 | <.0001 |
| Error | 90 | 15.0554188 | 0.1672824 | | |
| Corrected Total | 95 | 216.6966156 | | | |

| R-Square | Coeff Var | Root MSE | PH Mean |
|---|---|---|---|
| 0.930523 | 7.492161 | 0.409002 | 5.459062 |

| Source | DF | Anova SS | Mean Square | F Value | Pr > F |
|---|---|---|---|---|---|
| LIMING | 1 | 196.2818010 | 196.2818010 | 1173.36 | <.0001 |
| WATERING | 2 | 3.1753313 | 1.5876656 | 9.49 | 0.0002 |
| LIMING*WATERING | 2 | 2.1840646 | 1.0920323 | 6.53 | 0.0023 |

**Figure 5.2.4.** ANOVA table with two–factorial design for the pH data from Example 5.2.1; factors *watering/liming*.

```
***    Program 5_2_4    ***;
TITLE1 'Two-Factorial ANOVA Table';
TITLE2 'pH Data';
LIBNAME datalib 'c:\data';

PROC ANOVA DATA=datalib.ph;
    CLASS  liming watering;
    MODEL  ph=liming|watering;
RUN; QUIT;
```

As the pH data fit into a balanced design, one can use the ANOVA procedure. Specifications of a full factorial model can be shortened by using the bar in the MODEL statement, i.e., the cross effect is computed here as well.

At first it is tested, whether the six possible level combinations of *watering* and *liming* do have any influence at all on the pH values. This is done by means of a one–way analysis of variance for the single factor (watering–liming) with the 6 different levels *ac/n, no/n, w/n, ac/y, no/y, w/y* and 16 observations in each group as described in Section 5.1. The hypothesis that the pH means of the six groups coincide, is rejected by the value 241.08 of the $F$ statistic with the corresponding $p$–value $1 - F_{5,90}(241.08) < 0.0001$ (compare with Figure 5.1.3).

A more detailed analysis, based on a two–factorial design, follows. All hypotheses $H_{0,\gamma}$ (no cross effect), $H_{0,\beta}$ (no main effect of *watering*) and $H_{0,\alpha}$ (no main effect of *liming*) are rejected due to the $p$–values 0.0023, 0.0002 and 0.0001 of the corresponding test statistics $F_\gamma, F_\beta$ and $F_\alpha$ (see also Figure 5.2.2).

This detailed ANOVA does not reveal, however, *which* level combinations of the two factors have a significant influence on the pH means. This can be achieved by applying Tukey's studentized range test to the 6 different levels of the *single* factor (watering–liming).

## The Case $J = 1$

Suppose that we have only one observation per group $(i, k)$ in the two–way design, i.e., $J = 1$. This entails $Y_{ikj} = \bar{Y}_{ik}$. and, thus,

$$SS_R = 0.$$

The above $F$ statistics for testing the hypotheses $H_{0,\alpha}, H_{0,\beta}$ and $H_{0,\gamma}$, consequently, *cannot* be applied. A common approach in this case is to assume no cross effect $\gamma_{ik}$ i.e, one supposes the reduced model

$$Y_{ik} = \mu + \alpha_i + \beta_k + \varepsilon_{ik}, \qquad i = 1, \ldots, I, \quad k = 1, \ldots, K. \qquad (5.8)$$

The main effects $\alpha_i, \beta_k$ satisfy again the constraint

$$\sum_{i=1}^{I} \alpha_i = 0 = \sum_{k=1}^{K} \beta_k$$

and the $\varepsilon_{ik}$ are independent $N(0, \sigma^2)$ distributed measurement errors.

**5.2.4 Example** (Wheat Data; Rohatgi (1976), page 522). To study the effects of fertilization on various types of wheat, each of four different fertilizers (factor $A$) is applied to each of three different types of wheat (factor $B$). The following table lists the crop per combination in pounds per plot:

```
|                |       WHEAT                            |
|                |--------------------------------------|
|                |   1    |   2    |   3    |
|-------------+----------+----------+----------|
|FERTILIZ     |          |          |          |
|-------------|          |          |          |
|1            |    8.00|    3.00|    7.00|
|-------------+----------+----------+----------|
```

```
|2            |      10.00|      4.00|       8.00|
|-------------+-----------+-----------+-----------|
|3            |       6.00|      5.00|       6.00|
|-------------+-----------+-----------+-----------|
|4            |       8.00|      4.00|       7.00|
-------------------------------------------------
```

**Figure 5.2.5.** Wheat data; crop per combination.

```
***    Program 5_2_5    ***;
TITLE1 'Table of Crop';
TITLE2 'Wheat Data';
LIBNAME datalib 'c:\data';

DATA datalib.wheat;
    DO fertiliz=1 TO 4;
        DO wheat=1 TO 3;
            INPUT crop @@;
            OUTPUT;
    END; END;
    CARDS;
  8  3  7   10 4 8   6 5 6   8 4 7
  ;
PROC TABULATE DATA=datalib.wheat;
    CLASS  fertiliz wheat;
    VAR    crop;
    TABLE  fertiliz, wheat*(SUM*crop=' ');
    KEYLABEL SUM=' ';
RUN; QUIT;
```

The way of reading data in this DATA step was already used in Programs 4_1_7, 4_1_9 and 4_2_1. It is quite common for data from agricultural experiments and forestry.

The printed table is produced with PROC TABULATE, cf SAS Procedures Guide (SAS OnlineDoc).

To test the hypotheses

$$H_{0,\alpha} : \alpha_1 = \alpha_2 = \cdots = \alpha_I = 0, \quad H_{0,\beta} : \beta_1 = \beta_2 = \cdots = \beta_K = 0$$

that the two corresponding factors have no impact, we use the sums of squares $SS_A, SS_B$ and $SS_{AB}$. By Theorem 5.2.2, for model (5.8), we have

$$E(SS_A) = (I-1)\sigma^2 + K \sum_{i=1}^{I} \alpha_i^2,$$

$$E(SS_B) = (K-1)\sigma^2 + I\sum_{k=1}^{K} \beta_k^2,$$

$$E(SS_{AB}) = (I-1)(K-1)\sigma^2.$$

It is, therefore, plausible to reject the hypothesis $H_{0,\alpha}$ if

$$\tilde{F}_\alpha := \frac{SS_A/(I-1)}{SS_{AB}/((I-1)(K-1))}$$

is significantly large, and to reject $H_{0,\beta}$ if

$$\tilde{F}_\beta := \frac{SS_B/(K-1)}{SS_{AB}/((I-1)(K-1))}$$

is significantly large. Theorem 5.2.3 and the definition of the $F$ distribution in 2.1.9 provide the distributions of these $F$ statistics under the null hypotheses. Under $H_{0,\alpha}$, the sum of squares $SS_A/\sigma^2$ is $\chi_{I-1}^2$ distributed, under $H_{0,\beta}$, $SS_B/\sigma^2$ is $\chi_{K-1}^2$ distributed and $SS_{AB}/\sigma^2$ is for arbitrary $\alpha_i, \beta_k$, i.e., in the general model (5.8), $\chi_{(I-1)(K-1)}^2$ distributed. This is due to the fact that model (5.8) coincides with model (5.7) for the null hypothesis $\gamma_{ik} = 0$. Hence, under $H_{0,\alpha}$ we obtain that

$$\tilde{F}_\alpha \text{ is } F_{I-1,(I-1)(K-1)} \text{ distributed},$$

and under $H_{0,\beta}$ that

$$\tilde{F}_\beta \text{ is } F_{K-1,(I-1)(K-1)} \text{ distributed}.$$

The $p$–values are $p_\alpha = 1 - F_{I-1,(I-1)(K-1)}(\tilde{F}_\alpha)$, $p_\beta = 1 - F_{K-1,(I-1)(K-1)}(\tilde{F}_\beta)$. The ANOVA table for the case $J = 1$ is now

| | degrees of freedom (df) | sum of squares | test statistic |
|---|---|---|---|
| main effect of factor $A$ | $I-1$ | $SS_A$ | $\tilde{F}_\alpha = \dfrac{SS_A/(I-1)}{SS_{AB}/((I-1)(K-1))}$ |
| main effect of factor $B$ | $K-1$ | $SS_B$ | $\tilde{F}_\beta = \dfrac{SS_B/(K-1)}{SS_{AB}/((I-1)(K-1))}$ |
| cross effect | $(I-1)(K-1)$ | $SS_{AB}$ | |

Example 5.2.4 yields the following ANOVA table.

The ANOVA Procedure

Class Level Information

| Class | Levels | Values |
|-------|--------|--------|
| FERTILIZ | 4 | 1 2 3 4 |
| WHEAT | 3 | 1 2 3 |

Number of observations     12

Dependent Variable: CROP

| Source | DF | Sum of Squares | Mean Square | F Value | Pr > F |
|--------|----|----|----|----|----|
| Model | 5 | 39.33333333 | 7.86666667 | 6.44 | 0.0211 |
| Error | 6 | 7.33333333 | 1.22222222 | | |
| Corrected Total | 11 | 46.66666667 | | | |

| R-Square | Coeff Var | Root MSE | CROP Mean |
|----------|-----------|----------|-----------|
| 0.842857 | 17.45592 | 1.105542 | 6.333333 |

| Source | DF | Anova SS | Mean Square | F Value | Pr > F |
|--------|----|----|----|----|----|
| FERTILIZ | 3 | 4.66666667 | 1.55555556 | 1.27 | 0.3654 |
| WHEAT | 2 | 34.66666667 | 17.33333333 | 14.18 | 0.0053 |

**Figure 5.2.6.** ANOVA table for the wheat data in Example 5.2.4; factors *fertilizer/wheat*.

```
***    Program 5_2_6    ***;
TITLE1 'ANOVA Table';
TITLE2 'Wheat Data';
LIBNAME datalib 'c:\data';

PROC ANOVA DATA=datalib.wheat;
   CLASS   fertiliz wheat;
   MODEL  crop=fertiliz wheat;
RUN; QUIT;
```

This program is analogous to Program 5_2_4, except for the fact that in the case of only one observation for each factor combination no cross effect is computed.

The *global F* statistic

$$F_{\alpha,\beta} := \frac{(SS_A + SS_B)/(I + K - 2)}{SS_{AB}/((I - 1)(K - 1))}$$

tests firstly, whether $H_{0,\alpha}$ and $H_{0,\beta}$ are simultaneously true. By Theorem 5.2.3 (iv) $F_{\alpha,\beta}$ is in this case in model (5.8) $F_{I+K-2,(I-1)(K-1)}$ distributed. For the wheat data, with $I = 4$, $K = 3$, we obtain

$$SS_A + SS_B = 39.3333, \quad SS_{AB} = 7.3333, \quad F_{\alpha,\beta} = 6.44$$

and the $p$-value $1 - F_{5,6}(6.44) = 0.0211$. This result tends to reject the hypothesis that $H_{0,\alpha}$ and $H_{0,\beta}$ are simultaneously true, i.e., that neither the type of fertilizer nor the type of wheat have an impact on the crop.

The individual $F$ statistics $\tilde{F}_\alpha = 1.27$ and $\tilde{F}_\beta = 14.18$ with the corresponding $p$-values $p_\alpha = 0.3654$ and $p_\beta = 0.0053$ indicate that the fertilizers perform equally, the type of wheat, however, does have an impact on the crop. To determine the most effective type of wheat, one could now drop the factor fertilizer and use Tukey's approach for the multiple means comparison for the factor wheat, see Figure 5.1.5. It is obvious from the table of crops in Figure 5.2.5 that type 1 is the most productive one. But is it *significantly* the most productive one, or is type 2 *significantly* the least productive type?

## Exercises

1. Let $X_i$, $i = 1, \ldots, n$, be independent and square integrable variables with $E(X_i) = \mu_i$ and $Var(X_i) = \sigma^2$. Show that

$$E(X_i - \bar{X})^2 = (\mu_i - \bar{\mu})^2 + \frac{n-1}{n}\sigma^2,$$

where

$$\bar{X} = \frac{1}{n}\sum_{i=1}^{n} X_i, \quad \bar{\mu} = \frac{1}{n}\sum_{i=1}^{n} \mu_i.$$

2. Consider the one–way analysis of variance model (5.2) with unbalanced data. Put

$$\bar{Y}_{i.} := \frac{1}{J_i}\sum_{j=1}^{J_i} Y_{ij}, \quad \bar{Y}_{..} := \frac{1}{\sum_{i=1}^{I} J_i}\sum_{i=1}^{I}\sum_{j=1}^{J_i} Y_{ij}$$

$$SS_R := \sum_{i=1}^{I}\sum_{j=1}^{J_i}(Y_{ij} - \bar{Y}_{i.})^2, \quad SS_A := \sum_{i=1}^{I} J_i(\bar{Y}_{i.} - \bar{Y}_{..})^2.$$

Prove that:

(i) $SS_R/\left(\sum_{i=1}^{I} J_i - I\right)$ is an unbiased estimator of $\sigma^2$.

(ii) $E(SS_A) = \sum_{i=1}^{I} J_i \alpha_i^2 + (I-1)\sigma^2$, i.e., under $H_0$, $SS_A/(I-1)$ is an unbiased estimator of $\sigma^2$.

Hint: Use Exercise 1. (The $\varepsilon_{ij}$ do not need to be normally distributed; it suffices to require that the errors $\varepsilon_{ij}$ are independent with common mean 0 and variance $\sigma^2$.)

**3.** (Generalization of Theorem 5.1.2 to the unbalanced case) Consider the model (5.2). Let $SS_R$ and $SS_A$ be defined as in Exercise 2. Show that:

(i) $SS_R/\sigma^2$ is chi–square distributed with $\sum_{i=1}^{I} J_i - I$ degrees of freedom.

(ii) $SS_A/\sigma^2$ is chi–square distributed with $I-1$ degrees of freedom under $H_0$ and $SS_R$ and $SS_A$ are independent.

Hint to (ii): Suppose without loss of generality that $E(Y_{ij}) = 0$, $E(Y_{ij}^2) = 1$. Then $W_i := J_i^{1/2}\bar{Y}_{i.}$, $i = 1, \ldots, I$ are $N(0,1)$ distributed. Let $A$ be an orthogonal $I \times I$-matrix, whose first row is $(J_1^{1/2}/(\sum_{i=1}^{I} J_i)^{1/2}, \ldots, J_I^{1/2}/(\sum_{i=1}^{I} J_i)^{1/2})$. Consider the vector $(X_1, \ldots, X_I)^T := A(W_1, \ldots, W_I)^T$, show that $X_2^2 + \cdots + X_I^2 = SS_A$ and conclude as in the proof of Theorem 2.2.1.)

**4.** (Wheat data) Compute a Tukey test in order to determine the most effective type of wheat.

**5.** Generate plots of the distribution function of $T_{m,n}$ and of the density of the studentized range distribution for various values of $m$ and $n$.

**6.** Prove the representation (see Section 5.1)

$$\frac{12}{N(N+1)} SRS_A = \frac{12}{N(N+1)}\left(\sum_{i=1}^{I} J_i \bar{R}_{i.}^2\right) - 3(N+1).$$

**7.** (Course data) Use the Kruskal–Wallis test to decide, whether the way of teaching a course has a significant influence on performance in the final examination.

**8.** (i) Consider the one–way analysis of variance model (5.1). Show that the $i$-th sample mean $\bar{Y}_{i.}$ is the least squares estimator of $\mu_i$, $i = 1, \ldots, I$. (ii) Consider the two–way analysis of variance model (5.5). Show that $\bar{Y}_{ik.}$ is the least squares estimator of $\mu_{ik}$. Hint: (5.1) and (5.5) can be written in matrix notation as $Y = X\mu + \varepsilon$ with $\mu = (\mu_1, \ldots, \mu_I)^T$. The arguments in Section 5.3 then imply that $(X^T X)^{-1} X^T Y$ is the least squares estimator of $\mu$.

**9.** Show that the sum of squares $SS_A$ and $SS_B$ are independent in model (5.7). Hint: Use the projection matrices $C_1, C_2$ in the proof of Theorem 5.2.3 (iv).

**10.** Consider the two–way analysis of variance model (5.7). Prove that the maximum likelihood estimator $\hat{\Theta} = (\hat{\mu}, \hat{\alpha}, \hat{\beta}, \hat{\gamma})$, $\hat{\alpha} = (\hat{\alpha}_i)_{1 \leq i \leq I}$, $\hat{\beta} = (\hat{\beta}_k)_{1 \leq k \leq K}$, $\hat{\gamma} = (\hat{\gamma}_{ik})_{1 \leq i \leq I, 1 \leq k \leq K}$ has the representation given in Section 5.2.

**11.** Prove the equation

$$SS_T = SS_A + SS_B + SS_{AB} + SS_R.$$

Hint:

$$Y_{ikj} - \bar{Y}_{...}$$
$$= (Y_{ikj} - \bar{Y}_{ik.}) + (\bar{Y}_{i..} - \bar{Y}_{...}) + (\bar{Y}_{.k.} - \bar{Y}_{...}) + (\bar{Y}_{ik.} - \bar{Y}_{i..} - \bar{Y}_{.k.} + \bar{Y}_{...}).$$

**12.** Establish Theorem 5.2.2.

**13.** (Air pollution data) Compute a two–way analysis of variance of the ozone measurements ($o_3$) with the two factors *date* and *county*. Include the values of ozone, that were estimated by a regression analysis, see the SAS data set datalib.air1 in Program 3_3_2. Use the graphic test in Section 5.2 to check, whether the two factors interact.

The MEANS Procedure

| STATUS | N Obs | Variable | Mean | Std Dev |
|--------|-------|----------|------|---------|
| healthy | 40 | AN | 2.3660000 | 0.2906033 |
|  |  | ON | 1.0715000 | 0.2137282 |
|  |  | MN | 0.2162500 | 0.0392028 |
| ill | 58 | AN | 2.3256897 | 1.2596632 |
|  |  | ON | 1.1470690 | 0.8233628 |
|  |  | MN | 0.1884483 | 0.0582433 |

**Figure 6.1.1.** Table of means and standard deviations of cns data; variables $a/n$, $o/n$, $m/n$.

```
***    Program 6_1_1    ***;
TITLE1 'Means and Standard Deviations';
TITLE2 'CNS Data';
LIBNAME datalib 'c:\data';

PROC MEANS DATA=datalib.cns MEAN STD;
   CLASS status;
   VAR an on mn;
RUN; QUIT;
```

The MEANS procedure computes means, standard deviations and other descriptive statistics for stratified data also. This ability to compute the statistics for each distinct level of one or more variables is required by the statements CLASS or BY. The CLASS statement generates the above layout of the output, which is different from that one based on BY.

Both statements are in particular useful if they are used together with the OUTPUT statement. The statistics for the stratified data can be stored separately in a data file and can then be processed individually.

In order to get a first idea whether the objects can actually be classified by the *observed* variables, we evaluate scatterplots of *training samples* of objects, whose classes are known. A significant spatial separation of the classes by clearly separated clouds of associated vectors of observations, would indicate that the observed variables have a good capacity of discriminating.

# Chapter 6

# Discriminant Analysis

The discriminant analysis evaluates samples of events or objects in order to discriminate between or to classify them: Suppose we are given an object with an observable set of variables $x$. We want to classify this object as a member of one of $K$ classes. This kind of problem occurs in various fields; take, for example, the identification of a disease from its signs and symptoms on a patient, or the classification of prehistorical materials in archaeology as well as the *optical character recognition*. The detection of criminals on the basis of various observable characteristics of a person is a typical application of *pattern recognition*. It is obvious that the patterns are accompanied by stochastic variations and, hence, discriminant analysis is part of statistics.

## 6.1 Bayes' Approach

We consider a population $\Omega$ of objects, which is divided into $K \geq 2$ disjoint subsets $\Omega_1, \ldots, \Omega_K$. These subsets will be called *classes*. Let $\omega \in \Omega$ be an object, whose actual class is unknown. The object $\omega$ carries $p$ observable characteristics, which we file as a $p$ dimensional vector $x = x(\omega) \in S \subset \mathbb{R}^p$. The set $S$ is the totality of all possible vectors in this situation. Given this vector $x \in S$, the goal is to determine the actual class of $\omega \in \Omega$ in the $K$ disjoint classes $\Omega_1, \ldots, \Omega_K$. The number $K$ is known.

**6.1.1 Example.** We want to detect morphologial deviations of the central nervous system of a human being by the numbers of different types of nerve cells. This will be done by evaluating the cns data of Example 1.1.3.

The following table lists the means and standard deviations of the three ratios of cell types $a/n$, $o/n$ and $m/n$, separately for healthy and mentally ill subjects.

The following figure is a scatterplot of the 98 data points $(a/n, o/n, m/n)$ of the cns data in Example 1.1.3. A point representing an ill person is marked by a pyramid, a healthy person is marked by a star. The stars are visually surrounded by pyramids, which indicates that the variables $(a/n, o/n, m/n)$ have a certain capacity of discriminating between ill and healthy persons. The classification of a *new* object by means of these variables might, therefore, be successful with high probability. But which rule of classification or discriminant function shall we use?

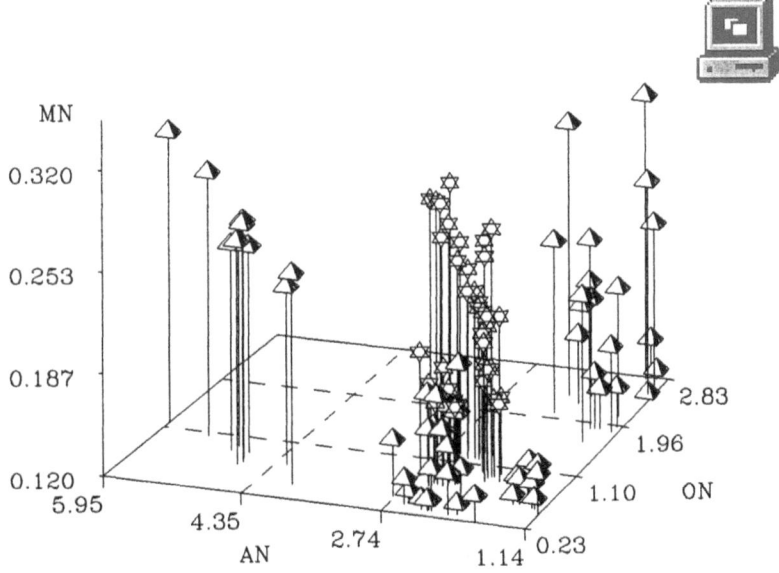

**Figure 6.1.2.** Scatterplot in 4 dimensions of cns data from Figure 6.1.1; healthy/ill is marked by a star/pyramid.

```
***    Program 6_1_2    ***;
TITLE1 'Fourdimensional Scatterplot';
TITLE2 'CNS Data';
LIBNAME datalib 'c:\data';

DATA data1;
   LENGTH shapev $10;
   SET datalib.cns;
   IF status='ill' THEN shapev='PYRAMID';
   ELSE shapev='STAR';
PROC G3D DATA=data1;
   SCATTER an*on=mn / SHAPE=shapev;
RUN; QUIT;
```

The procedure G3D produces three-dimensional graphs that plot one vertical variable z against two horizontal variables x, y, cf Program 6_4_2. The coordinates (x,y,z) are specified as y*x=z and must correspond to the values of three numeric variables in an observation of the input data set.

A scatter or needle plot is created using the SCATTER statement. A surface plot is created with the PLOT statement. The 'SHAPE=' option specifies the character variable in the input data set that contains the shape information for the plot. The shape variable 'shapev' is in this example generated in the DATA1 step. It has the values 'STAR' and 'PYRAMID'.

We divide the sample space $S$ into disjoint and nonempty subsets $G_1, \ldots, G_K$ and we estimate the class index of $\omega$ by the *decision rule*

$$\omega \text{ is classified to } \Omega_k \Leftrightarrow x \in G_k. \tag{6.1}$$

The subsets will be chosen in some optimal way. This optimality is based on a probabilistic reasoning. To this end we consider $x = x(\omega)$ and the *true* class index $k = k(\omega)$ to be realizations of random variables $X : \Omega \to S$ and $\kappa : \Omega \to \{1, \ldots, K\}$, where $\kappa(\omega) := k$ for $\omega \in \Omega_k$, $k = 1, \ldots, K$. The random variables $X$ and $\kappa$ will generally be dependent variables, as the vector $X(\omega)$ contains the observable information about the class index $\kappa(\omega)$.

## The Model

We assume that the sampling of an object from the population $\Omega$ is governed by a probability distribution $P$. This is called the *prior distribution*. Denote by $p(k) := P(\Omega_k)$ the prior probability of the event $\Omega_k = \{\kappa = k\}$, i.e., of the event that $\kappa$ attains the value $k$. We assume $p(k) > 0, k = 1, \ldots, K$. We assume further that the elementary conditional distribution of $X$, given that $\kappa = k$,

$$P(X \in \cdot | \kappa = k) = P\{X \in \cdot, \kappa = k\}/p(k)$$

has a $p$ dimensional density $f(x|k)$, $k = 1, \ldots, K$, i.e.,

$$P(X \in G | \kappa = k) = \int_G f(x|k)\, dx = \int_G f(x_1, \ldots, x_p|k)\, dx_1 \ldots dx_p$$

for any (Borel) subset $G$ of $S \subset \mathbb{R}^p$ and $k = 1, \ldots, K$. We suppose $S$ to be a Borel set as well. This conditional distribution $P(X \in \cdot | \kappa = k)$ is the *group-specific distribution* of $X$ on $\Omega_k$. The unconditional distribution of $X$ on $S$ has, consequently, the density

$$f(x) := \sum_{k=1}^{K} p(k) f(x|k), \quad x \in S,$$

(see Exercise 2). As our decision for a particular class has to be based on the observation $x$, the conditional distribution of $\kappa$, given that $X = x$,

$$P(\kappa \in \cdot | X = x).$$

will be crucial. This is the *posterior distribution* of $\kappa$ on $\Omega$. The function $(x, k) \mapsto p(k)f(x|k)$ is the joint density of $(X, \kappa)$. The posterior distribution $P(\kappa \in \cdot | X = x)$ has, consequently, the density

$$p(k|x) = \begin{cases} p(k)f(x|k)/f(x) & \text{, if } f(x) > 0, \\ g(k) & \text{elsewhere,} \end{cases}$$

with respect to the counting measure on $\{1, \ldots, K\}$. By $g \geq 0$ we denote an arbitrary probability density on this set, i.e., $g(1) + \cdots + g(K) = 1$ (Exercise 3). We choose in the sequel the particular values $g(1) = 1$, $g(k) = 0$, $k \geq 2$, cf (6.3). The prior distribution $p(k)$ and the conditional density $f(x|k)$ of $X$ are commonly unknown and have to be estimated by means of a *training sample*. This is a sample of objects $\omega_1, \ldots, \omega_n$, for which the actual class indices are known. For simplicity, we start with the case where both $p(k)$ and $f(x|k)$ are known.

Let $\{G_1, \ldots, G_K\}$ be an arbitrary partition of $S$. In the case of the event $\{X \in G_j, \kappa = i\}$ with $j \neq i$, our decision rule (6.1) leads to a *misclassification*. We assume that the pertaining loss of this misclassification is $C(j|i) > 0$, where

$$C(\cdot|\cdot) : \{1, \ldots, K\} \times \{1, \ldots, K\} \to [0, \infty)$$

with $C(k|k) = 0$, $k = 1, \ldots, K$, is a *loss function*; There is, consequently, no loss in the case of a correct classification $\{X \in G_k, \kappa = k\}$. Misclassifications can lead to various consequences, some of which might be quite serious; especially in medical diagnosis. The loss function enables us to take these differences into account by putting different weights on the misclassifications.

The probability of classifying an object from the $i$-th class to the $j$-th class is

$$P(X \in G_j | \kappa = i) = \int_{G_j} f(x|i)\, dx.$$

The *total error rate* is, consequently,

$$P\{\text{an object is misclassified}\} = \sum_{i=1}^{K} \sum_{j=1, j\neq i}^{K} P\{X \in G_j, \kappa = i\}$$

$$= \sum_{i=1}^{K} \sum_{j=1, j\neq i}^{K} p(i) \int_{G_j} f(x|i)\, dx.$$

The expected loss or *risk R* is, therefore, given by

$$R := R(G_1, \ldots, G_K) := E\left(C\left(\sum_{j=1}^{K} j \, 1_{G_j}(X) \Big| \kappa\right)\right)$$

$$= \sum_{j=1}^{K} \sum_{i=1}^{K} p(i) \, C(j|i) \int_{G_j} f(x|i) \, dx.$$

Our goal is now to find a partition $\{G_1^*, \ldots, G_K^*\}$ with minimum risk

$$R(G_1^*, \ldots, G_K^*) = \min_{\{G_1, \ldots, G_K\}} R(G_1, \ldots, G_K).$$

Such a partition entails the *optimal* decision rule

$$\omega \text{ is classified to } \Omega_k \Leftrightarrow x \in G_k^*.$$

## Optimal Classes

The following result provides optimal classes. It is based on the idea of mini-
mizing the integrands $\sum_{i=1}^{K} p(i)C(j|i)f(x|i)$ of $R$ with respect to $j$.

**6.1.2 Theorem.** *Put for* $x \in S$ *and* $k = 1, \ldots, K$

$$d_k(x) := \sum_{i=1}^{K} p(i)C(k|i)f(x|i).$$

*The risk R is minimized by the partition*

$$G_1^* = \{y \in S : d_1(y) = \min_{1 \le k \le K} d_k(y)\},$$

$$G_j^* = \{y \in S : d_j(y) = \min_{1 \le k \le K} d_k(y)\} \setminus \bigcup_{i=1}^{j-1} G_i^*, \qquad j = 2, \ldots, K.$$

*The functions* $d_k : S \to [0, \infty)$ *are called discriminant functions.*

**Proof.** Let $\{G_1, \ldots, G_K\}$ be an arbitrary partition of the sample space $S$. We
have

$$R(G_1, \ldots, G_K) = \sum_{j=1}^{K} \sum_{i=1}^{K} p(i) \, C(j|i) \int_{G_j} f(x|i) \, dx$$

$$= \int \sum_{j=1}^{K} d_j(x) 1_{G_j}(x) \, dx.$$

By $1_G(x)$ we denote again the indicator function of a set $G$, i.e., $1_G(x) = 1$ if $x \in G$ and $1_G(x) = 0$ if $x \notin G$. The definition of $G_j^*$ implies that

$$\sum_{j=1}^{K} d_j(x)1_{G_j^*}(x) = \min_{1 \le k \le K} d_k(x)$$

and, hence, we obtain

$$R(G_1, \ldots, G_K) - R(G_1^*, \ldots, G_K^*)$$

$$= \int \left\{ \sum_{j=1}^{K} d_j(x)1_{G_j}(x) - \sum_{j=1}^{K} d_j(x)1_{G_j^*}(x) \right\} dx$$

$$= \int \underbrace{\left\{ \sum_{j=1}^{K} d_j(x)1_{G_j}(x) - \min_{1 \le k \le K} d_k(x) \right\}}_{\ge 0} dx \ge 0. \quad \square$$

The above definition of optimal classes $G_k^*$ entails the classification of an observation $x \in S$ according to the rule:

Determine the *smallest* number $\hat{k} \in \{1, \ldots, K\}$ with $d_{\hat{k}}(x) = \min_{1 \le j \le K} d_j(x)$,

classify the object $\omega$ associated with $x$ to the class with index $\hat{k}$.        (6.2)

The number $\hat{k} = \hat{k}(x) = \hat{k}(X(\omega))$ is by Theorem 6.1.2 an estimator of the class index $\kappa(\omega)$ with minimum risk $R$. Note that the subsets $G_k^*$ are obviously invariant under strictly monotone transformations of the discriminant functions $d_k$, such as $\log(d_k)$, $k = 1, \ldots, K$.

## Bayes' Rule

If we take the *simple symmetric* loss function

$$C(j|i) = \begin{cases} 0 & \text{if } i = j, \\ C & \text{elsewhere;} \end{cases}$$

with a constant $C > 0$, then (6.2) is *Bayes' decision rule*: Take the smallest class index $\hat{k}$, which maximizes the posterior probability for a given vector of observations $x$:

$$p(\hat{k}|x) = \max_{1 \le k \le K} p(k|x). \qquad (6.3)$$

If $f(x) > 0$, this is obviously equivalent to

$$p(\hat{k})f(x|\hat{k}) = \max_{1 \le k \le K} p(k)f(x|k). \qquad (6.4)$$

Formula (6.3) can be seen as follows. In case $f(x) > 0$ we can write

$$d_k(x) = C \sum_{i=1, i \neq k}^{K} p(i)f(x|i)$$

$$= Cf(x) \sum_{i=1, i \neq k}^{K} p(i|x) = Cf(x)(1 - p(k|x)).$$

Minimizing $d_k(x)$ with respect to $k$ is, consequently, equivalent to maximizing $p(k|x)$ with respect to $k$:

$$d_{\hat{k}}(x) = \min_{1 \leq k \leq K} d_k(x) \Leftrightarrow p(\hat{k}|x) = \max_{1 \leq k \leq K} p(k|x).$$

Since, in addition, $0 = f(x) = \sum_{k=1}^{K} p(k)f(x|k)$ is equivalent to $f(x|k) = 0$, $k = 1, \ldots, K$, we obtain $d_1(x) = 0 = p(k|x)$, $k = 2, \ldots, K$ and, thus, $\hat{k} = 1$. For a simple symmetric loss function, Bayes' rule minimizes, therefore, the total error rate.

If the prior probabilities are $p(k) = 1/K$, $k = 1, \ldots, K$, then Bayes' rule in (6.4) becomes the *maximum–likelihood rule*: Take the smallest class index $\hat{k}$, which maximizes the group–specific density for a given vector of observations $x$

$$f(x|\hat{k}) = \max_{1 \leq k \leq K} f(x|k). \tag{6.5}$$

In the case $K = 2$ of two classes as in Example 6.1.1 we obtain from Theorem 6.1.2 the two discriminant functions

$$d_1(x) = p(2)C(1|2)f(x|2), \quad d_2(x) = p(1)C(2|1)f(x|1). \tag{6.6}$$

This entails the optimal partition (with the convention $a/0 := \infty$, $a \geq 0$)

$$G_1^* = \{y \in S : d_1(y) \leq d_2(y)\} = \left\{ y \in S : \frac{f(y|1)}{f(y|2)} \geq \frac{p(2)C(1|2)}{p(1)C(2|1)} \right\}$$

and

$$G_2^* = \{y \in S : d_2(y) < d_1(y)\} = \left\{ y \in S : \frac{f(y|1)}{f(y|2)} < \frac{p(2)C(1|2)}{p(1)C(2|1)} \right\}. \tag{6.7}$$

## Training Samples

Up to now we assumed the densities $p(k)$, $f(x|k)$ and $p(k|x)$ to be known. In applications, however, they are commonly unknown and have to be estimated.

This is either done in a *parametric* way, such as the assumption that $f(x|k) = f(x|\Theta_k)$ is a known function $f$ but with an unknown parameter $\Theta_k$, cf Section 6.2, or in a *nonparametric* way, where no special assumptions on the densities are made, cf Sections 6.3–6.5; but each case includes an estimation procedure, for which a training sample is considered.

# 6.2  Parametric Discriminant Analysis: Normal Observations

Traditional parametric discriminant analysis assumes normally distributed observations in each class with identical covariances. Precisely, we assume now that the group–specific densities are

$$f(x|k) = f(x|\Theta_k) = \frac{1}{(2\pi)^{p/2}(\det \Sigma)^{1/2}} \exp\left(-\frac{1}{2}(x - \mu_k)^T \Sigma^{-1}(x - \mu_k)\right),$$

$x \in S = \mathbb{R}^p$ and $k = 1, \ldots, K$. The parameter vector $\Theta_k$ consists of the vector of means $\mu_k \in \mathbb{R}^p$ of the observations $X$ on the $k$–th class and the common $p \times p$–covariance matrix $\Sigma$. We have, therefore, $\Theta_k = (\mu_k, \Sigma)$, cf Theorem 2.1.3.

### Linear Discriminant Functions

We consider in the following the case $K = 2$ of only two classes. Using the strictly monotone transformation $x \mapsto \log(x)$, $x > 0$, and neglecting the additive term $-(p\log(2\pi) + \log(\det \Sigma))/2$, which is independent of $x$ and $k$, we obtain from (6.6) the quadratic discriminant functions

$$d_1(x) = -\frac{1}{2}(x - \mu_2)^T \Sigma^{-1}(x - \mu_2) + \log(p(2)C(1|2)),$$

$$d_2(x) = -\frac{1}{2}(x - \mu_1)^T \Sigma^{-1}(x - \mu_1) + \log(p(1)C(2|1)). \tag{6.8}$$

Neglecting the term $(-1/2)x^T \Sigma^{-1} x$, which is independent of $k$, we then obtain the *linear discriminant functions*

$$\tilde{d}_1(x) = \mu_2^T \Sigma^{-1} x - \frac{1}{2}\mu_2^T \Sigma^{-1} \mu_2 + \log(p(2)C(1|2)),$$

$$\tilde{d}_2(x) = \mu_1^T \Sigma^{-1} x - \frac{1}{2}\mu_1^T \Sigma^{-1} \mu_1 + \log(p(1)C(2|1)). \tag{6.9}$$

These discriminant functions are built–in procedures in common statistical software packages. An object $\omega$ with associated vector of observations $x = X(\omega)$ is by (6.2) classified to $\Omega_1$ if

$$\tilde{d}_2(x) - \tilde{d}_1(x) \geq 0. \tag{6.10}$$

Otherwise it is classified to $\Omega_2$. The common border $\{\tilde{d}_1 = \tilde{d}_2\}$ of the two classes is a hyperplane in $I\!\!R^p$, i.e., it is a straight line in case $p = 2$ (see Exercise 8). For the maximum–likelihood decision rule, where the smallest class index $\hat{k}$ with

$$f(\boldsymbol{x}|\hat{k}) = \max_{k=1,2} f(\boldsymbol{x}|k),$$

is taken, the log–terms in (6.9) cancel, since we assume $p(2) = p(1) = 1/2$ and $C(1|2) = C(2|1) = C > 0$. The object $\omega$ with vector of observations $\boldsymbol{x} = \boldsymbol{X}(\omega)$ is then classified to $\Omega_1$ if

$$(\boldsymbol{\mu}_1 - \boldsymbol{\mu}_2)^T \boldsymbol{\Sigma}^{-1} \boldsymbol{x} \geq \frac{1}{2} (\boldsymbol{\mu}_1^T \boldsymbol{\Sigma}^{-1} \boldsymbol{\mu}_1 - \boldsymbol{\mu}_2^T \boldsymbol{\Sigma}^{-1} \boldsymbol{\mu}_2). \tag{6.11}$$

## Estimated Discriminant Functions

The optimal classification rules (6.10) or (6.11) are, however, applicable only if the parameters $\boldsymbol{\mu}_1, \boldsymbol{\mu}_2, \boldsymbol{\Sigma}$ as well as the prior probabilities $p(1), p(2)$ are known. The parameters $\boldsymbol{\mu}_1, \boldsymbol{\mu}_2$ and $\boldsymbol{\Sigma}$ can be estimated in a straightforward and unbiased manner from a training sample, which is stratified according to the classes. To this end, we denote by $\boldsymbol{x}_{ki} \in I\!\!R^p$, $i = 1, \ldots, n_k$, those observations in the training sample, which belong to the $k$–th class, $k = 1, 2$. The sample means

$$\bar{\boldsymbol{x}}_k = \frac{1}{n_k} \sum_{i=1}^{n_k} \boldsymbol{x}_{ki}, \qquad k = 1, 2,$$

and the pooled sample covariance matrix

$$S_p = \frac{1}{n_1 + n_2 - 2} \sum_{k=1}^{2} \sum_{i=1}^{n_k} (\boldsymbol{x}_{ki} - \bar{\boldsymbol{x}}_k)(\boldsymbol{x}_{ki} - \bar{\boldsymbol{x}}_k)^T$$

are unbiased estimates of $\boldsymbol{\mu}_k$, $k = 1, 2$, and $\boldsymbol{\Sigma}$ (Exercise 11). Substituting these estimates in (6.9), we obtain the estimated discriminant functions

$$\hat{d}_1(\boldsymbol{x}) = \bar{\boldsymbol{x}}_2^T S_p^{-1} \boldsymbol{x} - \frac{1}{2} \bar{\boldsymbol{x}}_2^T S_p^{-1} \bar{\boldsymbol{x}}_2 + \log(p(2)C(1|2)),$$

$$d_2(\boldsymbol{x}) = \bar{\boldsymbol{x}}_1^T S_p^{-1} \boldsymbol{x} - \frac{1}{2} \bar{\boldsymbol{x}}_1^T S_p^{-1} \bar{\boldsymbol{x}}_1 + \log(p(1)C(2|1)).$$

We assume the matrix $S_p$ to be invertible. These discriminant functions then yield the following classification rules with *known* prior probabilities: An object is classified to $\Omega_1$ if the associated vector of observations $\boldsymbol{x}$ satisfies the inequality $\hat{d}_2(\boldsymbol{x}) - \hat{d}_1(\boldsymbol{x}) \geq 0$ or, equivalently,

$$\left(\boldsymbol{x} - \frac{\bar{\boldsymbol{x}}_1 + \bar{\boldsymbol{x}}_2}{2}\right)^T a \geq \log\left(\frac{p(2)C(1|2)}{p(1)C(2|1)}\right) \tag{6.12}$$

with
$$a := S_p^{-1}(\bar{x}_1 - \bar{x}_2).$$

Note that the matrix $S_p^{-1}$ is symmetric, since $S_p$ is symmetric (Exercise 12), and, thus, $\bar{x}_k^T S_p^{-1} x = x^T S_p^{-1} \bar{x}_k$ for $k = 1, 2$. For the maximum–likelihood classification rule with estimated parameters, one has to put again the log-arithmic terms on the right–hand side of (6.12) equal to zero. Compared to this maximum–likelihood rule, the Bayes decision rule with $C(1|2) = C(2|1) = C > 0$ prefers, therefore, the classification to class 2 if $p(2)/p(1) > 1$. We will introduce in the next section a nonparametric alternative due to Fisher, which coincides with the maximum–likelihood classification rule.

**6.2.1 Example.** The cns data in Example 6.1.1 constitute a training sample for a classification problem with the two classes of healthy and ill persons. To classify a future person to one of these classes by means of its vector of ob-servations $(x_1, x_2, x_3) = (a/n, o/n, m/n)$, we have to estimate the discriminant function from the training sample. The sample means $\bar{x}_1$ and $\bar{x}_2$ within the groups of healthy and ill persons are provided in Figure 6.1.1. The pooled sam-ple covariance matrix $S_p$ is part of the following printout:

The DISCRIM Procedure

| Observations | 98 | DF Total | 97 |
| Variables | 3 | DF Within Classes | 96 |
| Classes | 2 | DF Between Classes | 1 |

Class Level Information

| STATUS | Variable Name | Frequency | Weight | Proportion | Prior Probability |
|--------|------|-----------|--------|------------|-------------------|
| healthy | healthy | 40 | 40.0000 | 0.408163 | 0.408163 |
| ill | ill | 58 | 58.0000 | 0.591837 | 0.591837 |

-----------------------------------------------------------------

Pooled Within-Class Covariance Matrix,    DF = 96

| Variable | AN | ON | MN |
|----------|----|----|----|
| AN | 0.9764414835 | -.1911676329 | 0.0287532507 |
| ON | -.1911676329 | 0.4210761638 | 0.0118079292 |
| MN | 0.0287532507 | 0.0118079292 | 0.0026385192 |

### Pooled Covariance Matrix Information

|                              | Natural Log of the                        |
|:----------------------------:|:-----------------------------------------:|
| Covariance<br>Matrix Rank    | Determinant of the<br>Covariance Matrix   |
| 3                            | -7.89034                                  |

--------------------------------------------------------------------------

### Linear Discriminant Function

$$\text{Constant} = -.5\ \bar{X}'_j\ COV^{-1}\ \bar{X}_j + \ln PRIOR_j \qquad \text{Coefficient} = COV^{-1}\ \bar{X}_j$$
$$\text{Vector}$$

#### Linear Discriminant Function for STATUS

| Variable | healthy  | ill      |
|:---------|:--------:|:--------:|
| Constant | -9.78535 | -7.75213 |
| AN       | 0.25830  | 1.39087  |
| ON       | 0.50607  | 2.03291  |
| MN       | 76.87924 | 47.16725 |

--------------------------------------------------------------------------

### Classification Summary for Calibration Data: DATALIB.CNS
Resubstitution Summary using Linear Discriminant Function

#### Generalized Squared Distance Function

$$D^2_j(X) = (X-\bar{X}_j)'\ COV^{-1}\ (X-\bar{X}_j) - 2 \ln PRIOR_j$$

#### Posterior Probability of Membership in Each STATUS

$$Pr(j|X) = \exp(-.5\ D^2_j(X))\ /\ \text{SUM}_k\ \exp(-.5\ D^2_k(X))$$

Number of Observations and Percent Classified into STATUS

| From<br>STATUS | healthy | ill | Total |
|---|---|---|---|
| healthy | 21<br>52.50 | 19<br>47.50 | 40<br>100.00 |
| ill | 9<br>15.52 | 49<br>84.48 | 58<br>100.00 |
| Total | 30<br>30.61 | 68<br>69.39 | 98<br>100.00 |
| Priors | 0.40816 | 0.59184 | |

Error Count Estimates for STATUS

| | healthy | ill | Total |
|---|---|---|---|
| Rate | 0.4750 | 0.1552 | 0.2857 |
| Priors | 0.4082 | 0.5918 | |

**Figure 6.2.1.** Linear discriminant analysis of the cns data from Example 6.1.1 using Bayes' classification rule. The variables are $a/n$, $o/n$, $m/n$.

```
***    Program 6_2_1    ***;
TITLE1 'Discriminant Analysis';
TITLE2 'CNS Data';
LIBNAME datalib 'c:\data';

PROC DISCRIM DATA=datalib.cns METHOD=NORMAL POOL=YES PCOV;
   CLASS status;
   VAR  an on mn;
   PRIORS PROP;
RUN; QUIT;
```

This program calls the procedure DISCRIM together with the options METHOD=NORMAL POOL =YES PCOV and the statement PRIORS PROP. It returns a Bayesian discriminant analysis for a multivariate normal distribution in each group that uses the pooled covariance matrix and estimated prior probabilities. Without the statement PRIORS PROP equal prior probabilites $p(1) = p(2) = 1/2$ are used, which is the maximum–likelihood approach.

The CLASS statement defines the clas-

sification variable, whose values define the
groups for analysis. The VAR statement
specifies the quantitative variables to be included in the analysis. The 'DATA=data1'
option (here datalib.cns) specifies the data
set to be analysed. If the additional option
'TESTDATA=data2' is used, a discriminant
function is computed from the data from the
data1 file (training sample) and applied to
the data in the file data2.

We estimate the prior probabilities $p(1), p(2)$ in this case from the initial cns data set:

$$\hat{p}(1) := 40/98 = 0.408163, \quad \hat{p}(2) := 58/98 = 0.591837.$$

Hence, we obtain the following Bayes' classification rule with estimated parameters, where $C(1|2) = C(2|1) = C > 0$ and $C(1|1) = C(2|2) = 0$: A person with the vector of observations $\boldsymbol{x} = (x_1, x_2, x_3)^T = (a/n, o/n, m/n)^T$ is classified to be healthy if

$$(\boldsymbol{x} - (\bar{\boldsymbol{x}}_1 + \bar{\boldsymbol{x}}_2)/2)^T S_p^{-1}(\bar{\boldsymbol{x}}_1 - \bar{\boldsymbol{x}}_2) \geq \log(\hat{p}(2)/\hat{p}(1))$$
$$\Leftrightarrow \boldsymbol{x}^T S_p^{-1}(\bar{\boldsymbol{x}}_1 - \bar{\boldsymbol{x}}_2) \geq 0.5(\bar{\boldsymbol{x}}_1 + \bar{\boldsymbol{x}}_2)^T S_p^{-1}(\bar{\boldsymbol{x}}_1 - \bar{\boldsymbol{x}}_2) + \log(\hat{p}(2)/\hat{p}(1))$$
$$\Leftrightarrow \boldsymbol{x}^T \left(S_p^{-1}\bar{\boldsymbol{x}}_1 - S_p^{-1}\bar{\boldsymbol{x}}_2\right)$$
$$\geq 0.5\bar{\boldsymbol{x}}_1^T S_p^{-1}\bar{\boldsymbol{x}}_1 - \log(\hat{p}(1)) - \left(0.5\bar{\boldsymbol{x}}_2 S_p^{-1}\bar{\boldsymbol{x}}_2 - \log(\hat{p}(2))\right).$$

This is equivalent to

$$\begin{pmatrix} x_1 \\ x_2 \\ x_3 \end{pmatrix}^T \begin{pmatrix} 0.25830 - 1.39087 \\ 0.50607 - 2.03291 \\ 76.87924 - 47.16725 \end{pmatrix} \geq 9.78535 - 7.75213$$

$$\Leftrightarrow -1.13257x_1 - 1.52684x_2 + 29.71199x_3 \geq 2.03322.$$

To get an idea of the performance of this classification rule with estimated parameters, it is reasonable to apply it to the training sample. It turns out that 19 of the 40 healthy persons are misclassified to be ill, while 9 of the 59 ill persons are misclassified to be healthy. The failure rate is estimated by

$$\frac{19}{40}\hat{p}(1) + \frac{9}{58}\hat{p}(2) = \frac{\text{total number of misclassified objects}}{\text{total number of objects}}$$

$$= 0.4750 \times 0.4082 + 0.1552 \times 0.5918 = 0.2857.$$

If, however, we apply the maximum–likelihood classification rule to this data set, then 14 of the 40 healthy persons are misclassified and 20 of the 58 ill ones. The failure rate is 0.3474. Since we have in this case $\hat{p}(2)/\hat{p}(1) = 58/40 =$

$1.45 > 1$, it is more likely that an observation is classified to group 2 of ill persons by Bayes' decision rule than by the maximum–likelihood rule. The high failure rates are due to the fact that a *linear* discriminant function, i.e., a plane in $\mathbb{R}^3$, cannot separate the twisted classes in Figure 6.1.2.

## Quadratic Discriminant Functions

If the covariances $\Sigma_k$ within each class differ, the quadratic term in (6.8) depends on $k$ and, thus, we obtain the *quadratic discriminant functions*

$$d_1(x) = x^T A_2 x + a_2^T x + c_1,$$
$$d_2(x) = x^T A_1 x + a_1^T x + c_2$$

with

$$A_k = \frac{1}{2}\Sigma_k^{-1}, \quad a_k = \Sigma_k^{-1}\mu_k$$

and

$$c_1 = -\frac{1}{2}\mu_2^T \Sigma_2^{-1}\mu_2 - \frac{1}{2}\log(\det \Sigma_2) + \log(p(2)C(1|2)),$$
$$c_2 = -\frac{1}{2}\mu_1^T \Sigma_1^{-1}\mu_1 - \frac{1}{2}\log(\det \Sigma_1) + \log(p(1)C(2|1)).$$

# 6.3 Fisher's Approach (Projection Pursuit)

Here and in the sequel we will introduce various nonparametric classification rules.

## Orthogonal Projection onto a Line

We consider the case of two classes and a training sample $x_{11}, \ldots, x_{1n_1}$, $x_{21}, \ldots, x_{2n_2}$ of data in $\mathbb{R}^p$, which is stratified according to these classes. Fisher's idea was to reduce the classification problem in $\mathbb{R}^p$ to a one dimensional problem by substituting the vector of observations $x_{kj}$ by the linear combination

$$y_{kj} = a^T x_{kj} \in \mathbb{R}, \qquad a = (a_1, \ldots, a_p)^T \in \mathbb{R}^p,$$

with $\|a\|^2 = \sum_{i=1}^p a_i^2 = 1$. The weights $a_1, \ldots, a_p$ will be chosen such that the $y_{kj}$ substitutes the $x_{kj}$ in some optimal way described below.

The transition from a $p$–dimensional vector $x$ to a linear combination $a^T x$ has the following geometric interpretation: The vector $a$ defines in $\mathbb{R}^p$ the line $sa$, $s \in \mathbb{R}$. If we project the vector $x$ orthogonally onto this line in $\mathbb{R}^p$, then $a^T x = s_0$ is the coordinate on the line $sa$, $s \in \mathbb{R}$, of this projection: The vector $a$ stands at right angles to the vector $x - s_0 a$, i.e., their inner product is zero $(x - s_0 a)^T a = 0$ or, equivalently,

$$x^T a = s_0 a^T a = s_0 ||a||^2 = s_0.$$

This implies $s_0 = x^T a = a^T x$.

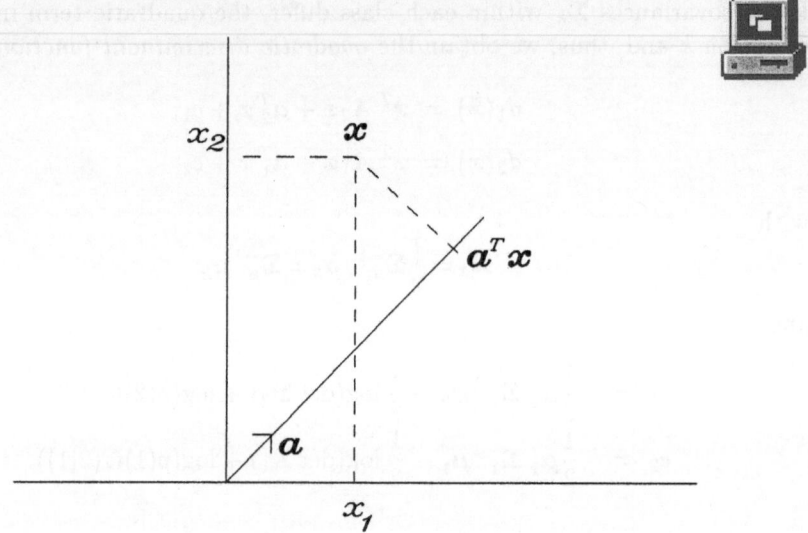

Figure 6.3.1. Orthogonal projection of $x$ onto the line $sa$.

This graph was created using SAS– 3_2_1, 6_4_3 and 7_1_1. ANNOTATE, cf the comments to Programs

## The Classification Rule

An object with the vector of observations $x$ is now classified according to the following rule. First we compute the vectors of sample means

$$\bar{x}_k = \frac{1}{n_k} \sum_{i=1}^{n_k} x_{ki}, \qquad k = 1, 2,$$

in each class. The information in the two samples is reduced to these two representations in $\mathbb{R}^p$. From now on we assume that $\bar{x}_1 \neq \bar{x}_2$. In a further step we compress the information by projecting the two vectors of means $\bar{x}_1, \bar{x}_2$ orthogonally onto the line $sa$, $s \in \mathbb{R}$. This results in a reduction of the dimension of the original data, where the two coordinates $s_1 = a^T \bar{x}_1$, $s_2 = a^T \bar{x}_2 \in \mathbb{R}$ on the line $s\boldsymbol{a}$ are one dimensional representations of our possibly high dimensional data in the two classes of the training sample. A future object with the vector of observations $\boldsymbol{x}$ is now classified to that class $\hat{k}$, whose projected vector of means $\boldsymbol{a}^T \bar{\boldsymbol{x}}_{\hat{k}}$ is the closest to the projection $\boldsymbol{a}^T \boldsymbol{x}$:

$$|\boldsymbol{a}^T \boldsymbol{x} - \boldsymbol{a}^T \bar{\boldsymbol{x}}_{\hat{k}}| = \min\{|\boldsymbol{a}^T \boldsymbol{x} - \boldsymbol{a}^T \bar{\boldsymbol{x}}_1|, |\boldsymbol{a}^T \boldsymbol{x} - \boldsymbol{a}^T \bar{\boldsymbol{x}}_2|\}.$$

## The Optimal Projection

We can choose an arbitrary vector $\boldsymbol{a} \in \mathbb{R}^p$ for the above projection. As we want to separate the two classes as clearly as possible, it seems reasonable to choose $\boldsymbol{a}$ such that the distance between the projected representatives $s_1 = \boldsymbol{a}^T \bar{\boldsymbol{x}}_1, s_2 = \boldsymbol{a}^T \bar{\boldsymbol{x}}_2$ of the two classes is maximized in a proper sense. To this end we project all data $\boldsymbol{x}_{ki}$ from the stratified training sample onto the line $s\boldsymbol{a}, \ s \in \mathbb{R}$. Then we compute the usual $t$ statistic for the comparison of the means of the projected data $\boldsymbol{a}^T \boldsymbol{x}_{ki}$ within the two classes

$$t = t_{\boldsymbol{a}} = \frac{\boldsymbol{a}^T \bar{\boldsymbol{x}}_1 - \boldsymbol{a}^T \bar{\boldsymbol{x}}_2}{s_p \sqrt{\frac{1}{n_1} + \frac{1}{n_2}}}.$$

By $s_p^2$ we denote the pooled sample variance, cf Section 2.3,

$$s_p^2 = \frac{1}{n_1 + n_2 - 2} \left( \sum_{i=1}^{n_1} (\boldsymbol{a}^T \boldsymbol{x}_{1i} - \boldsymbol{a}^T \bar{\boldsymbol{x}}_1)^2 + \sum_{j=1}^{n_2} (\boldsymbol{a}^T \boldsymbol{x}_{2j} - \boldsymbol{a}^T \bar{\boldsymbol{x}}_2)^2 \right) = \boldsymbol{a}^T S_p \boldsymbol{a},$$

with the pooled sample covariance matrix

$$S_p = \frac{1}{n_1 + n_2 - 2} \left( \sum_{i=1}^{n_1} (\boldsymbol{x}_{1i} - \bar{\boldsymbol{x}}_1)(\boldsymbol{x}_{1i} - \bar{\boldsymbol{x}}_1)^T + \sum_{j=1}^{n_2} (\boldsymbol{x}_{2j} - \bar{\boldsymbol{x}}_2)(\boldsymbol{x}_{2j} - \bar{\boldsymbol{x}}_2)^T \right).$$

We want to determine now that directional vector $\boldsymbol{a}$ with the corresponding $t$ statistic $t_{\boldsymbol{a}}$, which separates best the projected means $s_1 = \boldsymbol{a}^T \bar{\boldsymbol{x}}_1$ and $s_2 = \boldsymbol{a}^T \bar{\boldsymbol{x}}_2$, i.e., we want to determine that vector $\boldsymbol{a}$ with length $\|\boldsymbol{a}\| = 1$ such that

$$t_{\boldsymbol{a}}^2 = \frac{(\boldsymbol{a}^T \bar{\boldsymbol{x}}_1 - \boldsymbol{a}^T \bar{\boldsymbol{x}}_2)^2}{(\frac{1}{n_1} + \frac{1}{n_2})\boldsymbol{a}^T S_p \boldsymbol{a}}$$

is maximized. This is obviously equivalent to maximizing the function

$$Q(\boldsymbol{a}) := \frac{(\boldsymbol{a}^T \bar{\boldsymbol{x}}_1 - \boldsymbol{a}^T \bar{\boldsymbol{x}}_2)^2}{\boldsymbol{a}^T S_p \boldsymbol{a}}$$

$$= \frac{\text{squared distance of the means in the direction of } \boldsymbol{a}}{\text{pooled sample covariance in the direction of } \boldsymbol{a}}.$$

This is the measure of performance defined by Fisher (1936) for the separation of two samples in $\mathbb{R}^p$ in the direction of vector $a$. That vector $a$, which maximizes $Q(a)$, performs best. The matrix $S_p$ is positive semidefinite, i.e., we have $x^T S_p x \geq 0$ for arbitrary $x \in \mathbb{R}^p$. We assume in addition that $S_p$ is positive definite, i.e., $x^T S_p x > 0$ for any $x \neq 0$.

**6.3.1 Lemma.** *Suppose that the matrix $S_p$ is positive definite. If $\bar{x}_1 \neq \bar{x}_2$, then each vector $x \in \mathbb{R}^p$ can be written as*

$$x = s S_p^{-1}(\bar{x}_1 - \bar{x}_2) + b,$$

*where $s \in \mathbb{R}$ and the vectors $b \in \mathbb{R}^p$ and $\bar{x}_1 - \bar{x}_2$ are orthogonal, i.e., $b^T(\bar{x}_1 - \bar{x}_2) = 0$.*

**Proof:** The set $M := \{c \in \mathbb{R}^p : c^T(\bar{x}_1 - \bar{x}_2) = 0\}$ is a $(p-1)$–dimensional vector space in $\mathbb{R}^p$. The vector $S_p^{-1}(\bar{x}_1 - \bar{x}_2)$ is not an element of $M$, since

$$(S_p^{-1}(\bar{x}_1 - \bar{x}_2))^T(\bar{x}_1 - \bar{x}_2) = (\bar{x}_1 - \bar{x}_2)^T S_p^{-1}(\bar{x}_1 - \bar{x}_2) > 0$$

(cf Exercise 13). Hence, we obtain that $S_p^{-1}(\bar{x}_1 - \bar{x}_2)$ together with the basis of $M$ span $\mathbb{R}^p$. This yields the assertion.                                        □

By means of the above lemma we can easily show that $Q(a)$ is maximized by the vector

$$a^* := \frac{1}{\|S_p^{-1}(\bar{x}_1 - \bar{x}_2)\|} S_p^{-1}(\bar{x}_1 - \bar{x}_2).$$

**6.3.2 Theorem.** *Suppose that $\bar{x}_1 \neq \bar{x}_2$ and that $S_p$ is positive definite. Then we have*

$$Q(a^*) = \max_{a \in \mathbb{R}^p, \|a\|=1} Q(a) = (\bar{x}_1 - \bar{x}_2)^T S_p^{-1}(\bar{x}_1 - \bar{x}_2).$$

**Proof:** Choose $a \in \mathbb{R}^p$, $\|a\| = 1$. By Lemma 6.3.1 there exist $s \in \mathbb{R}$ and $b \in \mathbb{R}^p$ with $b^T(\bar{x}_1 - \bar{x}_2) = 0$, such that $a = s S_p^{-1}(\bar{x}_1 - \bar{x}_2) + b$. This implies that

$$Q(a) = \frac{(a^T(\bar{x}_1 - \bar{x}_2))^2}{a^T S_p a} = \frac{(s(\bar{x}_1 - \bar{x}_2)^T S_p^{-1}(\bar{x}_1 - \bar{x}_2))^2}{(s(\bar{x}_1 - \bar{x}_2)^T S_p^{-1} + b^T) S_p (s S_p^{-1}(\bar{x}_1 - \bar{x}_2) + b)}$$

$$= s^2 \frac{((\bar{x}_1 - \bar{x}_2)^T S_p^{-1}(\bar{x}_1 - \bar{x}_2))^2}{s^2(\bar{x}_1 - \bar{x}_2)^T S_p^{-1}(\bar{x}_1 - \bar{x}_2) + b^T S_p b}.$$

In the case $s = 0$ we have $Q(a) = 0$. As $S_p$ is positive definite, its inverse $S_p^{-1}$ is positive definite as well (cf Exercise 13), and, hence,

$$Q(a) \leq \frac{((\bar{x}_1 - \bar{x}_2)^T S_p^{-1} (\bar{x}_1 - \bar{x}_2))^2}{(\bar{x}_1 - \bar{x}_2)^T S_p^{-1} (\bar{x}_1 - \bar{x}_2)} = (\bar{x}_1 - \bar{x}_2)^T S_p^{-1} (\bar{x}_1 - \bar{x}_2) = Q(a^*).  \quad \square$$

### Fisher's Discriminant Function

We have obviously

$$a^{*T}(\bar{x}_1 - \bar{x}_2) = \frac{(\bar{x}_1 - \bar{x}_2)^T S_p^{-1}(\bar{x}_1 - \bar{x}_2)}{\|S_p^{-1}(\bar{x}_1 - \bar{x}_2)\|} > 0. \tag{6.13}$$

By the inequality

$$|y - y_1| \leq |y - y_2| \Leftrightarrow y \geq \frac{y_1 + y_2}{2},$$

which is true for arbitrary real numbers $y, y_1, y_2$ with $y_2 \leq y_1$, we obtain now Fisher's classification rule: An object $\omega \in \Omega$ with the associated vector of observations $x = X(\omega)$ is classified to the class $\Omega_1$ if the projection $a^{*T}x$ of $x$ onto the line $sa^*$ is closer to the projected mean $a^{*T}\bar{x}_1$ than to $a^{*T}\bar{x}_2$. This is equivalent to

$$|a^{*T}x - a^{*T}\bar{x}_1| \leq |a^{*T}x - a^{*T}\bar{x}_2|$$
$$\Leftrightarrow 0 \leq a^{*T}x - \frac{a^{*T}\bar{x}_1 + a^{*T}\bar{x}_2}{2} = a^{*T}\left(x - \frac{\bar{x}_1 + \bar{x}_2}{2}\right)$$
$$\Leftrightarrow 0 \leq \left(x - \frac{\bar{x}_1 + \bar{x}_2}{2}\right)^T S_p^{-1}(\bar{x}_1 - \bar{x}_2), \tag{6.14}$$

since we have $a^{*T}\bar{x}_2 < a^{*T}\bar{x}_1$ by (6.13). This nonparametric classification rule coincides, consequently, with the maximum–likelihood rule for normally distributed data, cf (6.12). The linear function

$$d_F(x) := x^T S_p^{-1}(\bar{x}_1 - \bar{x}_2), \qquad x \in \mathbb{R}^p,$$

is called *Fisher's discriminant function*, and an object with associated vector of observations $x$ is classified to $\Omega_1$ if

$$d_F(x) \geq \left(\frac{\bar{x}_1 + \bar{x}_2}{2}\right)^T S_p^{-1}(\bar{x}_1 - \bar{x}_2) = \frac{1}{2}\bar{x}_1^T S_p^{-1}\bar{x}_1 - \frac{1}{2}\bar{x}_2^T S_p^{-1}\bar{x}_2.$$

The function $Q$ is a particular example of a *projection index* within the framework of a *projection pursuit*, which searches for "interesting" low dimensional projections of high dimensional data sets by maximizing (or minimizing) a certain objective function (Huber (1985), Jones and Sibson (1987)). The projection index $Q(a)$ is, however, quite sensitive to possible outliers in the data.

Huber (1985), Section 6, recommends, therefore, to take as a robust alternative to $Q(a)$ the projection index

$$\left( \frac{med\,(z_1) - med\,(z_2)}{MAD\,((z_1 - med\,(z_1)),\,(z_2 - med\,(z_2)))} \right)^2 ,$$

where $z_k := (a^T x_{k1}, \ldots, a^T x_{kn_k})$, $k = 1, 2$. See Sections 1.2 and 1.3 for the definitions of $med$ and $MAD$. This robustified classification rule is, however, not available in SAS 8.

**6.3.3 Example** (Crystal Data). Six different physicochemical variables such as the pH level ($pH$), the concentration of calcium ($Ca$) and the specific gravity ($g$) were measured on 79 urine specimen, some of which contain certain crystals. The question is, whether the crystallization is related to these variables, cf Example 2.3.1.

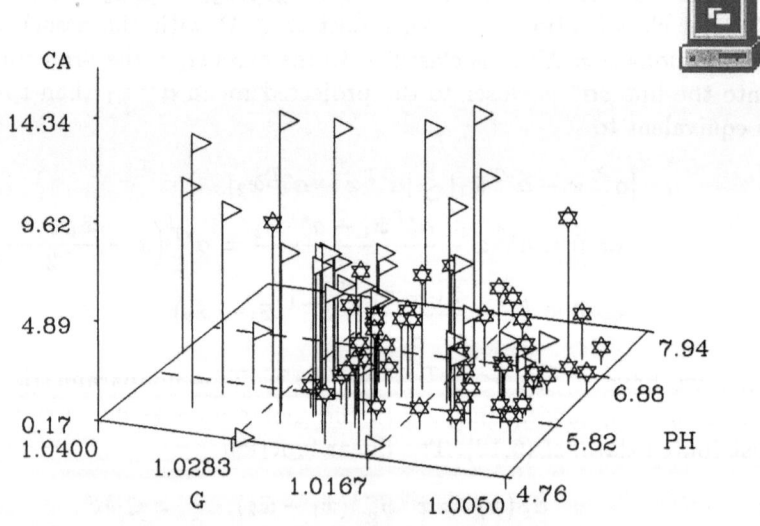

**Figure 6.3.2.** Scatterplot of crystal data in 4 dimensions, the variables are $Ca$, $g$, $pH$; crystallization/no crystallization in specimen corresponds to flag/star.

The subsequent discriminant analysis of the crystal data uses Fisher's classification rule. The specimen are classified only by means of the variables $pH$ level and specific weight. The number 1 indicates no crystallization , the number 2 indicates crystallization.

The DISCRIM Procedure

| Observations | 79 | DF Total | 78 |
|---|---|---|---|
| Variables | 2 | DF Within Classes | 77 |
| Classes | 2 | DF Between Classes | 1 |

Class Level Information

| CRYSTAL | Variable Name | Frequency | Weight | Proportion | Prior Probability |
|---|---|---|---|---|---|
| 1 | _1 | 45 | 45.0000 | 0.569620 | 0.500000 |
| 2 | _2 | 34 | 34.0000 | 0.430380 | 0.500000 |

Pooled Covariance Matrix Information

| Covariance Matrix Rank | Natural Log of the Determinant of the Covariance Matrix |
|---|---|
| 2 | -10.73595 |

Pairwise Generalized Squared Distances Between Groups

$$D^2(i|j) = (\bar{X}_i - \bar{X}_j)' \, COV^{-1} \, (\bar{X}_i - \bar{X}_j)$$

Generalized Squared Distance to CRYSTAL

| From CRYSTAL | 1 | 2 |
|---|---|---|
| 1 | 0 | 0.85087 |
| 2 | 0.85087 | 0 |

Linear Discriminant Function

$$\text{Constant} = -.5 \, \bar{X}_j' \, COV^{-1} \, \bar{X}_j \qquad \text{Coefficient Vector} = COV^{-1} \, \bar{X}_j$$

Linear Discriminant Function for CRYSTAL

| Variable | 1 | 2 |
|---|---|---|
| Constant | -12790 | -12931 |
| PH | 63.41349 | 63.39279 |
| G | 24808 | 24947 |

Classification Summary for Calibration Data: DATALIB.CRYSTAL
Resubstitution Summary using Linear Discriminant Function


Generalized Squared Distance Function

$$D_j^2(X) = (X-\bar{X}_j)' \; COV^{-1} \; (X-\bar{X}_j)$$


Posterior Probability of Membership in Each CRYSTAL

$$Pr(j|X) = \exp(-.5 \, D_j^2(X)) \; / \; \underset{k}{SUM} \; \exp(-.5 \, D_k^2(X))$$


Number of Observations and Percent Classified into CRYSTAL

| From CRYSTAL | 1 | 2 | Total |
|---|---|---|---|
| 1 | 28 | 17 | 45 |
|  | 62.22 | 37.78 | 100.00 |
| 2 | 12 | 22 | 34 |
|  | 35.29 | 64.71 | 100.00 |
| Total | 40 | 39 | 79 |
|  | 50.63 | 49.37 | 100.00 |
| Priors | 0.5 | 0.5 |  |


Error Count Estimates for CRYSTAL

|  | 1 | 2 | Total |
|---|---|---|---|
| Rate | 0.3778 | 0.3529 | 0.3654 |
| Priors | 0.5000 | 0.5000 |  |


**Figure 6.3.3.** Fisher's discriminant analysis of the crystal data by means of the pH level and specific weight.

```
***    Program 6_3_3    ***;
TITLE1 'Discriminant Analysis';
TITLE2 'Crystal Data';
LIBNAME datalib 'c:\data';

PROC DISCRIM DATA=datalib.crystal METHOD=NORMAL POOL=YES;
   CLASS crystal;
   VAR ph g;
RUN; QUIT;
```

This program is analogous to Program 6_2_1, but without the statement PRIORS PROP: In the case of two classes and normal group–specific distributions, Fisher's nonparametric approach coincides with the maximum–likelihood classification rule.

A discriminant analysis of the crystal data using Fisher's classification rule for the variables pH level ($x_1$) and specific weight ($x_2$) yields the following result. For easier reference to the above SAS output we use its notation as well. The maximum projection index is

$$Q(a^*) = (\bar{x}_1 - \bar{x}_2)^T S_p^{-1}(\bar{x}_1 - \bar{x}_2) = D^2(i|j) = 0.85087.$$

Fisher's discriminant function is with $x = (x_1, x_2)^T \in \mathbb{R}^2$ given by

$$d_F(x) = x^T S_p^{-1}(\bar{x}_1 - \bar{x}_2) = x^T(Cov^{-1}\bar{x}_1 - Cov^{-1}\bar{x}_2)$$

$$= x^T \left( \begin{pmatrix} 63.41349 \\ 24808 \end{pmatrix} - \begin{pmatrix} 63.39279 \\ 24947 \end{pmatrix} \right) = 0.0207x_1 - 139x_2.$$

This function yields, therefore, the following decision rule: a specimen with the vector of variables $x = (x_1, x_2)^T$ of pH level and specific weight $g$ is classified to class 1 (no crystallization) if

$$d_F(x) \geq \frac{1}{2}\bar{x}_1^T S_p^{-1}\bar{x}_1 - \frac{1}{2}\bar{x}_2^T S_p^{-1}\bar{x}_2$$

$$= 0.5\,\bar{x}_1^T Cov^{-1}\bar{x}_1 - 0.5\,\bar{x}_2^T Cov^{-1}\bar{x}_2 = 12790 - 12931 = -141,$$

i.e., if

$$0.0207x_1 - 139x_2 + 141 \geq 0.$$

Applying this classification rule to the crystal data, we obtain the following estimate of its failure rate

$P\{$an object is misclassified$\}$

$$= P\{X \in G_2, \kappa = 1\} + P\{X \in G_1, \kappa = 2\}$$

$$= p(1)P(X \in G_2|\kappa = 1) + p(2)P(X \in G_1|\kappa = 2)$$

$$\approx p(1)\frac{\text{number of objects in class 1, which are classified to class 2}}{\text{number of observations in class 1}}$$

$$+ p(2)\frac{\text{number of objects in class 2, which are classified to class 1}}{\text{number of observations in class 2}}$$

$$= p(1)\frac{17}{45} + p(2)\frac{12}{34} = p(1)0.3778 + p(2)0.3529 = 0.3654,$$

where it is assumed that $p(1) = p(2) = 0.5$. If the 79 sets of data were sampled at random in such a way, that their division into classes was at random, we can estimate $p(1)$ and $p(2)$ by their frequencies

$$p(1) \approx \hat{p}(1) = \frac{45}{79} = 0.5696, \quad p(2) \approx \hat{p}(2) = \frac{34}{79} = 0.4304.$$

An estimate of the failure rate is now given by

$$\hat{p}(1)\frac{17}{45} + \hat{p}(2)\frac{12}{34} = \frac{17 + 12}{79}$$

$$= \frac{\text{total number of misclassified objects}}{\text{total number of objects}} = 0.3671.$$

## 6.4   Density Estimators

Bayes' decision rule (6.4) classified an object with the vector of observations $x$ to that particular class with (least) index $\hat{k}$ such that the posterior probability is maximized, i.e.,

$$p(\hat{k})f(x|\hat{k}) = \max_{1 \leq k \leq K} p(k)f(x|k).$$

The densities $f(\cdot|k)$, $k = 1, \ldots, K$, of the group–specific distributions within the classes were assumed to be known. The particular case of normal densities was investigated in Section 6.2. If the densities $f(\cdot|k)$ are, however, unknown, the idea is natural to replace $f(\cdot|k)$ by a density estimator $\hat{f}(\cdot|k)$. In complete analogy to Bayes' decision rule, an object with the vector of observations $x$ is then classified to that class with (least) index $\hat{k}$ such that the estimated posterior probability is maximized, i.e.,

$$p(\hat{k})\hat{f}(x|\hat{k}) = \max_{1 \leq k \leq K} p(k)\hat{f}(x|k).$$

But now we have to use density estimators based on multivariate data. We will introduce in the following two different approaches: On the one hand, we will generalize the kernel density estimator of Section 1.1 to higher dimensions, on the other hand, we will discuss in Section 6.5 a nearest neighbor estimator.

### The Kernel Density Estimator Approach

Let $Y_1, \ldots, Y_{n_k}$ be independent random vectors in $\mathbb{R}^p$ with common distribution $P\{Y_i \in \cdot\} = P(X \in \cdot | \kappa = k)$. The *multivariate kernel density estimator* of the group–specific density $f(x|k)$ with kernel $\mathcal{K} : \mathbb{R}^p \to \mathbb{R}$ and bandwidth $h_k > 0$ is defined by

$$\hat{f}_{n_k}(x|k) = \int \frac{1}{h_k^p} \mathcal{K}\left(\frac{x-y}{h_k}\right) F_{n_k}(dy) = \frac{1}{n_k h_k^p} \sum_{i=1}^{n_k} \mathcal{K}\left(\frac{x-Y_i}{h_k}\right),$$

where

$$F_{n_k}(y) = \frac{1}{n_k} \sum_{i=1}^{n_k} 1_{(-\infty, y]}(Y_i), \qquad y = (y_1, \ldots, y_p)^T \in \mathbb{R}^p,$$

denotes the $p$–dimensional sample distribution function of $Y_1, \ldots, Y_{n_k}$. The division $(x - Y_i)/h_k$ is meant componentwise for all $p$ components of $x - Y_i$. An obvious estimator of the posterior probability $p(k|x)$ is then

$$\hat{p}(k|x) = p(k)\hat{f}_{n_k}(x|k) \Big/ \Big( \sum_{j=1}^{K} p(j)\hat{f}_{n_j}(x|j) \Big).$$

The prior probabilities $p(k)$ are again assumed to be known. Bayes' decision rule (6.3) with estimated posterior probabilities now becomes: Classify to that class with least index $\hat{k}$, which satisfies

$$\hat{p}(\hat{k}|x) = \max_{1 \leq k \leq K} \hat{p}(k|x).$$

Suppose the priors $p(k)$ are unknown. Then we can estimate them by the empirical frequencies $n_k/n$ of the associated classes, provided that the data were sampled with random stratification, $n = n_1 + \cdots + n_k$. We will discuss in the sequel general properties of the above multivariate kernel density estimator.

## Multivariate Kernel Density Estimators

We will generalize the results in Section 1.1 for univariate kernel density estimators to multivariate data now.

Let $Y$ be a $p$–dimensional random vector with density $f$ and denote by $Y_1, \ldots, Y_n$ independent copies of $Y$. The function

$$\hat{f}_n(x) = \frac{1}{nh^p} \sum_{i=1}^{n} \mathcal{K}\Big(\frac{x - Y_i}{h}\Big), \qquad x \in \mathbb{R}^p, \tag{6.15}$$

is the kernel density estimator of $f(x)$ with kernel $\mathcal{K} : \mathbb{R}^p \to \mathbb{R}$ and bandwidth $h > 0$. We require $\mathcal{K}$ to be the density of a probability measure, i.e., we require

$$\mathcal{K}(y) \geq 0 \quad \text{and} \quad \int \mathcal{K}(y)\, dy = 1. \tag{6.16}$$

We require in addition that $\mathcal{K}$ is radially symmetric and invariant under permutations, i.e.,

$$\mathcal{K}((y_1, \ldots, y_p)^T) = \mathcal{K}((-y_1, y_2, \ldots, y_p)^T) = \cdots = \mathcal{K}((y_1, \ldots, y_{p-1}, -y_p)^T)$$

$$\mathcal{K}((y_1, \ldots, y_p)^T) = \mathcal{K}((y_{\sigma(1)}, \ldots, y_{\sigma(p)})^T) \tag{6.17}$$

for arbitrary $y_1, \ldots, y_p \in \mathbb{R}$ and an arbitrary permutation $(\sigma(1), \ldots, \sigma(p))$ of the vector $(1, \ldots, p)$. The radial symmetry implies $\mathcal{K}(y) = \mathcal{K}(-y)$. The kernels listed below satisfy the conditions (6.16) and (6.17).

**6.4.1 Examples.** Denote by $v_p$ the volume of the $p$–dimensional unit ball

$$v_p = \mathrm{Vol}(\{y \in \mathbb{R}^p : y^T y \leq 1\}) = 2\pi^{p/2}/(p\Gamma(p/2)),$$

where $\Gamma$ is the gamma function, as defined in Theorem 2.1.7. For arbitrary $y \in \mathbb{R}^p$

(i)
$$\mathcal{K}_u(y) = \begin{cases} 1/v_p, & y^T y \leq 1, \\ 0 & \text{elsewhere,} \end{cases}$$

is the *uniform kernel*,

(ii)
$$\mathcal{K}_E(y) = \begin{cases} c_1(p)(1 - y^T y), & y^T y \leq 1, \\ 0 & \text{elsewhere,} \end{cases}$$

with $c_1(p) = (1 + p/2)/v_p$ is the *Epanechnikov kernel*, whose support is the $p$–dimensional unit ball,

(iii)
$$\mathcal{K}_2(y) = \begin{cases} c_2(p)(1 - y^T y)^2, & y^T y \leq 1, \\ 0 & \text{elsewhere,} \end{cases} \quad \text{with}$$

$c_2(p) = (1 + p/4)c_1(p)$ is the *quadratic kernel*,

(iv)
$$\mathcal{K}_3(y) = \begin{cases} c_3(p)(1 - y^T y)^3, & y^T y \leq 1, \\ 0 & \text{elsewhere,} \end{cases}$$

with $c_3(p) = (1 + p/6)c_2(p)$ is the *cubic kernel*,

(v)
$$\mathcal{K}_\varphi(y) = (2\pi)^{-p/2} \exp\left(-\frac{1}{2}y^T y\right)$$

is the *normal kernel*.

(vi) Any function $k : \mathbb{R} \to [0, \infty)$ with $\int k(x)\,dx = 1$, and $k(x) = k(-x)$ for $x \in \mathbb{R}$ can serve as a basis of the *tensor kernel*

$$\mathcal{K}_T(y) := \prod_{i=1}^{p} k(y_i), \qquad y = (y_1, \ldots, y_p)^T,$$

which obviously satisfies conditions (6.16) and (6.17).

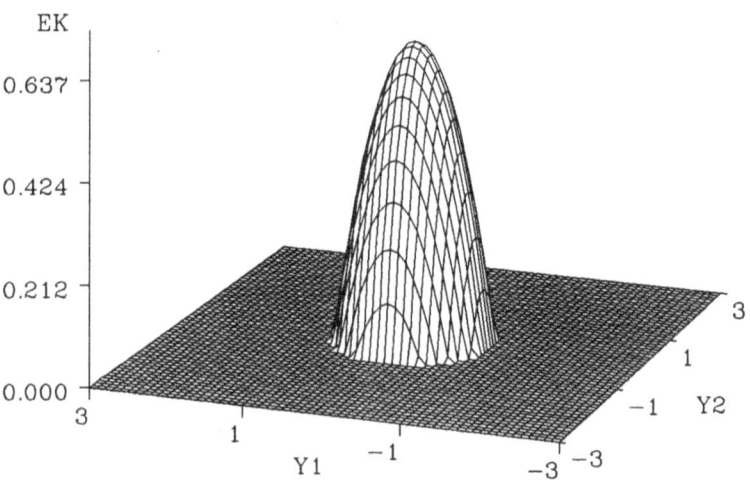

**Figure 6.4.1.** Epanechnikov kernel, $p = 2$.

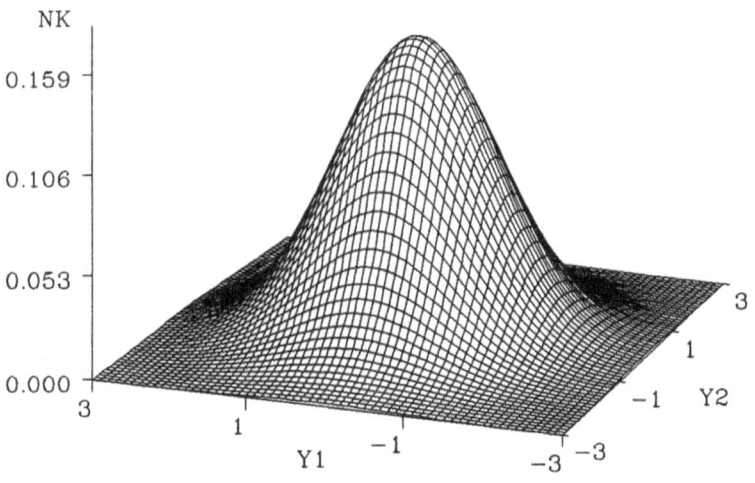

**Figure 6.4.2.** Normal kernel, $p = 2$.

```
***     Program 6_4_2    ***;
TITLE1 'Normal Kernel, p=2';

DATA data1;
    DO y1=-3 TO 3 BY 0.1;
        DO y2=-3 TO 3 BY 0.1;
            yy=y1**2+y2**2;
            nk=0.159155*EXP(-0.5*yy);
            OUTPUT;
        END;
    END;
PROC G3D DATA=data1;
    PLOT y1*y2=nk;
RUN; QUIT;
```

A function z=f(y1,y2) can be plotted using the procedure G3D. It requires data sets that include the values of three variables y1, y2, z. This data set (here 'data1') is produced here in the DATA step. The nested DO loops generate the grid of values (y1,y2) on the plane and the vertical z (here 'nk', the value of the normal kernel). The result of each loop is written into the data set data1 by using the OUTPUT statement. Each DO loop is terminated with an END statement. The three–dimensional graph is then produced by the G3D procedure, cf the comments on Programs 2_1_1 and 6_1_2.

The plot of the Epanechnikov kernel is generated in an analogous way.

## Bias and Variance

Next we compute the expectation and the variance of the kernel density estimator and, thus, its mean squared error.

**6.4.2 Lemma.** *Suppose that the density f has continuous partial derivatives of second order in a neighborhood of $x$. We assume the kernel $K$ to have a bounded support and to satisfy conditions (6.16) and (6.17). Then we have the expansion*

$$Bias\,(\hat{f}_n(x)) := E(\hat{f}_n(x)) - f(x) = \frac{h^2}{2}\Delta f(x) \int y_1^2 K(y)\,dy + o(h^2), \quad as \ \ h \to 0,$$

*where*

$$\Delta f(x) := \sum_{i=1}^{p} \frac{\partial^2 f(x)}{\partial^2 x_i}$$

*is the Laplace operator.*

**Proof:** By substituting $y \mapsto x - hy$ we obtain

$$E(\hat{f}_n(x)) = h^{-p} E(K((x-Y)/h))$$

$$= h^{-p} \int K((x-y)/h))f(y)\,dy = \int K(y)f(x-hy)\,dy.$$

The multivariate Taylor expansion of $f(x - hy)$ at $x$ (see, for example, Section 12.4 in Smith (1983))

$$f(x - hy) = f(x) - h \sum_{i=1}^{p} \frac{\partial f(x)}{\partial x_i} y_i + \frac{h^2}{2} \sum_{i=1}^{p} \sum_{j=1}^{p} \frac{\partial^2 f(x)}{\partial x_i \partial x_j} y_i y_j + o(h^2)$$

yields

$$E(\hat{f}_n(x)) = f(x) \int K(y)\, dy - h \sum_{i=1}^{p} \frac{\partial f(x)}{\partial x_i} \int y_i K(y)\, dy$$

$$+ \frac{h^2}{2} \sum_{i=1}^{p} \sum_{j=1}^{p} \frac{\partial^2 f(x)}{\partial x_i \partial x_j} \int y_i y_j K(y)\, dy + o(h^2)$$

$$= f(x) + \frac{h^2}{2} \Delta f(x) \int y_1^2 K(y)\, dy + o(h^2)$$

by (6.16), (6.17) and the fact that $K$ has a bounded support. Note that condition (6.17) then implies

$$\int y_i K(y)\, dy = 0, \quad \int y_i y_j K(y)\, dy = 0, \quad i \neq j,$$

and

$$\int y_i^2 K(y)\, dy = \int y_1^2 K(y)\, dy, \quad i = 2, \ldots, p.$$

$\square$

**6.4.3 Lemma.** *We assume that $f$ is continuous at $x$. If $K$ has a bounded support and finite second moment, i.e., $\int K^2(y)\, dy < \infty$, the variance of $\hat{f}_n(x)$ has the expansion*

$$Var(\hat{f}_n(x)) = \frac{1}{nh^p} f(x) \int K^2(y)\, dy + o\left(\frac{1}{nh^p}\right), \quad as \ h \to 0.$$

**Proof:** Exercise 15. $\square$

**6.4.4 Corollary.** *We assume that the density $f$ has continuous partial derivatives of second order in a neighborhood of $x$. Suppose that the kernel $K$ has a bounded support and that it satisfies the conditions (6.16), (6.17) as well as $\int K^2(y)\, dy < \infty$. Then we obtain the following expansion for the mean squared error*

$$MSE\left(\hat{f}_n(\boldsymbol{x})\right) := E\left((\hat{f}_n(\boldsymbol{x}) - f(\boldsymbol{x}))^2\right)$$

$$= Var(\hat{f}_n(\boldsymbol{x})) + (Bias\,(\hat{f}_n(\boldsymbol{x})))^2$$

$$= \frac{1}{nh^p} f(\boldsymbol{x}) \int \mathcal{K}^2(\boldsymbol{y})\,d\boldsymbol{y} + \frac{h^4}{4}\left\{\Delta f(\boldsymbol{x}) \int y_1^2 \mathcal{K}(\boldsymbol{y})\,d\boldsymbol{y}\right\}^2$$

$$+\,o\left(\frac{1}{nh^p} + h^4\right),\ as\ h \to 0.$$

## The Optimal Bandwidth

Choose the *sequence* of bandwidths $h = h(n)$ such that $h(n) \to 0$ and $nh(n)^p \to \infty$ as $n \to \infty$. Then under the conditions of Corollary 6.4.4, we obtain that the mean squared error of $\hat{f}_n(\boldsymbol{x})$ converges to zero as the sample size $n$ increases to infinity, i.e.,

$$MSE\left(\hat{f}_n(\boldsymbol{x})\right) \to 0, \qquad n \to \infty.$$

The weak consistency of $\hat{f}_n(\boldsymbol{x})$ is now immediate from Chebyshev's inequality (see Exercise 17, Chapter 1):

$$P\{|\hat{f}_n(\boldsymbol{x}) - f(\boldsymbol{x})| \geq \varepsilon\} \leq \frac{MSE(\hat{f}_n(\boldsymbol{x}))}{\varepsilon^2} \longrightarrow 0, \qquad n \to \infty,$$

for arbitrary $\varepsilon > 0$. The optimal bandwidth for the sample size $n$, which minimizes the sum of the leading terms in the expansion of $MSE(\hat{f}_n(\boldsymbol{x}))$, is now

$$h^*(n) = \frac{1}{n^{1/(4+p)}} \frac{(p\,f(\boldsymbol{x}) \int \mathcal{K}^2(\boldsymbol{y})\,d\boldsymbol{y})^{1/(4+p)}}{(\Delta f(\boldsymbol{x}) \int y_1^2 \mathcal{K}(\boldsymbol{y})\,d\boldsymbol{y})^{2/(4+p)}}.$$

We have, obviously, $h^*(n) \to 0$ and $nh^*(n)^p \to \infty$ as $n \to \infty$. This bandwidth, however, depends again on the unknown density and has therefore only minor practical impact.

## Sphering the Data

The only smoothing parameter in the kernel density estimator of the form (6.15)

$$\hat{f}_n(\boldsymbol{x}) = \frac{1}{nh^p} \sum_{i=1}^{n} \mathcal{K}\left(\frac{\boldsymbol{x} - \boldsymbol{Y}_i}{h}\right)$$

is $h > 0$, by which all components of the vectors $\boldsymbol{Y}_i$ are equally scaled. This means that this kernel density estimator puts equal weights to each of the $p$ directions of the data cloud $\{\boldsymbol{Y}_1, \ldots, \boldsymbol{Y}_n\}$ in $\mathbb{R}^p$, given by the $p$ unit vectors $(1, 0, \ldots, 0)^T, \ldots, (0, \ldots, 0, 1)^T$. A different principal direction of the data cloud would, however, be ignored. Take, for example, the case of a data cloud in $\mathbb{R}^2$ which is almost parallel to the $x$ axis. The variation of the data is in this case much larger in the first coordinate than in the second one. The bandwidth $h$ should, therefore, be large for the first component in order to catch a sufficient number of data by the corresponding band. The bandwidth for the second component could, however, be small, since a sufficient number of data would already be caught by a small band. The choice of only one smoothing parameter $h$ is, therefore, often not adequate to the shape of the data cloud.

This deficiency of a kernel density estimator of the form (6.15) can be overcome by using the tensor kernel from Example 6.4.1 (vi) with possibly different bandwidths in each component. The corresponding density estimator with the vector $h := (h_1, \ldots, h_p)^T$ of bandwidths $h_j > 0$, $j = 1, \ldots, p$, is defined by

$$\hat{f}_{n,h}(x) := \frac{1}{n \prod_{j=1}^p h_j} \sum_{i=1}^n \mathcal{K}_T\left(\frac{x - Y_i}{h}\right)$$

$$:= \frac{1}{n \prod_{j=1}^p h_j} \sum_{i=1}^n \prod_{j=1}^p k\left(\frac{x_j - Y_{ij}}{h_j}\right),$$

where $x = (x_1, \ldots, x_p)^T \in \mathbb{R}^p$ and $Y_i = (Y_{i1}, \ldots, Y_{ip})^T$, $i = 1, \ldots, n$. This density estimator can take into account some principal directions of the data cloud by properly choosing the bandwidth $h_i$ in each component. The results 6.4.2 up to 6.4.4 can be generalized to this estimator, cf Exercise 17.

But it is by no means obvious to evaluate the shape of a data cloud in higher dimensions. The approach to this problem via principal component analysis is developed in Chapter 8. Another solution to this problem is the idea to remove principal directions by applying a linear transformation to the data such that the transformed cloud is ball–shaped. The transformed data are evaluated and the result obtained is then transformed back.

To this end we transform the data $Y_i$ by the inverse of the *symmetric root* $S_n^{1/2}$ of the sample $p \times p$–covariance matrix

$$S_n = \frac{1}{n-1} \sum_{i=1}^n (Y_i - \bar{Y})(Y_i - \bar{Y})^T.$$

We assume that the covariance matrix $S_n$ is positive definite, i.e., its rank is $p$ (Exercise 13). The inverse matrix $S_n^{-1/2}$ satisfies in particular the equation $S_n^{-1/2} S_n S_n^{-1/2} = I_p$, cf Exercise 18, which is crucial for our purposes. The sample covariance matrix of the transformed data $Z_i := S_n^{-1/2} Y_i$, $i = 1, \ldots, n$, is then the unity matrix:

$$\frac{1}{n-1}\sum_{i=1}^{n}(Z_i - \bar{Z})(Z_i - \bar{Z})^T$$

$$= \frac{1}{n-1}\sum_{i=1}^{n}(S_n^{-1/2}Y_i - S_n^{-1/2}\bar{Y})(S_n^{-1/2}Y_i - S_n^{-1/2}\bar{Y})^T$$

$$= \frac{1}{n-1}\sum_{i=1}^{n}S_n^{-1/2}(Y_i - \bar{Y})(Y_i - \bar{Y})^T S_n^{-1/2}$$

$$= S_n^{-1/2} S_n S_n^{-1/2} = I_p,$$

where $\bar{Z} = (Z_1 + \cdots + Z_n)/n = S_n^{-1/2}\bar{Y}$ is the vector of sample means of $Z_1, \ldots, Z_n$. We use the fact that $S_n^{-1/2}$ is a symmetric matrix, i.e., it coincides with its transpose, cf Exercise 12.

By the transformation $Z_i = S_n^{-1/2}Y_i$, which wastes no information contained in $Y_i$, we remove principal directions of the data cloud $Y_1, \ldots, Y_n$ in $\mathbb{R}^p$ by transforming them into a ball–shaped set, see Figure 6.4.3. For this reason, the transformation is called *sphering the data* . If we subtract, in addition, the vector of means $\bar{Z} = S_n^{-1/2}\bar{Y}$ and consider the data $Z_i - \bar{Z}$, $i = 1, \ldots, n$, we move the center of the ball–shaped data cloud to the origin of $\mathbb{R}^p$.

To these transformed data $Z_i = S_n^{-1/2}Y_i$ with equal variation in each direction we apply the density estimator

$$\frac{1}{nh^p}\sum_{i=1}^{n}\mathcal{K}\left(\frac{x - S_n^{-1/2}Y_i}{h}\right)$$

from (6.15). As these data have equal variation in each direction, the above estimator with the same bandwidth $h$ for all components is adequate. Note, however, that in general it does not estimate the density $f(x)$ of $Y$. If we denote by

$$S = E((Y - E(Y))(Y - E(Y))^T)$$

the covariance matrix of $Y$, then $S^{-1/2}Y$ has the density $(\det S)^{1/2}f(S^{1/2}x)$, cf Section 1.4 in Reiss (1989), note that $\det(S^{-1/2}) = (\det S)^{-1/2}$. The density estimator

$$\frac{1}{nh^p}\sum_{i=1}^{n}\mathcal{K}\left(\frac{x - S^{-1/2}Y_i}{h}\right),$$

where we have replaced $S_n$ by $S$, is consequently an estimator of $(\det S)^{1/2}$ $f(S^{1/2}x)$. The following modified version $\hat{f}_{S_n}(x)$ will then satisfy

$$\hat{f}_{S_n}(x) := \frac{1}{(\det S_n)^{1/2}nh^p} \sum_{i=1}^{n} K\left(\frac{S_n^{-1/2}(x - Y_i)}{h}\right)$$

$$\approx (\det S)^{-1/2}\frac{1}{nh^p} \sum_{i=1}^{n} K\left(\frac{S^{-1/2}x - S^{-1/2}Y_i}{h}\right)$$

$$\approx f(S^{1/2}S^{-1/2}x) = f(x). \tag{6.18}$$

## The Mahalanobis Distance

For the normal kernel from Example 6.4.1 (v) we obtain

$$\hat{f}_{S_n}(x)$$

$$= \frac{1}{(2\pi)^{p/2}(\det S_n)^{1/2}nh^p} \sum_{i=1}^{n} \exp\left(-\frac{1}{2h^2}(S_n^{-1/2}(x - Y_i))^T(S_n^{-1/2}(x - Y_i))\right)$$

$$= \frac{1}{(2\pi)^{p/2}(\det S_n)^{1/2}nh^p} \sum_{i=1}^{n} \exp\left(-\frac{1}{2h^2}\Delta^2_{S_n}(x, Y_i)\right),$$

where
$$\Delta^2_B(x, y) := (x - y)^T B^{-1}(x - y)$$

denotes the generalized squared distance between $x$ and $y \in \mathbb{R}^p$ and $B$ is an arbitrary positive definite $p \times p$–matrix. With the particular choice $B = S_n$, $\Delta_{S_n}(x, y)$ is called *Mahalanobis distance* between $x$ and $y$, cf Theorem 6.3.2 and Section 7.2.

For the Epanechnikov kernel from Example 6.4.1 (ii) we obtain

$$\hat{f}_{S_n}(x) = \frac{(1 + p/2)}{v_p(\det S_n)^{1/2}nh^p} \sum_{i=1}^{n} \max\left\{1 - \frac{1}{h^2}\Delta^2_{S_n}(x, Y_i), 0\right\}.$$

The following figures visualize the sphering of data that have an obvious principal direction. The vector of the sample means is $0$. The plots are based on the variables $mOsm$ and $urea$ in a subsample of size twenty of the crystal data in Example 6.3.3. Using the same bandwidth $h = 2$ in each case, the density estimator $\hat{f}_{S_n}$ with data sphering reveals in Figure 6.4.4 this principal direction of the data, whereas $\hat{f}_n$ fails.

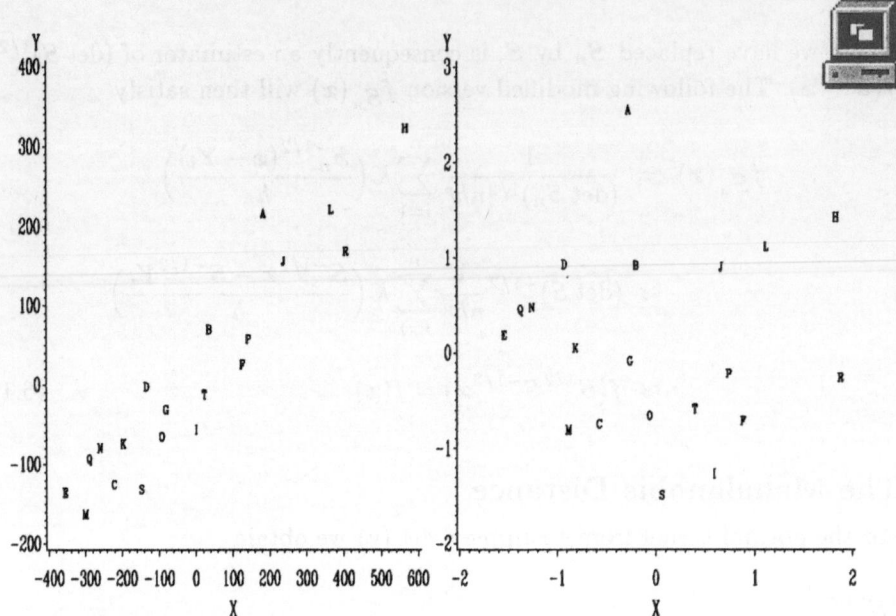

**Figure 6.4.3.** Scatterplot of size twenty observations from crystal data for the variables *mOsm* and *urea* with sample means 0, and the data transformed by sphering them.

```
***    Program 6_4_3    ***;
TITLE1 'Original and Transformed Data'; TITLE2 'Crystal Data';
LIBNAME datalib 'c:\data';

DATA letter(KEEP=mosm urea text); SET datalib.crystal;
    INPUT text $ @@; OUTPUT;
    CARDS;
A B C D E F G H I J K L M N O P Q R S T
;
PROC STANDARD DATA=letter M=0 OUT=data1;
PROC IML;
    USE data1 VAR{mosm urea}; * reading standardized data;
    READ ALL INTO x;
    covm=(x'*x)*(1/19);        * covariance matrix;
    is=INV(covm);              * inverse of covariance matrix;
    CALL SVD(u,q,v,is);        * decomposition for computation of;
    isr=u*DIAG(SQRT(q))*u';    * inverse symmetric root (isr);
    t_x=x*isr;
    vn={t_mosm t_urea};
    CREATE new1 FROM t_x [COLNAME=vn]; APPEND FROM t_x;
QUIT;                          * quit IML;
                               ↓
```

↑
```
DATA datalib.o_anno t_anno; MERGE data1 new1;
    FUNCTION='TEXT'; POSITION='2'; XSYS='2'; YSYS='2'; SIZE=1.1;
    X=mosm; Y=urea; OUTPUT datalib.o_anno;
    X=t_mosm; Y=t_urea; OUTPUT t_anno;
GOPTIONS NODISPLAY; SYMBOL1    V=NONE    I=NONE;
PROC GPLOT DATA=datlib.o_anno GOUT=figure;
    TITLE1 'Scatter Plot of Original Data';
    PLOT y*x /ANNO=datalib.o_anno NAME='original';
PROC GPLOT DATA=t_anno GOUT=figure;
    TITLE1 'Scatter Plot of Transformed Data';
    PLOT y*x /ANNO=t_anno NAME='transfor'; RUN; QUIT;
GOPTIONS DISPLAY;
PROC GREPLAY IGOUT=figure NOFS TC=SASHELP.TEMPLT TEMPLATE=H2;
    TREPLAY 1:original    2:transfor; RUN; DELETE _ALL_; RUN; QUIT;
```

This program gives a first impression of the capacities of the SAS/IML module (=Interactive Matrix Language). At first the procedure STANDARD with the option M=0 returns the data now having mean vector 0.

The first two lines of PROC IML (USE, READ) transform the SAS data file data1 into the matrix x. Matrix manipulation is then done using the IML functions INV, SVD and DIAG. The final two lines (CREATE, APPEND) then transform the manipulated matrix t_x into the SAS data set new1.

The subsequent DATA step generates two annotated data sets, cf Program 3_2_1, which contain the marks for the original data file o_anno and the transformed one t_anno.

The procedures GPLOT and GREPLAY are then used for plotting the data as in Program 3_1_1.

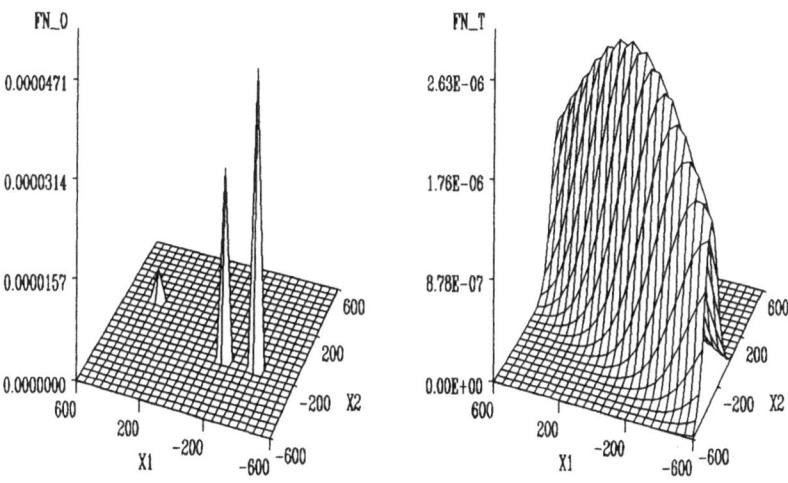

**Figure 6.4.4.** Density estimators without and with data sphering for the original data in Figure 6.4.3; normal kernel, $h = 2$.

```
***    Program 6_4_4    ***;
TITLE1 'Density Estimator';
TITLE2 'Crystal Data';
LIBNAME datalib 'c:\data';

DATA data1;
    a= 0.0078622; b=-0.008428;  * inverse of symmetric root;
    c=-0.008428;  d= 0.019606;  * of covariance matrix;
    sdet=12031.507;             * determinant of symmmetric root;
    n=20;                       * sample size;
    h=2;                        * choice of bandwidth;
    DO x1=-600 TO 600 BY 50;    * computation of density estimator;
    DO x2=-600 TO 600 BY 50;    * using the data;
      fn_o=0; fn_t=0;           * from Program 6_4_1;
      DO i=1 TO n;
          SET datalib.o_anno POINT=i;
          fn_o=fn_o+EXP(-0.5*(((x1-mosm)/h)**2+((x2-urea)/h)**2));
          fn_t=fn_t+EXP(-0.5*(((a*(x1-mosm))/h +(b*(x2-urea))/h )**2
                        +((c*(x1-mosm))/h +(d*(x2-urea))/h)**2));
      END;
    twopi=6.2831853;
    fn_o=fn_o/(twopi*n*h**2);
    fn_t=fn_t/(twopi*sdet*n*h**2);
    OUTPUT; END; END; STOP;

GOPTIONS NODISPLAY;
PROC G3D DATA=data1 GOUT=figure;
    PLOT x1*x2=fn_o / NAME='original';
RUN; QUIT;

TITLE1 'Density Estimator with Data Sphering';
PROC G3D DATA=data1 GOUT=figure;
    PLOT x1*x2=fn_t / NAME='transfor';
RUN; QUIT;

GOPTIONS DISPLAY;
PROC GREPLAY IGOUT=figure TC=SASHELP.TEMPLT NOFS TEMPLATE=H2;
    TREPLAY 1:original   2:transfor;
RUN;
    DELETE _ALL_;
RUN; QUIT;
```

This program is a combination of the computation of a kernel density estimator as in Program 1_1_4 and the visualization of its results by means of the procedures G3D and GREPLAY, cf the commands on Program 6_1_2 and 3_1_1.

In the following example we compute a kernel density estimator based discriminant analysis of the crystal data from Example 6.3.3 with the variables $pH$ value and specific weight as in Figure 6.3.3. We use the Epanechnikov kernel with data sphering. The bandwidth is $h = 1.2$ and the prior probabilities of the two classes are assumed to be equal, i.e., $p(1) = p(2) = 0.5$. The estimated failure rate 0.3281 is only a little bit smaller than that of Fisher's approach in Figure 6.3.3.

The DISCRIM Procedure

| Observations | 79 | DF Total | 78 |
|---|---|---|---|
| Variables | 2 | DF Within Classes | 77 |
| Classes | 2 | DF Between Classes | 1 |

Class Level Information

| CRYSTAL | Variable Name | Frequency | Weight | Proportion | Prior Probability |
|---|---|---|---|---|---|
| 1 | _1 | 45 | 45.0000 | 0.569620 | 0.500000 |
| 2 | _2 | 34 | 34.0000 | 0.430380 | 0.500000 |

--------------------------------------------------------------------

Classification Summary for Calibration Data: DATALIB.CRYSTAL
Resubstitution Summary using Epanechnikov Kernel Density

Squared Distance Function

$$D^2(X,Y) = (X-Y)' \, COV^{-1} \, (X-Y)$$

Posterior Probability of Membership in Each CRYSTAL

$$F(X|j) = n_j^{-1} \, \underset{i}{SUM} \, ( \, 1.0 - D^2(X,Y_{ji}) \, / \, R^2 \, )$$

$$Pr(j|X) = PRIOR_j \, F(X|j) \, / \, \underset{k}{SUM} \, PRIOR_k \, F(X|k)$$

Number of Observations and Percent Classified into CRYSTAL

| From CRYSTAL | 1 | 2 | Total |
|---|---|---|---|
| 1 | 34 | 11 | 45 |
| | 75.56 | 24.44 | 100.00 |
| 2 | 14 | 20 | 34 |
| | 41.18 | 58.82 | 100.00 |
| Total | 48 | 31 | 79 |
| | 60.76 | 39.24 | 100.00 |
| Priors | 0.5 | 0.5 | |

Error Count Estimates for CRYSTAL

| | 1 | 2 | Total |
|---|---|---|---|
| Rate | 0.2444 | 0.4118 | 0.3281 |
| Priors | 0.5000 | 0.5000 | |

**Figure 6.4.5.** Kernel density estimator based discriminant analysis of crystal data, variables $pH$, $g$; Epanechnikov kernel, $h = 1.2$. Compare with Figure 6.3.3.

```
***    Program 6_4_5    ***;
TITLE1 'Discriminant Analysis using Kernel Density Estimators';
TITLE2 'Crystal Data';
LIBNAME datalib 'c:\data';

PROC DISCRIM DATA=datalib.crystal METHOD=NPAR KERNEL=EPA R=1.2;
    CLASS crystal;
    VAR   ph g;
RUN; QUIT;
```

The nonparametric and kernel density estimator based discriminant analysis is computed if the options METHOD=NPAR and KERNEL=EPA are added to PROC DIS-CRIM. Other kernels are UNI, BI, TRI, NORMAL. The option R=1.2 determines the bandwidth $h$.

# 6.5  The Nearest Neighbor Approach

In the final section of this chapter we introduce the nearest neighbor approach in nonparametric discriminant analysis. Its basic idea is as follows: Suppose we have a training sample with random stratification into several classes. A future object $\omega$ will be classified to that particular class, which has the highest number of elements of the training sample in a neighborhood of the associated vector of observations $x = x(\omega)$.

## The Uniform Kernel

Denote by

$$B_h(x) := \{y \in \mathbb{R}^p : (y - x)^T(y - x) \leq h^2\}$$

the ball in $\mathbb{R}^p$ with center $x$ and radius $h > 0$. Using the uniform kernel

$$\mathcal{K}_u(y) = \begin{cases} 1/v_p, & y \in B_1(0), \\ 0 & \text{elsewhere}; \end{cases}$$

and the bandwidth $h > 0$, the kernel density estimator in (6.15) becomes

$$\hat{f}_n(x) = \frac{r/n}{h^p v_p} = \frac{r/n}{v_p(h)}.$$

By $r$ we denote the *random* number of those observations $Y_i$, which fall into the ball $B_h(x)$, and $v_p(h)$ is the volume of $B_h(x)$, cf Example 6.4.1.

## Nearest Neighbor Estimators

Data sphering was introduced in the preceding Section 6.4 to overcome the disadvantage of the ordinary density estimator, which ignores a possible principal direction of the data cloud by the use of a single smoothing parameter. An alternative approach to overcome this disadvantage is the idea to let the data themselves determine the bandwidth $h$ in the above density estimator with uniform kernel. The radius $h$ is, consequently, a *random variable*. The number $r$ of data will, however, be fixed. Precisely, we take as the bandwidth the Euclidean distance $h_r(x)$ between $x$ and its $r$-nearest neighbor $Y_l$ among $Y_1, \ldots, Y_n$.

For the uniform kernel with this random bandwidth $h_r(x)$, the ordinary density estimator becomes the *r-nearest neighbor estimator*

$$\hat{f}_n(x) = \frac{r/n}{h_r(x)^p v_p} = \frac{r/n}{v_p(h_r(x))}.$$

Note that *we* choose the parameter $r \in \{1, \ldots, n\}$, the volume $v_p(h_r(x)) = h_r(x)^p v_p$, however, is a random variable.

## Estimators of Parameters within the Classes

Suppose we have a training sample $(x_1, k_1), \ldots, (x_n, k_n)$. Then, within each class $k$, we define the $r$–nearest neighbor estimator of the density $f(x|k)$ of the group–specific distribution within the $k$-th class by

$$\hat{f}_{n_k}(x|k) = \frac{r_k/n_k}{v_p(h_r(x))}.$$

By $n_k$ we denote the number of observations in the $k$-th class, and by $r_k$ the random number of those among them, which fall into the ball $B_{h_r(x)}(x)$. We have, consequently, $\sum_{k=1}^{K} n_k = n$, $\sum_{k=1}^{K} r_k = r$.

We now estimate the prior probability $p(k)$ by

$$\hat{p}(k) = n_k/n$$

and the unconditional density $f(x) = \sum_{k=1}^{K} p(k)f(x|k)$ of the vector $X$ of observations by

$$\hat{f}_n(x) = \sum_{k=1}^{K} \hat{p}(k)\hat{f}_{n_k}(x|k) = \frac{r}{nv_p(h_r(x))}.$$

An estimator of the posterior probability $p(k|x) = p(k)f(x|k)/f(x)$ is then given by

$$\hat{p}_n(k|x) := \hat{p}(k)\hat{f}_{n_k}(x|k)/\hat{f}_n(x) = \frac{n_k}{n} \frac{r_k/n_k}{v_p(h_r(x))} \Big/ \frac{r}{nv_p(h_r(x))} = r_k/r.$$

## The Bayes' Rule, the Maximum–Likelihood Rule

We thus arrive at Bayes' classification rule with estimated posterior probabilities: An object with the vector of observations $x$ is classified to that class with the particular index $\hat{k}$, which maximizes the estimated posterior probabilities $\hat{p}_n(k|x)$, i.e., $\hat{k}$ is the (least) index with

$$r_{\hat{k}} = \max_{1 \leq k \leq K} r_k.$$

This is the $r$–nearest neighbor classification rule: An object is classified to that index $\hat{k}$, such that the total number of observations $(x_i, k)$, which fall into the ball with center $x$ and radius $h_r(x)$, is maximized. This rule aims at maximizing the posterior probabilities. For the particular choice $r = 1$, this is simply called the *nearest neighbor rule*.

If we assume identical prior probabilities $p(1) = \cdots = p(K) = 1/K$, the estimator of the unconditional density $f(x)$ becomes

$$\tilde{f}_n(x) = \frac{1}{K} \sum_{k=1}^{K} \hat{f}_{n_k}(x|k).$$

This yields the estimator

$$\tilde{p}_n(k|\boldsymbol{x}) = \frac{p(k)\hat{f}_{n_k}(\boldsymbol{x}|k)}{\tilde{f}_n(\boldsymbol{x})} = \frac{r_k/n_k}{\sum_{j=1}^{K} r_j/n_j}$$

of the posterior probability $p(k|\boldsymbol{x})$. The maximum–likelihood classification rule is, therefore: An object with the vector of observations $\boldsymbol{x}$ is classified to that class with least index $\tilde{k}$, which maximizes $\tilde{p}_n(k|\boldsymbol{x})$ or, equivalently, $\hat{f}_{n_k}(\boldsymbol{x}|k)$, i.e.,

$$r_{\tilde{k}}/n_{\tilde{k}} = \max_{1\le k\le K} r_k/n_k.$$

If we put $m_k := r_k/n_k$, $k = 1,\ldots,K$, the Bayes' rule and the maximum–likelihood rule can be unified as follows: An object is classified to that class with least index $\bar{k}$, such that the estimated posterior probabilities

$$\bar{p}_n(k|\boldsymbol{x}) = \frac{m_k(\boldsymbol{x})\,\mathrm{prior}(k)}{\sum_{j=1}^{K} m_j(\boldsymbol{x})\,\mathrm{prior}(j)}$$

are maximized, i.e.,

$$\bar{p}_n(\bar{k}|\boldsymbol{x}) = \max_{1\le k\le K} \bar{p}_n(k|\boldsymbol{x}). \tag{6.19}$$

By putting the $j$–th prior probability equal to $\mathrm{prior}(j) = \hat{p}(j) = n_j/n$ we obtain $\bar{p}_n(k|\boldsymbol{x}) = \hat{p}_n(k|\boldsymbol{x})$, i.e., (6.19) yields Bayes' rule. With the $j$–th prior probability equal to $\mathrm{prior}(j) = 1/K$, $j = 1,\ldots,K$, (6.19) is the maximum–likelihood rule, since we have in this case $\bar{p}_n(k|\boldsymbol{x}) = \tilde{p}_n(k|\boldsymbol{x})$.

In the following we continue Example 6.2.1 by applying the nearest neighbor classification rule with $r = 6$ to the cns data and the variables $a/n$, $o/n$, $m/n$. Figure 6.1.2 already indicated that a *linear* discriminant analysis would not be a proper approach for these data. The failure rates were 0.3474 and 0.2857 for the maximum–likelihood rule and for Bayes' rule, respectively, in the parametric model of Example 6.2.1. The above nonparametric approach yields, however, the considerably smaller failure rates 0.0431 for the maximum–likelihood rule with identical prior probabilities $p(1) = p(2) = 0.5$, and 0.0714 for Bayes' rule with estimated priors $\hat{p}(1) = 0.4082$ and $\hat{p}(2) = 0.5918$.

The DISCRIM Procedure

| Observations | 98 | DF Total | 97 |
|---|---|---|---|
| Variables | 3 | DF Within Classes | 96 |
| Classes | 2 | DF Between Classes | 1 |

Class Level Information

| STATUS | Variable Name | Frequency | Weight | Proportion | Prior Probability |
|--------|---------------|-----------|--------|------------|-------------------|
| healthy | healthy | 40 | 40.0000 | 0.408163 | 0.500000 |
| ill | ill | 58 | 58.0000 | 0.591837 | 0.500000 |

-------------------------------------------------------------------

Classification Summary for Calibration Data: DATALIB.CNS
Resubstitution Summary using 6 Nearest Neighbors

Squared Distance Function

$$D^2 (X,Y) = (X-Y)' (X-Y)$$

Posterior Probability of Membership in Each STATUS

$m_k (X)$ = Proportion of obs in group k in 6
nearest neighbors of X

$$Pr(j|X) = m_j (X) PRIOR_j / SUM_k ( m_k (X) PRIOR_k )$$

Number of Observations and Percent Classified into STATUS

| From STATUS | healthy | ill | Total |
|-------------|---------|-----|-------|
| healthy | 40 | 0 | 40 |
|  | 100.00 | 0.00 | 100.00 |
| ill | 5 | 53 | 58 |
|  | 8.62 | 91.38 | 100.00 |
| Total | 45 | 53 | 98 |
|  | 45.92 | 54.08 | 100.00 |
| Priors | 0.5 | 0.5 | |

Error Count Estimates for STATUS

|  | healthy | ill | Total |
|---|---|---|---|
| Rate | 0.0000 | 0.0862 | 0.0431 |
| Priors | 0.5000 | 0.5000 | |

**Figure 6.5.1.** $r$–nearest neighbor discriminant analysis of cns data in Example 6.1.1; variables $a/n$, $o/n$, $m/n$; $r = 6$; maximum–likelihood rule.

```
***    Program 6_5_1    ***;
TITLE1 'Nearest Neighbor Discriminant Analysis';
TITLE2 'CNS Data';
LIBNAME datalib 'c:\data';

PROC DISCRIM DATA=datalib.cns METHOD=NPAR K=6 METRIC=IDENTITY;
    CLASS status;
    VAR an on mn;
RUN; QUIT;
```

The DISCRIM procedure together with the option METHOD=NPAR computes a kernel density estimator based nonparametric discriminant analysis with the uniform kernel and identical priors. The additional option K=6 invokes the nearest neighbor approach with 6 nearest neighbors. The additional statement PRIORS PROP in the DISCRIM procedure would invoke the Bayes rule with estimated priors in place of the maximum–likelihood rule with identical priors, cf Program 6_2_1. METRIC specifies the metric in which squared distances are computed. METRIC=IDENTITY uses Euclidean distance. The default is the Mahalanobis distance.

The classification rule derived from the training sample 'DATA=data1' (here DATA= datalib.cns) can be applied to objects in some data set data2 by adding the option 'TESTDATA=data2'.

Exercises

1. Let $(\Omega, \mathcal{A}, P)$ be a probability space and let $(B_i)_{i \in I}$ $(I = \{1, \ldots, n\}$ or $I = \mathbb{N})$ be a division of $\Omega$ into disjoint subsets with $P(B_i) > 0$, $i \in I$. Prove Bayes' rule,

$$P(B_i|A) = \frac{P(B_i)\,P(A|B_i)}{\sum_{j \in I} P(B_j)\,P(A|B_j)}, \quad i \in I, \text{ whenever } P(A) > 0.$$

Hint: Use the law of total probability in Exercise 4, Chapter 4).

**2.** Suppose that the random variable $X$ has the group–specific density $f(\cdot|k)$, $k = 1, \ldots, K$. Show that $X$ has the density

$$f(x) = \sum_{k=1}^{K} p(k) f(x|k)$$

and that $(X, \kappa)$ has the joint density

$$(x, k) \mapsto p(k) f(x|k).$$

Hint: Fubini's theorem.

**3.** (Conditional distribution) A function $(k, x) \mapsto P(\kappa = k|X = x)$, $k \in \{1, \ldots, K\}$, $x \in S \subset \mathbb{R}^p$, $S$ Borel measurable, is called *conditional distribution* of $\kappa$ given $X = x$ if it satisfies the following three conditions:

(1) The function $S \ni x \mapsto P(\kappa = k|X = x)$ is Borel measurable for each $k = 1, \ldots, K$.

(2) The function $k \mapsto P(\kappa = k|X = x)$ defines for each $x \in \mathbb{R}^p$ a probability measure on $\{1, \ldots, K\}$.

(3) $P\{X \in B, \kappa = k\} = \displaystyle\int_B P(\kappa = k|X = x) \, d\mathcal{L}(X|P)(x)$ for each Borel subset $B \subset \mathbb{R}^p$ and $k \in \{1, \ldots, K\ \}$.

(i) We continue Exercise 2 and put $p(k|x) := p(k) f(x|k)/f(x)$ if $f(x) > 0$, and $p(k|x) = g(k)$ elsewhere, where $g(1) + \cdots + g(K) = 1$, $g \geq 0$. Prove that the conditional distribution of $\kappa$ given $X = x$ has the density $p(k|x)$, i.e., $P(\kappa = k|X = x) := p(k|x)$ satisfies (1)–(3).

(ii) Suppose that the density $f$ of $X$ is positive on the interval $[x - \varepsilon, x + \varepsilon]$ and that $p(k|\cdot)$ is continuous at $x$. Prove the convergence

$$P(\kappa = k|X \in [x - \varepsilon, x + \varepsilon]) \to P(\kappa = k|X = x) \text{ if } \|\varepsilon\| \to 0.$$

**4.** Put

$$\tilde{G}_k^* := \{y \in S : d_k(y) < d_j(y), \ j \neq k, \ j = 1, \ldots, K\}, \quad k = 1, \ldots, K.$$

Show that:

(i) $\tilde{G}_1^*, \ldots, \tilde{G}_K^*$ are disjoint.

(ii) $\tilde{G}_k^* \subset G_k^*$, $1 \leq k \leq K$.

(iii) $S \setminus \bigcup_{k=1}^{K} \tilde{G}_k^* \subset \bigcup_{1 \leq j < k \leq K} \{y \in S : d_j(y) = d_k(y)\}$.

**5.** (i) Consider a division of the population into 2 classes and suppose that the group-specific distributions are binomial distributions, i.e., $f(x|j) = B(n, p_j)(\{x\})$, $p_j \in (0, 1)$, $j = 1, 2$, $x \in \{0, \ldots, n\}$. Prove that Bayes' rule yields the following classification rule: An object with the observation $x$ is assigned to class 1 if

$$x \geq \frac{\ln(p(2)/p(1)) - n\ln((1-p_1)/(1-p_2))}{\ln(p_1(1-p_2)/(p_2(1-p_1)))}, \quad \text{whenever} \quad \frac{p_1(1-p_2)}{p_2(1-p_1)} > 1$$

$$x \leq \frac{\ln(p(2)/p(1)) - n\ln((1-p_1)/(1-p_2))}{\ln(p_1(1-p_2)/(p_2(1-p_1)))}, \quad \text{whenever} \quad \frac{p_1(1-p_2)}{p_2(1-p_1)} < 1$$

(ii) Suppose that $f(x|j) = P_{\lambda_j}(\{x\})$, $j = 1, 2$, where $P_\lambda(\{x\}) = e^{-\lambda}\lambda^x/x!$, $x \in \mathbb{N} \cup \{0\}$ is the Poisson distribution with parameter $\lambda > 0$. Prove that Bayes' criterion yields the rule: $x$ is assigned to class 1 if

$$x \geq \frac{\ln(p(2)/p(1)) + \lambda_1 - \lambda_2}{\ln(\lambda_1/\lambda_2)}, \quad \text{in case} \quad \lambda_1/\lambda_2 > 1$$

$$x \leq \frac{\ln(p(2)/p(1)) + \lambda_1 - \lambda_2}{\ln(\lambda_1/\lambda_2)}, \quad \text{in case} \quad \lambda_1/\lambda_2 < 1$$

**6.** (CNS data) Generate scatterplots of the values of each pair of variables *an, on, mn, gn, ao* of mentally ill subjects (classes 1–3) and healthy subjects (classes 4 and 5). Can the points representing the two groups be separated by a line?

**7.** Show that the linear discriminant rule (6.10) is invariant under nonsingular transformations $x \mapsto Ax + b$, i.e., $x$ and $Ax + b$ assign to the same class $k \in \{1, 2\}$. Hint: Theorem 3.3.6 and Theorem 3.3.7.

**8.** Let $\tilde{d}_1$ and $\tilde{d}_2$ be the linear discriminant functions defined in (6.9). Show that the hyperplane $\{\tilde{d}_1 = \tilde{d}_2\}$ has the representation

$$n^T(x - x_0) = 0,$$

where

$$n = \Sigma^{-1}(\mu_1 - \mu_2)$$

and

$$x_0 = \frac{1}{2}(\mu_1 + \mu_2) + (\mu_1 - \mu_2)\,c/\Delta^2, \quad c := \ln\left(\frac{p(2)C(1|2)}{p(1)C(2|1)}\right).$$

By $\Delta$ we denote the Mahalanobis distance

$$\Delta = \Delta_\Sigma(\mu_1, \mu_2) := \left((\mu_1 - \mu_2)^T \Sigma^{-1}(\mu_1 - \mu_2)\right)^{1/2}.$$

**9.** Consider the classification problem in $\mathbb{R}^p$ with 2 classes. Let $X$ be a random vector of observations in $\mathbb{R}^p$ with a positive definite covariance matrix $\Sigma$. Let $c$ and $\Delta = \Delta_\Sigma$ be defined as in Exercise 8.

(i) Show that
$$\tilde{d}_2(X) - \tilde{d}_1(X) \text{ is } N(-c + \Delta^2/2, \Delta^2) \text{ distributed}$$
if $X$ is $N(\mu_1, \Sigma)$ distributed and that
$$\tilde{d}_2(X) - \tilde{d}_1(X) \text{ is } N(-c - \Delta^2/2, \Delta^2) \text{ distributed}$$

if $X$ is $N(\mu_2, \Sigma)$ distributed. By $\tilde{d}_i$, $i = 1, 2$, we denote the linear discriminant functions defined in (6.9). Hint: Exercise 8, Theorem 3.3.6 and Theorem 3.3.7.

(ii) Let $P(X \in \cdot | \kappa = k) = N(\mu_k, \Sigma)(\cdot)$, $k = 1, 2$. Prove that Bayes' rule has the risk
$$R(G_1^*, G_2^*) = p(1)C(2|1)\Phi((c - \Delta^2/2)/\Delta) + p(2)C(1|2)\Phi(-(c + \Delta^2/2)/\Delta),$$

where $\Phi$ is the standard normal distribution function.

**10.** Consider the two group–specific distributions $N(\mu_1, \Sigma)$ and $N(\mu_2, \Sigma)$ with
$$\mu_1 = \begin{pmatrix} \mu_{11} \\ \mu_{12} \end{pmatrix}, \quad \mu_2 = \begin{pmatrix} \mu_{21} \\ \mu_{22} \end{pmatrix}, \quad \Sigma = \begin{pmatrix} \sigma_1^2 & \sigma_1\sigma_2\varrho \\ \sigma_1\sigma_2\varrho & \sigma_2^2 \end{pmatrix}.$$

Denote as in Exercise 8 by $\Delta = \Delta_\Sigma(\mu_1, \mu_2)$ the Mahalanobis distance between the two group–specific distributions. Prove the following result

(i)
$$\Delta^2 = \frac{(\mu_{11} - \mu_{21})\sigma_2^2 + (\mu_{12} - \mu_{22})\sigma_1^2 - 2\sigma_1\sigma_2\varrho(\mu_{11} - \mu_{21})(\mu_{12} - \mu_{22})}{\sigma_1\sigma_2(1 - \varrho^2)}.$$

(ii) $\Delta^2$ has the representation
$$\Delta^2 = \Delta^2(\varrho) = \frac{\delta_1^2 + \delta_2^2 - 2\delta_1\delta_2\varrho}{(1 - \varrho^2)}$$

with $\delta_1 = (\mu_{11} - \mu_{21})/\sigma_1$ and $\delta_2 = (\mu_{12} - \mu_{22})/\sigma_2$.

(iii) The probability of a correct classification of $x$ increases with $\Delta^2$. If $\varrho = 0$, we have $\Delta^2 = \delta_1^2 + \delta_2^2$. Determine the set $\{\varrho \in (-1, 1) : \Delta(\varrho) > \delta_1^2 + \delta_2^2\}$. What conditions have to be satisfied by $\delta_1$ and $\delta_2$ in order to improve the classification? Give a graphic interpretation of the results.

**11.** Let $X_i = (X_{i1}, \ldots, X_{ip})^T$, $i = 1, \ldots, n$, be independent $p$–dimensional random vectors with mean vector $\mu \in \mathbb{R}^p$ and covariance matrix $\Sigma = (\sigma_{ij})$. Show that the empirical covariance matrix
$$S_n := \frac{1}{n-1}\sum_{i=1}^n (X_i - \bar{X})(X_i - \bar{X})^T$$

is an unbiased estimator of $\Sigma$, i.e., with $S_n = (S_{ij})$ we have $E(S_n) := (E(S_{ij})) = \Sigma$. Hint: Consider $Y_i := X_i - \mu$ and $\bar{Y} := \bar{X} - \mu$.

**12.** Let $A$ be an invertible $p \times p$-matrix. Show that $(A^T)^{-1} = (A^{-1})^T$. If $A$, in addition, is symmetric, show that $A^{-1}$ is symmetric as well.

**13.** Let $A$ be a positive semidefinite $p \times p$-matrix. Prove the following two facts: $A$ is invertible if and only if $A$ is positive definite. The inverse matrix $A^{-1}$ is positive definite as well. Hint: The principal axes transformation in (7.4) and Section 8.2 implies the representation $\Lambda = R^T A R$, where $R$ is an orthogonal matrix and $\Lambda$ is the diagonal matrix of the nonnegative eigenvalues of $A$. Note further that $\det(AB) = \det A \det B$.

**14.** (CNS data) Compute a discriminant analysis (healthy/ill) under the assumption of normal group–specific distributions. Divide the observations properly in two parts and compute a discriminant analysis with a training sample and a test sample (see the comments on Programs 6_2_1 and 6_5_1). How can the results be displayed graphically?

**15.** Prove Lemma 6.4.3.

**16.** Plot the cubic kernel.

**17.** Let $\hat{f}_{n,h}$ be the tensor kernel density estimator with bandwidth $h = (h_1, \ldots, h_p)^T$, $h_i > 0$, $i = 1, \ldots, p$, and kernel $k : \mathbb{R} \to [0, \infty)$.

(i) Under the assumption of Lemma 6.4.2 we have

$$Bias\left(\hat{f}_{n,h}(x)\right) = \frac{1}{2} \sum_{j=1}^{p} h_j^2 \frac{\partial f(x)}{\partial x_j \partial x_j} \int y_1^2 k(y_1) \, dy_1 + o(\|h\|^2).$$

(ii) Under the assumption of Lemma 6.4.3 we have

$$Var\left(\hat{f}_{n,h}(x)\right) = \frac{1}{n \prod_{i=1}^{p} h_i} f(x) \int K_T^2(y) \, dy + o\left(\frac{1}{n \prod_{i=1}^{p} h_i}\right).$$

(iii) Under the assumption of Corollary 6.4.4 we have

$$MSE\left(\hat{f}_{n,h}(x)\right)$$

$$= \frac{1}{n \prod_{i=1}^{p} h_i} f(x) \int K_T^2(y) \, dy$$

$$+ \frac{1}{4} \left\{ \sum_{i=1}^{p} h_i^2 \frac{\partial f(x)}{\partial x_i \partial x_i} \int y_1^2 k(y_1) \, dy_1 \right\}^2$$

$$+ o\left(\frac{1}{n \prod_{i=1}^{p} h_i} + \|h\|^4\right).$$

**18.** For any positive definite $p \times p$-matrix $A$ there exists an orthogonal matrix $R$, i.e., $R^T = R^{-1}$, such that $\Lambda = R^T A R$ is a diagonal matrix (see the hint in Exercise 13). The entries $d_{ii}$ of $\Lambda$ are the $p$ eigenvalues of $A$; these are positive. Put $s \in \mathbb{R}$

$$A^s := R\Lambda^s R^T,$$

where $\Lambda^s$ is the diagonal matrix with the entries $d_{ii}^s$. Show that for any $s, t \in \mathbb{R}$,

$$A^s A^t = A^{s+t}, \quad A^{-s} = (A^{-1})^s.$$

$A^{1/2}$, in particular, is called the *symmetric root* of $A$; it satisfies the equations $A^{1/2} A^{1/2} = A$, $A^{-1/2} A A^{-1/2} = I_p$.

**19.** (CNS data) Perform a nonparametric discriminant analysis in the situation of Exercise 14.

**20.** (Crystal data) (i) Compute a parametric and nonparametric discriminant analysis with the two classes crystallization / no crystallization and the variables pH level ($pH$), calcium concentration ($Ca$) and specific weight ($g$). (ii) Suppose that the class membership of the last 5 objects (75–79) is unknown. Compute a discriminant analysis as in (i) of the first 74 observations and apply the derived discriminant functions to classify the last 5 objects (see the comments on Programs 6_2_1 and 6_5_1).

# Chapter 7

# Cluster Analysis

The purpose of cluster analysis is to place objects into groups or clusters suggested by data, such that objects in a given cluster tend to be similar in some sense, and objects in different clusters tend to be dissimilar. Different to discriminant analysis, the clusters or groups are not defined a priori. The partitioning of a set of objects has to be done a posteriori, instead. Cluster analysis, therefore, is part of exploratory data analysis.

## 7.1 The Art of Clustering

The crucial problem when partitioning a set of objects into clusters is the fact that *cluster* is a rather vague term and, consequently, the partition usually cannot be uniquely defined. This problem and our approach towards clustering is illustrated by the following example.

**7.1.1 Example** (Air Pollution II Data; Gibbons et al. (1987)). The following measurements were taken in 1960 in 80 American cities. The variables *smin* up to *pmax* are taken in $\mu g/m^3 \times 10$:

| | | |
|---|---|---|
| *t* | = | total mortality rate |
| *smin* | = | smallest biweekly sulfate reading |
| *smean* | = | arithmetic mean of biweekly sulfate reading |
| *smax* | = | largest biweekly sulfate reading |
| *pmin* | = | smallest biweekly suspended particle reading |
| *pmean* | = | arithmetic mean of biweekly suspended particle reading |
| *pmax* | = | largest biweekly suspended particle reading |
| *pm2* | = | population density per square mile $\times 0.1$ |
| *lpop* | = | logarithm (base 10) of population $\times 10$ |
| *perwh* | = | percentage of whites in population |
| *nonpoor* | = | percentage of families with income above poverty level |
| *ge65* | = | percentage of population at least 65 years old $\times 10$. |

A detailed analysis of this data set is given in Chapters 9 and 10 by Jobson (1992). The following scatterplot displays the values of *smean* and *pmean* of

15 cities. Des Moines, Johnstown, Milwaukee, York and Providence seem to be arranged in a belt–shaped cluster with respect to these measurements. Jersey City might be a group of its own; it could, however, also belong to the above cluster. The remaining 9 cities seem to be one cluster.

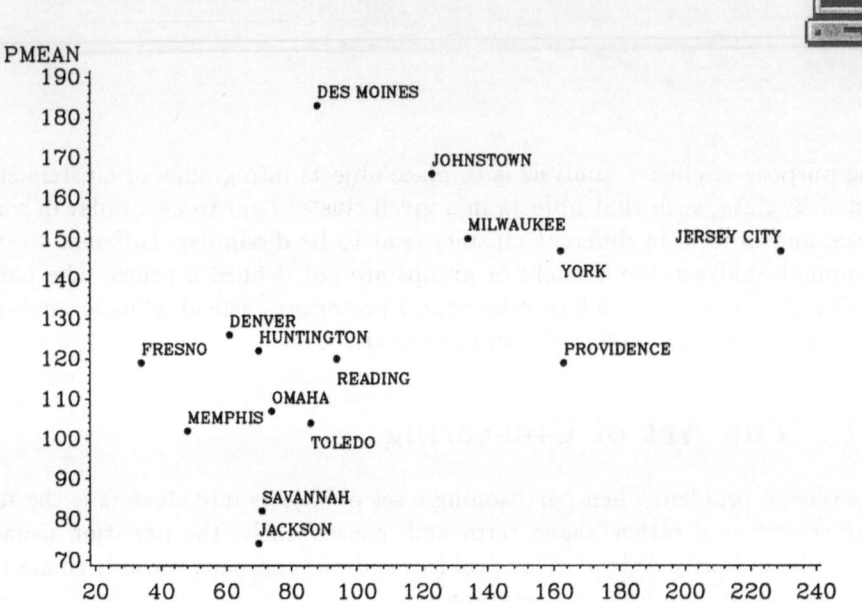

**Figure 7.1.1.** Scatterplot of 15 cities of Air Data; variables *smean* and *pmean*.

```
***     Program 7_1_1     ***;
TITLE1 'Scatterplot of 15 Cities';
TITLE2 'Air Pollution II Data';
LIBNAME datalib 'c:\data';

DATA anno;
    SET datalib.polution(OBS=15);
    FUNCTION='LABEL'; TEXT=city; XSYS='2'; YSYS='2';
    HSYS='4'; POSITION='3'; SIZE=1.5;
    X=smean; Y=pmean;
    IF city IN ('TOLEDO' 'YORK' 'READING') THEN POSITION='9';
    IF city IN ('JERSEY CITY') THEN POSITION='1';

SYMBOL1 V=DOT C=GREEN H=0.8;
PROC GPLOT DATA=datalib.polution(OBS=15);
    PLOT pmean*smean / ANNOTATE=anno;
RUN; QUIT;
```

This program gives a further illustration of the use of ANNOTATE data sets, which were already introduced in Program 3_2_1.

Different to Program 3_2_1 we do not want to draw lines here but text at prescribed locations in the graphics output.

FUNCTION='LABEL' instructs the AN-NOTATE facility to draw text. The text string is specified by the variable TEXT. Since the names of the cities are contained in the variable 'city' of the data set, the statement 'TEXT=city' writes these names also to the variable TEXT.

The statements XSYS='2'and YSYS='2' specify that the coordinate system of the ANNOTATE graphics coincides with that of the GPLOT procedure. The option POSI-TION='3' specifies the test string's orientation around (X,Y). The value 3 specifies that the text is to begin one cell above and left aligned at point (X,Y). The IF condition requires the names ('TOLEDO' etc.) to be written in capital letters, since this is the way they are automatically stored by the system.

To study the locations in $I\!\!R^{12}$ of the cities with respect to all 12 variables, we can use the matrix of their Euclidean distances. Those cities, whose corresponding distances in $I\!\!R^{12}$ are small, have a similar profile with respect to the 12 variables. This similarity will be quantified by a *similarity measure*.

Next we will aggregate similar cities to a cluster and we will determine a similarity measure *between the clusters*. Similar groups will then be aggregated to one cluster etc. If this agglomeration of clusters is stopped according to some stopping rule, the process of clustering the objects is terminated with the final configuration. The choice of the similarity measure as well as of the stopping rule can, however, heavily influence the final clusters.

## 7.2   Distance and Similarity Measures

In this section we introduce popular similarity and distance measures. As the distance or dissimilarity of two objects decreases, the corresponding similarity measure increases. The most common measure of similarity is the correlation coefficient of two objects based on the corresponding measurements, cf Section 3.1. The Euclidean distance between two objects is a common distance measure. Both similarity and distance or dissimilarity measures are called *proximity measures*. In the sequel we identify $n$ different objects with the $n$ numbers in the set $I = \{1, \ldots, n\}$.

### Distance Measures

**7.2.1 Definition.** A function $d : I \times I \to [0, \infty)$ is a *distance measure* if it satisfies

(i)  $d(i, i) = 0$

(ii) $d(i, j) = d(j, i)$

for $i, j \in I$. The symmetric $n \times n$–matrix $(d(i, j))$ is the *distance matrix*. A distance measure $d$ is called a *metric distance measure* if it satisfies the triangular inequality

(iii) $d(i, j) \leq d(i, k) + d(k, j),$        $i, j, k \in I.$

If the $n$ objects are measurements with values in $\mathbb{R}^p$, i.e., $\boldsymbol{x}_i \in \mathbb{R}^p$, $i = 1, \ldots, n$, one usually defines the distance of two objects by the distance of the vectors of measurements

$$d(i, j) := d_M(\boldsymbol{x}_i, \boldsymbol{x}_j),$$

where $d_M$ is a *metric* on $\mathbb{R}^p$. A function $d_M : \mathbb{R}^p \times \mathbb{R}^p \to \mathbb{R}$ is called a metric if it satisfies for $\boldsymbol{x}, \boldsymbol{y}, \boldsymbol{z} \in \mathbb{R}^p$

(i)  $d_M(\boldsymbol{x}, \boldsymbol{y}) = 0 \Leftrightarrow \boldsymbol{x} = \boldsymbol{y},$

(ii)  $d_M(\boldsymbol{x}, \boldsymbol{y}) \leq d_M(\boldsymbol{x}, \boldsymbol{z}) + d_M(\boldsymbol{y}, \boldsymbol{z}).$                    (triangular inequality).

Note that (i) and (ii) imply the nonnegativity and the symmetry of the function $d_M$, i.e., $d_M(\boldsymbol{x}, \boldsymbol{y}) \geq 0$ and $d_M(\boldsymbol{x}, \boldsymbol{y}) = d_M(\boldsymbol{y}, \boldsymbol{x})$, $\boldsymbol{x}, \boldsymbol{y} \in \mathbb{R}^p$ (Exercise 3).

A popular example of a distance matrix is a table of driving mileages between cities.

| El Paso | Flag staff | Los An ge les | Las Ve gas | Phoe nix | Re no | Salt Lake City | San Die go | San Fran cis co | Tuc son | |
|---|---|---|---|---|---|---|---|---|---|---|
| 0 | 544 | 870 | 686 | 401 | 1132 | 919 | 724 | 1270 | 310 | El Paso |
| | 0 | 478 | 246 | 137 | 692 | 509 | 490 | 773 | 255 | Flagstaff |
| | | 0 | 276 | 376 | 470 | 707 | 127 | 387 | 524 | Los Angeles |
| | | | 0 | 285 | 446 | 431 | 331 | 571 | 403 | Las Vegas |
| | | | | 0 | 731 | 646 | 353 | 763 | 118 | Phoenix |
| | | | | | 0 | 520 | 571 | 225 | 849 | Reno |
| | | | | | | 0 | 762 | 745 | 764 | Salt Lake City |
| | | | | | | | 0 | 514 | 414 | San Diego |
| | | | | | | | | 0 | 911 | San Francisco |
| | | | | | | | | | 0 | Tucson |

**Figure 7.2.1.** Table of driving mileages in statute miles between 10 U.S. cities.

**7.2.2 Examples.** Popular metric distance measures are provided by the *Minkowski metrics* or $L_q$*-norms* on $I\!R^p$:

$$d_q(\boldsymbol{x}, \boldsymbol{y}) := \Big( \sum_{i=1}^{p} |x_i - y_i|^q \Big)^{1/q},$$

where $\boldsymbol{x} = (x_1, \ldots, x_p)^T$, $\boldsymbol{y} = (y_1, \ldots, y_p)^T \in I\!R^p$ and $q \in [1, \infty]$. With $q = 2$ we obtain the *Euclidean* distance

$$d_2(\boldsymbol{x}, \boldsymbol{y}) = ||\boldsymbol{x} - \boldsymbol{y}|| = \Big( \sum_{i=1}^{p} |x_i - y_i|^2 \Big)^{1/2},$$

whereas $d_1$ is the *Manhattan* or *city-block metric*

$$d_1(\boldsymbol{x}, \boldsymbol{y}) = \sum_{i=1}^{p} |x_i - y_i|.$$

The case $q = \infty$ yields the maximum distance of the coordinates

$$d_\infty(\boldsymbol{x}, \boldsymbol{y}) = \max_{1 \le i \le p} |x_i - y_i|$$

(see Exercise 9). The distance measure $d_q$ satisfies the triangular inequality

$$d_q(\boldsymbol{x}, \boldsymbol{y}) \le d_q(\boldsymbol{x}, \boldsymbol{z}) + d_q(\boldsymbol{y}, \boldsymbol{z}), \qquad \boldsymbol{x}, \boldsymbol{y}, \boldsymbol{z} \in I\!R^p$$

which is known as Minkowski's inequality. It is, therefore, a metric on $I\!R^p$, see e.g. Section 3.1 in Berberian (1999).

It seems to be desirable that the distance of two observations in $I\!R^p$ does not depend on the particular measuring unit. An example is the population density *pm2* in Example 7.1.1. The distances between the data points in $I\!R^{12}$ representing the cities should not be affected if we used square kilometers in the definition of *pm2* instead of square miles. This condition is satisfied by scale invariant measures on $I\!R^p$.

**7.2.3 Definition.** A distance measure $d$ on a set of points $\{\boldsymbol{x}_1, \ldots, \boldsymbol{x}_n\}$ in $I\!R^p$ is called *scale invariant* if it satisfies

$$d(\boldsymbol{x}_i, \boldsymbol{x}_j) = d(\boldsymbol{C}\boldsymbol{x}_i, \boldsymbol{C}\boldsymbol{x}_j), \qquad i, j = 1, \ldots, n$$

for any $p \times p$ diagonal matrix

$$\boldsymbol{C} = \begin{pmatrix} c_1 & & 0 \\ & \ddots & \\ 0 & & c_p \end{pmatrix}, \qquad c_k > 0, \quad k = 1, \ldots, p.$$

A distance measure $d$ is called *translation invariant* if for any $b \in \mathbb{R}^p$

$$d(x_i + b, x_j + b) = d(x_i, x_j), \qquad i, j = 1, \ldots, n.$$

The Minkowski metric $d_q$ is obviously translation invariant but not scale invariant. The Euclidean metric $d_2$ is invariant under orthogonal transformations: Let $A$ be a $p \times p$–matrix with $A^T A = I_p$, then we have

$$
\begin{aligned}
d_2^2(Ax, Ay) &= \|Ax - Ay\|^2 = \|A(x - y)\|^2 = (x - y)^T A^T A(x - y) \\
&= (x - y)^T (x - y) = \|x - y\|^2 = d_2^2(x, y).
\end{aligned}
$$

This property of $d_2$ means that the Euclidean distances of points in $\mathbb{R}^p$ are not affected by a rotation or a reflection of the points.

The disadvantage of not being scale invariant can be overcome for the Euclidean distance by considering the *standardized Euclidean distance* $d_{2,st}$ instead. Let $x_1, \ldots, x_n$ be arbitrary points in $\mathbb{R}^p$. Then $d_{2,st}$ is defined on the set $\{x_1, \ldots, x_n\}$ by

$$d_{2,st}^2(x_i, x_j) := \Delta_D^2(x_i, x_j) = (x_i - x_j)^T D^{-1}(x_i - x_j) = \sum_{k=1}^p \frac{1}{s_k^2}(x_{ik} - x_{jk})^2,$$

where $s_k^2$ is the empirical variance of the $k$–th coordinates of $x_1, \ldots, x_n \in \mathbb{R}^p$, which we assume to be positive, and

$$D = \begin{pmatrix} s_1^2 & & 0 \\ & \ddots & \\ 0 & & s_p^2 \end{pmatrix}.$$

The Mahalanobis distance has further invariance properties, see also Section 6.4.

**7.2.4 Definition.** Let $x_1, \ldots, x_n$ be points in $\mathbb{R}^p$ and denote the mean vector by $\bar{x}_n := n^{-1} \sum_{i=1}^n x_i$. If the empirical covariance matrix

$$S = \frac{1}{n-1} \sum_{i=1}^n (x_i - \bar{x}_n)(x_i - \bar{x}_n)^T$$

is invertible, then

$$\Delta_S(x_i, x_j) := \left( (x_i - x_j)^T S^{-1}(x_i - x_j) \right)^{1/2}, \qquad i, j = 1, \ldots, n$$

is called the *Mahalanobis distance* on $\{x_1, \ldots, x_n\}$.

**7.2.5 Lemma.** *The Mahalanobis distance pertaining to $x_1, \ldots, x_n \in \mathbb{R}^p$ is invariant under affine transformations: Put $y_i := Ax_i + b$, $i = 1, \ldots, n$. Then we have*

$$\Delta_{S_y}(y_i, y_j) = \Delta_S(x_i, x_j), \qquad i, j = 1, \ldots, n,$$

*where $A$ is any invertible $p \times p$-matrix, $b$ is any vector in $\mathbb{R}^p$ and $S_y$ is the empirical covariance matrix of $y_1, \ldots, y_n$. The Mahalanobis distance is, therefore, in particular scale and translation invariant.*

**Proof:** From Theorem 3.3.7 we obtain the representation $S_y = ASA^T$ and, thus, for any $i, j = 1, \ldots, n$

$$
\begin{aligned}
\Delta^2_{S_y}(y_i, y_j) &= (y_i - y_j)^T S_y^{-1}(y_i - y_j) \\
&= (x_i - x_j)^T A^T (ASA^T)^{-1} A(x_i - x_j) \\
&= (x_i - x_j)^T A^T (A^T)^{-1} S^{-1} A^{-1} A(x_i - x_j) \\
&= (x_i - x_j)^T S^{-1}(x_i - x_j) = \Delta^2_S(x_i, x_j). \qquad \square
\end{aligned}
$$

The following table is an excerpt from the Mahalanobis distance matrix of 15 cities from Example 7.1.1. It is based on six variables *smean, pmean, pm2, perwh, nonpoor* and *ge65*.

| CITY | COL1 | COL2 | COL3 | COL4 | COL5 | COL6 | COL7 |
|------|------|------|------|------|------|------|------|
| PROVIDENCE | 0.00 | . | . | . | . | . | . |
| JACKSON | 3.25 | 0.00 | . | . | . | . | . |
| JOHNSTOWN | 3.33 | 3.77 | 0.00 | . | . | . | . |
| JERSEY CITY | 4.39 | 4.55 | 4.67 | 0.00 | . | . | . |
| HUNTINGTON | 3.37 | 3.51 | 2.43 | 4.45 | 0.00 | . | . |
| DES MOINES | 4.40 | 4.32 | 3.43 | 4.83 | 4.32 | 0.00 | . |
| DENVER | 3.72 | 3.51 | 4.07 | 4.40 | 3.02 | 3.47 | 0 |

**Figure 7.2.2.** Excerpt from the Mahalanobis distance matrix of the cities in Example 7.1.1, based on the variables *smean, pmean, pm2, perwh, nonpoor, ge65*.

```
***    Program 7_2_2    ***;
TITLE1 'Mahalanobis Distance Matrix';
TITLE2 'Air Pollution II Data';
LIBNAME datalib 'c:\data';

DATA sub15;
    SET datalib.polution(OBS=15);
PROC CORR DATA=sub15 OUTP=covmat COV NOPRINT;
    VAR smean pmean pm2 perwh nonpoor ge65;
PROC IML;
*** import data from data sets sub15 and covmat;
    USE sub15;
    READ ALL VAR {smean pmean pm2 perwh nonpoor ge65} INTO x;
    USE covmat WHERE(_TYPE_='COV');
    READ ALL VAR {smean pmean pm2 perwh nonpoor ge65} INTO s;
*** computation of distance matrix 'dist';
*** as lower triangular matrix;
    dist=J(15,15,.);
    s_inv=INV(s);                        * s_inv=inverse matrix of s;
    DO k=1 TO 15;
        DO l=1 TO k;
        d1=(x[k, ]-x[l, ]);
        dist[k,l]=SQRT(d1 * s_inv * d1'); * see Definition 7.2.4;
    END;
        END;
    CREATE mahal FROM dist;              * export matrix to data set;
    APPEND FROM dist;
QUIT;                                    * quit IML;
*** create data set of type 'DISTANCE';
DATA datalib.dist1(TYPE=DISTANCE);
    MERGE sub15 mahal;
PROC PRINT DATA=datalib.dist1(OBS=7) ROUND;
    VAR city COL1-COL7;
RUN; QUIT;
```

In this program the Mahalanobis distances are computed by means of the SAS/IML–module, see Program 6_4_3. If the data set is 'TYPE=DISTANCE', the data are interpreted as a distance matrix. That is the input we need for PROC CLUSTER. If the data set is not 'TYPE=DISTANCE' in the second DATA step, the data are interpreted as coordinates in a Euclidean space, and Euclidean distances are computed (see Figures 7.4.3 and 7.4.4). 'VAR city COL1-COL7' restricts the printed output to the first 7 cities. The option COV in the CORR procedure computes the covariance matrix.

The Mahalanobis distance pertaining to $x_1, \ldots, x_n \in \mathbb{R}^p$ coincides with the Euclidean distance of empirically uncorrelated points $y_1, \ldots, y_n \in \mathbb{R}^p$. Denote the symmetric root of $S^{-1/2}$ by $S^{-1}$ as in Section 6.4 or in Exercise 18 of Chapter 6. The covariance matrix of $y_i := S^{-1/2} x_i \in \mathbb{R}^p$, $i = 1, \ldots, n$, is the unity matrix $I_p$ and the Mahalanobis distance of $x_1, \ldots, x_n$ satisfies

$$
\begin{aligned}
\Delta_S^2(x_i, x_j) &= (x_i - x_j)^T S^{-1}(x_i - x_j) \\
&= (y_i - y_j)^T (y_i - y_j) = d_2^2(y_i, y_j), \qquad i, j = 1, \ldots, n.
\end{aligned}
$$

Different from the standardized Euclidean distance $d_{2,st}$, the Mahalanobis distance removes the correlations of the coordinates of the $x_i$.

## Similarity Measures

Points in $\mathbb{R}^p$ that are close with respect to some distance measure will be called *similar*. The following definition provides a precise description of *similarity* in mathematical terms. Again we identify $n$ different objects by the set of integers $I = \{1, \ldots, n\}$.

**7.2.6 Definition.** A function $s : I \times I \to [0, 1]$ is called a *similarity measure* on $I$ if it satisfies

(i) $s(i, i) = 1$,

(ii) $s(i, j) = s(j, i)$

for $i, j \in I$. The number $s(i, j)$ is called *similarity coefficient* of $i$ and $j$. The symmetric $n \times n$–matrix $(s(i, j))$ is a *similarity matrix*.

The idea that two objects are similar iff their distance is small, suggests the following transformation of an arbitrary similarity measure on $I$ into a distance measure: Let $s$ be a similarity measure and let $f : [0, 1] \to [0, \infty)$ be a strictly monotone decreasing function with $f(1) = 0$. An example is the function $f(x) := \sqrt{2(1 - x)}$, $x \in [0, 1]$. A distance measure is then defined by

$$
d(i, j) := f(s(i, j)).
$$

Let, on the other hand, $d$ be a distance measure on $I$ with the property $d(i, j) = 0 \Leftrightarrow i = j$. Choose a strictly monotone decreasing function $g : [0, 1] \to [0, 1]$ with $g(0) = 1$ such as $g(x) = 1 - x$, $x \in [0, 1]$. Put $a := \max\{d(i, j) : i, j \in I\}$. Then

$$
s(i, j) := g(d(i, j)/a)
$$

defines a similarity measure on $I$.

## Binary Variables

Similarity coefficients are of particular importance for categorical data. Suppose we have $n$ data, each consisting of $p$ binary variables. We can code these observations by $x_i \in \{0,1\}^p, i = 1, \ldots, n$, which is the dummy coding from Section 4.2. Let now $f : \{0,1\}^p \times \{0,1\}^p \to [0,1]$ be a function with the properties

(i) $f(y, z) = 1 \Leftrightarrow y = z$

(ii) $f(y, z) = f(z, y)$

for $y, z \in \{0,1\}^p$, then

$$s(i, j) := f(x_i, x_j), \qquad i, j = 1, \ldots, n,$$

defines a similarity measure on $I = \{1, \ldots, n\}$.

The most popular similarity coefficients $s(i, j)$ for binary observations $x_i, x_j \in \{0,1\}^p$ can be deduced from the following $2 \times 2$ table, where $\mathbf{1} = (1, 1, \ldots, 1)^T \in \{0,1\}^p$.

|         |   | $x_j$ | | |
|---------|---|-------|---|---|
|         |   | 1 | 0 | |
| $x_i$ | 1 | $\alpha_{ij} := x_i^T x_j$ | $\beta_{ij} := x_i^T(1 - x_j)$ | $\alpha_{ij} + \beta_{ij}$ |
|         | 0 | $\gamma_{ij} := (1 - x_i)^T x_j$ | $\delta_{ij} := (1 - x_i)^T(1 - x_j)$ | $\gamma_{ij} + \delta_{ij}$ |
|         |   | $\alpha_{ij} + \gamma_{ij}$ | $\beta_{ij} + \delta_{ij}$ | $\alpha_{ij} + \beta_{ij} + \gamma_{ij} + \delta_{ij} = p$ |

The number of coordinates, where $x_i$ and $x_j$ both have the entry 1, is given by $\alpha_{ij}$. $\beta_{ij}$ counts the number of coordinates, where $x_i$ has the entry 1 and $x_j$ has a 0, and vice versa for $\gamma_{ij}$. Finally, $\delta_{ij}$ is the number of coordinates, where both $x_i$ and $x_j$ have a 0. The total sum $\alpha_{ij} + \beta_{ij} + \gamma_{ij} + \delta_{ij}$ must, consequently, coincide with the number $p$ of coordinates.

The most popular similarity coefficients are the *Jaccard Coefficient*

$$s_J(i, j) := \begin{cases} \dfrac{\alpha_{ij}}{\alpha_{ij} + \beta_{ij} + \gamma_{ij}}, & \text{if } \delta_{ij} < p, \\ 1, & \text{if } \delta_{ij} = p, \end{cases}$$

the *Czekanowski Coefficient*

$$s_C(i, j) := \begin{cases} \dfrac{2\alpha_{ij}}{2\alpha_{ij} + \beta_{ij} + \gamma_{ij}}, & \text{if } \delta_{ij} < p, \\ 1, & \text{if } \delta_{ij} = p, \end{cases}$$

and the $M$ or *simple matching coefficient*

$$s_M(i,j) := \frac{\alpha_{ij} + \delta_{ij}}{p}.$$

The $M$ coefficient is the relative frequency of matching coordinates. Note that we have $\beta_{ij} = \gamma_{ij} = 0$ if $i = j$ and, thus, the above coefficients are actually similarity measures.

The most popular similarity measure in case of categorical data with $K \geq 3$ different levels is the generalized matching or $M$ coefficient

$$s(i,j) = \frac{u_{ij}}{p},$$

where $u_{ij}$ is the number of matching coordinates of $x_i, x_j \in \{0, 1, \ldots, K-1\}^p$. For a detailed discussion of various similarity coefficients we refer to Jobson (1992), Section 10.1.1.

**7.2.7 Example** (Grounds for Divorce Data). The following tables lists the available grounds for divorce in the year 1982 from 20 selected U.S. states. For each of the 9 categories a state is coded 1 if that ground is available and 0 otherwise. The abbreviations mean *marriage breakdown, cruelty, desertion, nonsupport, alcohol/drug addiction, felony, impotency, insanity, (a period of) separation*. This data set is analyzed in Chapter 10 of Jobson (1992). Figure 7.2.4 shows an excerpt of the similarity matrix of the 20 states, based on the Jaccard coefficient. From this table we conclude that New Hampshire and Louisiana are quite similar, whereas Nebraska and Louisiana are quite dissimilar.

|   |   |   |   | N |   |   |   | I |   | S |
|---|---|---|---|---|---|---|---|---|---|---|
|   |   |   |   | 0 | A |   |   | M |   | E |
|   |   |   | D | S | L | F | P | I | P |
| S | B | C | E | U | C | E | 0 | N | A |
| T | R | R | S | P | 0 | L | T | S | R |
| A | E | U | E | P | H | 0 | E | A | A |
| T | A | E | R | 0 | 0 | N | N | N | T |
| E | K | L | T | R | L | Y | C | E | E |

| STATE | BREAK | CRUEL | DESERT | NOSUPPOR | ALCOHOL | FELONY | IMPOTENC | INSANE | SEPARATE |
|---|---|---|---|---|---|---|---|---|---|
| FLORIDA | 1 | 0 | 0 | 0 | 0 | 0 | 0 | 1 | 0 |
| LOUISIANA | 0 | 1 | 1 | 1 | 1 | 1 | 0 | 0 | 1 |
| MAINE | 1 | 1 | 1 | 1 | 1 | 0 | 1 | 1 | 0 |
| MARYLAND | 0 | 1 | 1 | 0 | 0 | 1 | 1 | 1 | 1 |
| MASSACHUSETTS | 1 | 1 | 1 | 1 | 1 | 1 | 1 | 0 | 1 |
| MONTANA | 1 | 0 | 0 | 0 | 0 | 0 | 0 | 0 | 0 |
| NEBRASKA | 1 | 0 | 0 | 0 | 0 | 0 | 0 | 0 | 0 |
| NEW HAMPSHIRE | 1 | 1 | 1 | 1 | 1 | 1 | 1 | 0 | 1 |

| NEW YORK      | 0 | 1 | 1 | 0 | 0 | 1 | 0 | 0 | 1 |
|---------------|---|---|---|---|---|---|---|---|---|
| NORTH DAKOTA  | 1 | 1 | 1 | 1 | 1 | 1 | 1 | 1 | 0 |
| OKLAHOMA      | 1 | 1 | 1 | 1 | 1 | 1 | 1 | 1 | 0 |
| OREGON        | 1 | 0 | 0 | 0 | 0 | 0 | 0 | 0 | 0 |
| RHODE ISLAND  | 1 | 1 | 1 | 1 | 1 | 1 | 1 | 0 | 1 |
| SOUTH CAROLINA| 0 | 1 | 1 | 0 | 1 | 0 | 0 | 0 | 1 |
| SOUTH DAKOTA  | 0 | 1 | 1 | 1 | 1 | 1 | 0 | 0 | 0 |
| TEXAS         | 1 | 1 | 1 | 0 | 0 | 1 | 0 | 1 | 1 |
| VERMONT       | 0 | 1 | 1 | 1 | 0 | 1 | 0 | 1 | 1 |
| VIRGINIA      | 0 | 1 | 1 | 0 | 0 | 1 | 0 | 0 | 1 |
| WASHINGTON    | 1 | 0 | 0 | 0 | 0 | 0 | 0 | 0 | 0 |
| WEST VIRGINIA | 1 | 1 | 1 | 0 | 1 | 1 | 0 | 1 | 1 |

**Figure 7.2.3.** Table of available grounds of divorce in 20 U.S. states.

| STATE          | COL1 | COL2 | COL3 | COL4 | COL5 | COL6 | COL7 | COL8 |
|----------------|------|------|------|------|------|------|------|------|
| FLORIDA        | 1.00 | .    | .    | .    | .    | .    | .    | .    |
| LOUISIANA      | 0.00 | 1.00 | .    | .    | .    | .    | .    | .    |
| MAINE          | 0.29 | 0.44 | 1.00 | .    | .    | .    | .    | .    |
| MARYLAND       | 0.14 | 0.50 | 0.44 | 1.00 | .    | .    | .    | .    |
| MASSACHUSETTS  | 0.11 | 0.75 | 0.67 | 0.56 | 1.00 | .    | .    | .    |
| MONTANA        | 0.50 | 0.00 | 0.14 | 0.00 | 0.13 | 1.00 | .    | .    |
| NEBRASKA       | 0.50 | 0.00 | 0.14 | 0.00 | 0.13 | 1.00 | 1.00 | .    |
| NEW HAMPSHIRE  | 0.11 | 0.75 | 0.67 | 0.56 | 1.00 | 0.13 | 0.13 | 1.0  |

**Figure 7.2.4.** Excerpt of the similarity matrix of the U.S. states in Figure 7.2.3; Jaccard coefficient.

```
***    Program 7_2_4    ***;
TITLE1 'Similarity Matrix of Binary Variables';
TITLE2 'Grounds for Divorce Data';
LIBNAME datalib 'c:\data';

PROC IML;
*** import data ;
   USE datalib.grounds WHERE(sub='J');
   READ ALL VAR _NUM_ INTO x;
   p=NCOL(x);                      * p = number of variables;
   n=NROW(x);                      * n = number of observations;
                         ↓
```

```
                                   ↑
*** computation of distance matrix;
*** as lower triangular matrix;
*** selection of coefficient by removing asterisk '*';
    dist=J(n,n,.);
    DO k=1 TO n;
       DO l=1 TO k;
          a=x[k,] * x[l,]';
          b=x[k,] * (1-x[l,]');
          c=x[l,] * (1-x[k,]');
          d=(1-x[k,]) * (1-x[l,]');
          IF d=p THEN dist[k,l] =1; ELSE
          dist[k,l]=a/(a+b+c);        * Jaccard;
* dist[k,l]=2*a/(2*a+b+c);           * Czekanowski;
* dist[k,l]=(a+d)/p;                 * Matching;
       END; END;
    CREATE distb FROM dist;           * export data;
    APPEND FROM dist;
QUIT;                                 * quit IML;

DATA distb(TYPE=DISTANCE);
    MERGE datalib.grounds(WHERE=(sub='J')) distb;

PROC PRINT DATA=distb(OBS=8) ROUND;
    VAR COL1-COL8;
    ID state;
RUN; QUIT;
```

Similarity matrices can quite easily be derived from binary variables using the IML procedure. The above program computes all three similarity coefficients discussed in the text. To select a particular coefficient, remove the asterisk '*' at the beginning of the corresponding line (here 'Jaccard').

The data set 'grounds' contains the grounds for divorce from 51 states. Those states, which are used here and in Figure 7.2.3, have to this end the value 'J' of the variable 'sub'. This variable was added just for this particular reason.

# 7.3   Multidimensional Scaling

*Multidimensional scaling (MDS)* is a collection of techniques that aim at the graphical visualization of an arbitrary $n \times n$–distance matrix $(d(i,j))$ of $n$ objects. *Metric MDS* seeks for a low dimension $p$ and $n$ points $x_1, \ldots, x_n \in \mathbb{R}^p$ representing the objects, such that the distance $d(i,j)$ between the $i$–th and the $j$–th object coincide with the Euclidean distance $\|x_i - x_j\|$ of the representatives $x_i, x_j \in \mathbb{R}^p$:

$$d(i, j) = \|x_i - x_j\|, \qquad i, j = 1, \ldots, n.$$

A scatterplot of the points $x_1, \ldots, x_n$ visualizes the distance matrix if $p$ is 2 or 3. While metric *MDS* attempts to preserve the distances between objects, the spatial representation in *nonmetric MDS* preserves only the rank order among the distances.

Suppose, for example, that different types of cars are compared with respect to various variables such as reliability, prestige, comfort etc. and that the results are written to a distance matrix. An *MDS* plot of this distance matrix might then help the marketing researchers of a car manufacturer to detect new opportunities for selling by finding regions in $\mathbb{R}^p$ with only sparse representatives. The following figure is an *MDS* plot of the table of distances in Figure 7.2.1. This is a map from the car driver's world. See also the remarks at the end of this section.

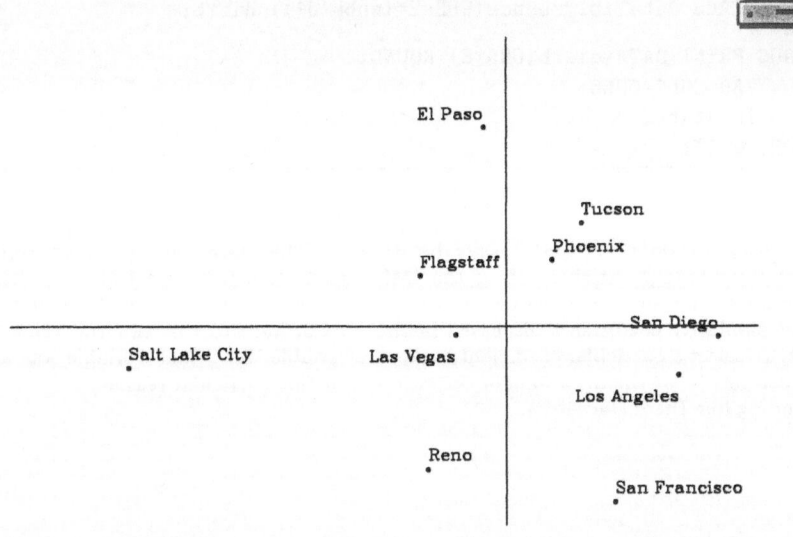

**Figure 7.3.1.** *MDS* plot of the distance matrix in Figure 7.2.1.

## The Doubly–Centered Matrix $A$

The following symmetric $n \times n$–matrix $A = (a_{ij})$ plays a crucial role in the metric *MDS* problem. Put for $i, j = 1, \ldots, n$

$$a_{ij} := -\frac{1}{2}\left(d^2(i, j) - d^2(i, .) - d^2(., j) + d^2(., .)\right),$$

where

$$d^2(i, .) := \frac{1}{n}\sum_{k=1}^{n} d^2(i, k),$$

$$d^2(., j) := \frac{1}{n}\sum_{k=1}^{n} d^2(k, j)$$

and

$$d^2(., .) := \frac{1}{n^2}\sum_{k=1}^{n}\sum_{l=1}^{n} d^2(k, l).$$

The columns and the rows of the matrix $A$ sum up to 0:

$$\sum_{k=1}^{n} a_{ik} = 0 = \sum_{k=1}^{n} a_{kj}, \qquad i, j = 1, \ldots, n,$$

i.e., the rank of $A$ cannot be larger than $n - 1$. The matrix $A$ is generated from the matrix $(d^2(i, j))$ by double centering. Denote by $E_n$ that $n \times n$–matrix, whose entries are all equal to 1. Then we have the representation

$$A = -\frac{1}{2}\left(I_n - \frac{1}{n}E_n\right)(d^2(i, j))\left(I_n - \frac{1}{n}E_n\right). \tag{7.1}$$

We suppose in the following that not all entries of $A$ are zero, which is equivalent to the assumption that not all $d(i, j)$ are 0.

## Diagonalizing $A$

Since $A$ is symmetric, all its eigenvalues are real numbers $\lambda_1 \geq \cdots \geq \lambda_n$, see, for example, Section VIII.3 in Lang (1987). If $A$, in addition, is positive semidefinite with rank $(A) = p \in \{1, \ldots, n-1\}$, then $p$ eigenvalues are positive, $\lambda_1 \geq \cdots \geq \lambda_p > 0$, whereas the others vanish, $\lambda_{p+1} = \cdots = \lambda_n = 0$ (Exercise 13). Choose orthonormal eigenvectors $r_1, \ldots, r_n \in \mathbb{R}^n$, corresponding to $\lambda_1, \ldots, \lambda_n$, i.e.,

$$r_j^T r_k = \begin{cases} 1 & \text{if } j = k \\ 0 & \text{if } j \neq k, \end{cases} \tag{7.2}$$

and

$$Ar_i = \lambda_i r_i, \quad i = 1, \ldots, n. \tag{7.3}$$

Denote by $R$ that $n \times n$-matrix, whose columns are the $n$ vectors $r_1, \ldots, r_n$. Then we can write (7.3) in closed form as

$$AR = R\Lambda, \tag{7.4}$$

where $\Lambda$ is the $n \times n$-diagonal matrix of the eigenvalues $\lambda, \ldots, \lambda_n$:

$$\Lambda = \begin{pmatrix} \lambda_1 & & 0 \\ & \ddots & \\ 0 & & \lambda_n \end{pmatrix}.$$

$R$ is by (7.2) an orthogonal matrix, i.e., it satisfies $R^T R = I_n$ and, thus, $R^T = R^{-1}$ which yields $RR^T = I_n$ as well. Multiplying equation (7.4) with $R^T$ from the left, we *diagonalize* the matrix $A$:

$$R^T A R = \Lambda \quad \text{viz.} \quad A = R\Lambda R^T = (R\Lambda^{1/2})(R\Lambda^{1/2})^T,$$

where $\Lambda^{1/2}$ is the diagonal matrix with the entries $\lambda_{ii}^{1/2}$, $i = 1, \ldots, n$, see also Exercise 18 in Chapter 6.

## The Solution of the Metric *MDS* Problem

If the eigenvalues $\lambda_{p+1}, \ldots, \lambda_n$ of $A$ are equal to 0, then we obtain

$$A = (R\Lambda^{1/2})(R\Lambda^{1/2})^T$$

$$= (\sqrt{\lambda_1} r_1, \ldots, \sqrt{\lambda_n} r_n) \begin{pmatrix} \sqrt{\lambda_1} r_1^T \\ \vdots \\ \sqrt{\lambda_n} r_n^T \end{pmatrix}$$

$$= (\sqrt{\lambda_1} r_1, \ldots, \sqrt{\lambda_p} r_p, 0, \ldots, 0) \begin{pmatrix} \sqrt{\lambda_1} r_1^T \\ \vdots \\ \sqrt{\lambda_p} r_p^T \\ 0^T \\ \vdots \\ 0^T \end{pmatrix}$$

$$= (\sqrt{\lambda_1} r_1, \ldots, \sqrt{\lambda_p} r_p) \begin{pmatrix} \sqrt{\lambda_1} r_1^T \\ \vdots \\ \sqrt{\lambda_p} r_p^T \end{pmatrix} = XX^T, \tag{7.5}$$

where the $j$-th column of the $n \times p$-matrix

$$X = (x_{ij}) = (\sqrt{\lambda_1}r_1, \ldots, \sqrt{\lambda_p}r_p)$$

is the vector $\sqrt{\lambda_j}\, r_j, j = 1, \ldots, p$. Now we take the number $p$ of the positive eigenvalues $\lambda_i$ of $A$ as the dimension of the vector space for the solution of the metric $MDS$ problem and we take the $i$-th row of the matrix $X$ as the vector $x_i \in I\!\!R^p$, i.e., we put

$$x_i := (x_{i1}, \ldots, x_{ip})^T, \qquad i = 1, \ldots, n.$$

This actually solves the metric $MDS$ problem. The following result is the fundamental theorem of the metric $MDS$.

**7.3.1 Theorem.** *Let $D = (d(i, j))$ be an arbitrary $n \times n$-distance matrix. The metric MDS problem has a solution iff the matrix $A$ is positive semidefinite. If the matrix $A$ is positive semidefinite with rank $(A) = p \in \{1, \ldots, n-1\}$, then we obtain with the above notations*

$$\|x_i - x_j\| = d(i, j), \qquad i, j = 1, \ldots, n.$$

**Proof:** Suppose first that the matrix $A$ is positive semidefinite with rank $p \in \{1, \ldots, n-1\}$. From (7.5) we obtain for $i, j = 1, \ldots, n$

$$\begin{aligned}
\|x_i - x_j\|^2 &= (x_i - x_j)^T(x_i - x_j) \\
&= x_i^T x_i - x_j^T x_i - x_i^T x_j + x_j^T x_j \\
&= a_{ii} - a_{ji} - a_{ij} + a_{jj} = d^2(i, j).
\end{aligned}$$

Suppose next that the metric $MDS$ problem can be solved for $D = (d(i, j))$ in $I\!\!R^p$, i.e., suppose that there exist $x_1, \ldots, x_n \in I\!\!R^p$ such that

$$(d^2(i, j)) = ((x_i - x_j)^T(x_i - x_j)).$$

Choose an arbitrary $z \in I\!\!R^n$ and put $y := (I_n - n^{-1}E_n)z = (y_1, \ldots, y_n)^T$. Then we obtain

$$\sum_{j=1}^n y_j = y^T(1, \ldots, 1)^T = z^T(I_n - \frac{1}{n}E_n)(1, \ldots, 1)^T = z^T 0 = 0.$$

Now we can conclude that the matrix $A$ is positive semidefinite:

$$\begin{aligned}
z^T A z &= -\frac{1}{2}y^T(d^2(i, j))y \\
&= -\frac{1}{2}\sum_{i,j=1}^n y_i(x_i - x_j)^T(x_i - x_j)y_j
\end{aligned}$$

$$= - \sum_{i,j=1}^{n} y_i x_i^T x_i y_j + \sum_{i,j=1}^{n} (y_i x_i)^T (y_j x_j)$$

$$= - \left( \sum_{j=1}^{n} y_j \right) \left( \sum_{i=1}^{n} y_i x_i^T x_i \right) + \left( \sum_{i=1}^{n} y_i x_i \right)^T \left( \sum_{j=1}^{n} y_j x_j \right)$$

$$= \left( \sum_{i=1}^{n} y_j x_i \right)^T \left( \sum_{i=1}^{n} y_i x_i \right) \geq 0. \qquad\qquad \square$$

Since the Euclidean distance of vectors is not affected by a translation $I\!\!R^p \ni x \mapsto x + b$ with an arbitrary $b \in I\!\!R^p$, we can assume without loss of generality that the mean vector of $x_1, \ldots, x_n$ is $0$.

The following figure visualizes the distance matrix of the industrial nations in Example 3.3.1 based on the vectors of studentized variables (*invest, inflatn, gnp, nukes, unempld*)$^T \in I\!\!R^5$. This *MDS* plot aims at visualizing economical differences. While France and Germany are in the center of the plot, Spain and Ireland are close to each other and clearly separated from the rest. Japan is an obvious outlier; see also the factor analysis in Figure 8.3.8 together with the subsequent comments and the comments on erroneous *MDS* plots below. To make sure that all variables have equal influence on an *MDS* plot, the variables are studentized by their individual standard deviations before computing the distance matrix. The covariance matrix of the studentized variables is then the correlation matrix of the original variables.

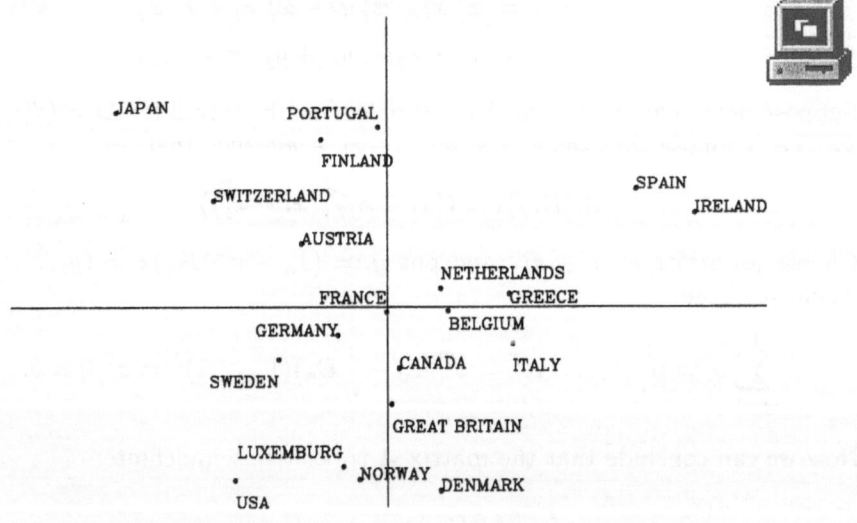

**Figure 7.3.2.** *MDS* plot of the distance matrix for the standardized economy data in Example 3.3.1; variables *invest, inflatn, gnp, nukes, unempld*.

```
***    Program 7_3_2    ***;
TITLE1 'MDS Plot;
TITLE2 'Economy Data';
LIBNAME datalib 'c:\data';

PROC STANDARD DATA=datalib.economy OUT=economy STD=1;
    VAR invest inflatn gnp nukes unempld;
PROC IML;
*** import data set economy;
  USE economy;
  READ ALL VAR{invest inflatn gnp nukes unempld} INTO x;
*** computation of distance matrix 'dist';
  dist=J(20,20,.);
  DO k=1 TO 20;  DO l=1 TO 20;
   d1=(x[k,]-x[l,]);
   dist[k,l]=d1*d1';
  END; END;
i20=I(20); e20=J(20,20,1); p=(i20-(1/20)*e20);
a1=-.5*XMULT(p,dist);             * computation of matrix A;
a2=XMULT(a1,p);                   * from (7.1);
CALL EIGEN(l,r,a2); x=J(20,5,.);
  DO k=1 TO 5;
   x[,k]=(l[k]##.5)*r[,k]; END;   * solution of MDS problem;
CREATE mds1 FROM x; APPEND FROM x;
QUIT;                            * quit IML;

DATA mds2;
    MERGE economy mds1;
DATA annomds;   SET mds2;
  FUNCTION='LABEL'; TEXT=country; XSYS='2'; YSYS='2'; HSYS='4';
  POSITION='C'; SIZE=1.5; X=COL2; Y=COL1;
   IF country IN('NETHERLANDS') THEN POSITION='3';
   IF country IN('GREECE') THEN POSITION='6';
   IF country IN('ITALY' 'GREAT BRITAIN' 'USA') THEN POSITION='9';
   IF country IN('PORTUGAL' 'LUXEMBURG' 'FRANCE') THEN POSITION='1';
   IF country IN('FINLAND' 'SWEDEN') THEN POSITION='7';
   IF country IN('BELGIUM') THEN POSITION='F';
   IF country IN('GERMANY') THEN POSITION='A'; RUN;

SYMBOL1 V=DOT C=GREEN H=0.6;
PROC GPLOT DATA=mds2;
  PLOT COL1*COL2 / ANNOTATE=annomds NOAXIS HREF=0 VREF=0;
RUN; QUIT;
```

See Programs 6_4_3 and 7_2_2 for comments on the procedures IML and CORR, and Program 7_1_1 for comments on the ANNOTATE functions.

## The Additive Constant Problem

If the doubly–centered matrix $A$ belonging to some $n \times n$–distance matrix $D = (d(i,j))$ is *not* positive semidefinite, then we cannot solve the metric *MDS* problem by Theorem 7.3.1. An appropriate solution can nevertheless be derived by adding a constant $c \in I\!\!R$ to each off–diagonal distance $d(i,j), i \neq j$. If $c$ is properly chosen, then the distance matrix

$$D_c := (d(i,j) + c\, l(i,j))$$

with $l(i,j) = 1$ if $i \neq j$ and 0 elsewhere, yields the positive semidefinite matrix $A_c$. This is the content of the following result. For extensions we refer to Cailliez (1983).

**7.3.2 Theorem.** *Suppose that the matrix $A$ is not positive semidefinite, i.e., we have $\lambda_n < 0$. Denote by $\mu_n$ the smallest eigenvalue of the matrix*

$$B := -(1/2)(I_n - n^{-1}E_n)(d(i,j))(I_n - n^{-1}E_n).$$

*Then $A_c$ is positive semidefinite for any $c \geq \sqrt{4\mu_n^2 - 2\lambda_n} - 2\mu_n \ (> 0)$.*

The proof of Theorem 7.3.2 uses the following auxiliary result.

**7.3.3 Lemma.** *Let $M$ be a symmetric $n \times n$–matrix with eigenvalues $\gamma_1 \geq \cdots \geq \gamma_n$. Then we have for arbitrary $x \in I\!\!R^n$ the inequalities*

$$\gamma_n x^T x \leq x^T M x \leq \gamma_1 x^T x.$$

**Proof:** Repeating the arguments in diagonalizing the matrix $A$ in (7.4), we can find for $M$ an orthogonal $n \times n$–matrix $R$ such that

$$M = R^T \begin{pmatrix} \gamma_1 & & 0 \\ & \ddots & \\ 0 & & \gamma_n \end{pmatrix} R.$$

Choose an arbitrary $x \in I\!\!R^n$ and put $y := Rx = (y_1, \ldots, y_n)^T$. Then we obtain

$$x^T M x = y^T \begin{pmatrix} \gamma_1 & & 0 \\ & \ddots & \\ 0 & & \gamma_n \end{pmatrix} y = \sum_{i=1}^n \gamma_i y_i^2 \begin{cases} \leq \gamma_1 \sum_{i=1}^n y_i^2 = \gamma_1 y^T y \\ \geq \gamma_n \sum_{i=1}^n y_i^2 = \gamma_n y^T y, \end{cases}$$

which implies the assertion by the equation $y^T y = x^T R^T R x = x^T x$.          □

**Proof of Theorem 7.3.2:** The matrix $A_c$ corresponding to $D_c$ has by (7.1) the representation

$$A_c = -\frac{1}{2}\left(I_n - \frac{1}{n}E_n\right)\left(d^2(i,j) + 2c\,d(i,j)l(i,j) + c^2 l(i,j)\right)\left(I_n - \frac{1}{n}E_n\right)$$

$$= -\frac{1}{2}\left(I_n - \frac{1}{n}E_n\right)\left(d^2(i,j) + 2c\,d(i,j) + c^2 l(i,j)\right)\left(I_n - \frac{1}{n}E_n\right).$$

The matrix $(l(i,j)) = E_n - I_n$ satisfies

$$(l(i,j))(I_n - n^{-1}E_n) = -(I_n - n^{-1}E_n).$$

Since the matrix $I_n - n^{-1}E_n$ is idempotent, we obtain for $A_c$ the representation

$$A_c = -\frac{1}{2}\left(I_n - \frac{1}{n}E_n\right)(d^2(i,j))\left(I_n - \frac{1}{n}E_n\right) - c\left(I_n - \frac{1}{n}E_n\right)(d(i,j))$$

$$\times\left(I_n - \frac{1}{n}E_n\right) + \frac{c^2}{2}\left(I_n - \frac{1}{n}E_n\right)\left(I_n - \frac{1}{n}E_n\right)$$

$$= \left(I_n - \frac{1}{n}E_n\right)\left(A + 2cB + \frac{c^2}{2}I_n\right)\left(I_n - \frac{1}{n}E_n\right).$$

Choose now an arbitrary $z \in \mathbb{R}^n$ and put $y := (I_n - n^{-1}E_n)z \in \mathbb{R}^n$. Then we obtain from Lemma 7.3.3

$$z^T A_c z = y^T\left(A + 2cB + \frac{c^2}{2}I_n\right)y \geq \left(\lambda_n + 2c\mu_n + \frac{c^2}{2}\right)y^T y \geq 0,$$

if $c \geq \sqrt{4\mu_n^2 - 2\lambda_n} - 2\mu_n (> 0)$, which is the larger one of the two solutions of the equation $c^2/2 + 2c\mu_n + \lambda_n = 0$. □

**7.3.4 Example** (Habitude Data). A student was asked to evaluate the way of teaching of the lecturers Becker, Falk, Femir, Marohn, Retel and Winter. To this end, all $\binom{6}{2}$ possible pairs that can be built out of the 6 lecturers were considered and the student was asked to rank the pairs according to the similarity in teaching habitudes of the members of each pair: The pair of those lecturers who are most similar in teaching habitude is ranked number 1, the second is ranked number 2 , ..., and the least similar is ranked $\binom{6}{2} = 15$. Thus we obtain a $6 \times 6$–distance matrix, which represents the similarities in teaching habitudes:

|         | Becker | Falk | Femir | Marohn | Retel | Winter |
|---------|--------|------|-------|--------|-------|--------|
| Becker  | 0      | 7    | 10    | 8      | 6     | 2      |
| Falk    |        | 0    | 1     | 4      | 11    | 3      |
| Femir   |        |      | 0     | 13     | 15    | 14     |
| Marohn  |        |      |       | 0      | 5     | 12     |
| Retel   |        |      |       |        | 0     | 9      |
| Winter  |        |      |       |        |       | 0      |

We visualize the above matrix $D$ by metric *MDS*. But now we have to add a constant to the off–diagonal elements of $D$ in order to get a positive semidefinite matrix $A_c$. Here we can take $c^* = 90.040866$, which results from Lemma 7.3.2 with the eigenvalues $\lambda_6 = -304.0895$ and $\mu_6 = -20.8216$. Figure 7.3.3 readily visualizes close similarites of teaching habitudes of Femir and Falk as well as Becker and Winter. They visually build two clusters. Marohn and Retel are quite far away from these two clusters and might be viewed as the third cluster. This result suggests the problem: *How to interpret a cluster*. To answer this question, however, more information has to be delivered by the student.

We can now interview a sample of students in the above way and by taking the averages of the pertaining matrices of ranks elementwise, we obtain a distance matrix which reflects the average similarities of teaching habitudes. This approach can in general be applied to those problems, where an interviewer comments on similarities or dissimilarities. A possible application is the detection of selling opportunities in the framework of marketing research, where specified goods or services already on the market are analyzed.

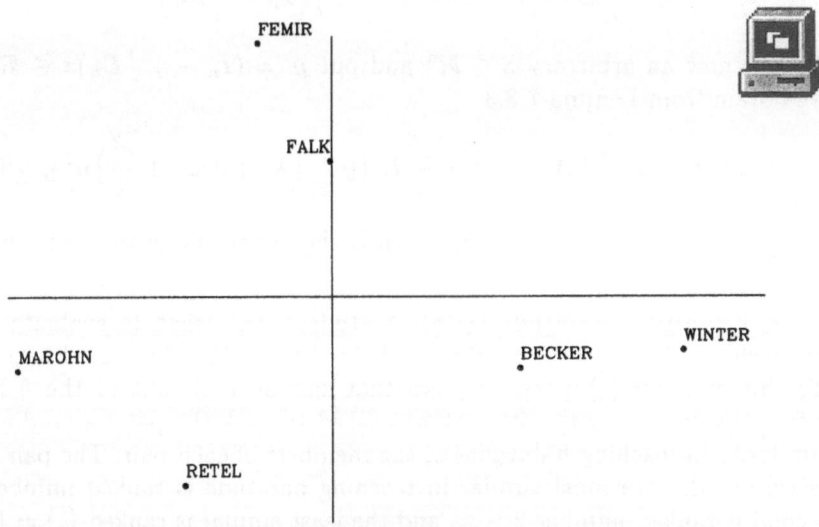

**Figure 7.3.3.** The additive constant problem; *MDS* plot of the distance matrix $D_1 = D + 90.040866\ (E_n - I_n)$ in Example 7.3.4.

## Solution of the *MDS* Problem for Similarity Matrices

**7.3.5 Theorem.** *Let $(s(i,j))$ be an arbitrary $n \times n$ similarity matrix. The matrix*

$$d(i,j) := \sqrt{2(1 - s(i,j))}, \qquad i,j = 1,\ldots,n,$$

*is a distance matrix, whose corresponding doubly–centered matrix $A$ satisfies*

$$A = \left(I_n - \frac{1}{n}E_n\right)(s(i,j))\left(I_n - \frac{1}{n}E_n\right).$$

*If, in addition, the similarity matrix $(s(i,j))$ is positive semidefinite, this property carries over to the matrix $A$.*

**Proof:** The symmetric matrix $I_n - n^{-1}E_n$ satisfies

$$\left(I_n - \frac{1}{n}E_n\right)E_n = 0.$$

For the doubly–centered matrix $A$ in (7.1) corresponding to $(d(i,j))$ we obtain the representation

$$
\begin{aligned}
A &= -\frac{1}{2}\left(I_n - \frac{1}{n}E_n\right)(2(1 - s(i,j)))\left(I_n - \frac{1}{n}E_n\right) \\
&= -\left(I_n - \frac{1}{n}E_n\right)E_n\left(I_n - \frac{1}{n}E_n\right) + \left(I_n - \frac{1}{n}E_n\right)(s(i,j))\left(I_n - \frac{1}{n}E_n\right) \\
&= \left(I_n - \frac{1}{n}E_n\right)(s(i,j))\left(I_n - \frac{1}{n}E_n\right).
\end{aligned}
$$

This representation immediately implies that positive semidefiniteness of $(s(i,j))$ carries over to $A$: Choose an arbitrary $z \in \mathbb{R}^n$ and put $y := (I_n - n^{-1}E_n)z \in \mathbb{R}^n$. Then we have

$$z^T A z = z^T\left(I_n - \frac{1}{n}E_n\right)(s(i,j))\left(I_n - \frac{1}{n}E_n\right)z = y^T(s(i,j))y \geq 0. \qquad \square$$

If the similarity $(s(i,j))$ is positive semidefinite, we can immediately solve the metric *MDS* problem for the distance matrix $(d(i,j)) = (\sqrt{2(1 - s(i,j))})$ in $\mathbb{R}^p$, where $p$ is the rank of the matrix $(s(i,j))$. Note that rank $A \leq$ rank $(s(i,j))$. If $(s(i,j))$ is not positive semidefinite, we can derive an appropriate solution by adding an appropriate constant $c$ to the off–diagonal elements of $(d(i,j))$ as described in Theorem 7.3.2. This yields a positive semidefinite matrix $A_c$.

## Approximate Solutions in Lower Dimensions

An *exact* solution of the metric *MDS* problem in $\mathbb{R}^p$ with a *large* $p$ is obviously not quite helpful, since the spatial representation of the objects cannot be displayed. One approach to overcome this problem is to take only the first $q$ coordinates of the exact solutions $x_1, \ldots, x_n \in \mathbb{R}^p$ in (7.5), i.e., put

$$x_{i(q)} := (x_{i1}, \ldots, x_{iq})^T \in \mathbb{R}^q, \qquad i = 1, \ldots, n.$$

The Euclidean distance between $x_{i(q)}$ and $x_{j(q)}$ in $I\!\!R^q$ is

$$
\begin{aligned}
\|x_{i(q)} - x_{j(q)}\|^2 &= (x_{i(q)} - x_{j(q)})^T (x_{i(q)} - x_{j(q)}) \\
&= (x_i^T - x_j^T - \underbrace{(0, \ldots, 0}_{q \text{ zeros}}, x_{iq+1} - x_{jq+1}, \ldots, x_{ip} - x_{jp}))(x_i - x_j)
\end{aligned}
$$

$$
= (x_i - x_j)^T (x_i - x_j) - \sum_{k=q+1}^{n} (x_{ik} - x_{jk})^2
$$

$$
= d^2(i,j) - \sum_{k=q+1}^{p} \lambda_k (r_{ki} - r_{kj})^2, \tag{7.6}
$$

where $r_i = (r_{i1}, \ldots, r_{in})^T$, $i = 1, \ldots, n$, are the orthonormal eigenvectors in (7.2) corresponding to the eigenvalues $\lambda_1 \geq \cdots \geq \lambda_p > 0 = \lambda_{p+1} = \cdots = \lambda_n$ of the matrix $A$.

Replacing the exact solution $x_i \in I\!\!R^p$ by the approximate solution $x_{i(q)} \in I\!\!R^q$, $i = 1, \ldots, n$, we obtain from (7.6), the general inequality $\sqrt{b} - \sqrt{a} \leq \sqrt{b-a}$, $0 \leq a \leq b$, and the triangular inequality for the Euclidean distance the following bound for the error $d(i,j) - \|x_{i(q)} - x_{j(q)}\|$:

$$
\begin{aligned}
0 \leq &\ d(i,j) - \|x_{i(q)} - x_{j(q)}\| \\
\leq &\ \Big( \sum_{k=q+1}^{p} \lambda_k (r_{ki} - r_{kj})^2 \Big)^{1/2} \\
\leq &\ \Big( \sum_{k=q+1}^{p} \lambda_k r_{ki}^2 \Big)^{1/2} + \Big( \sum_{k=q+1}^{p} \lambda_k r_{kj}^2 \Big)^{1/2} \\
\leq &\ \lambda_{q+1}^{1/2} \left( \Big( \sum_{k=q+1}^{p} r_{ki}^2 \Big)^{1/2} + \Big( \sum_{k=q+1}^{p} r_{kj}^2 \Big)^{1/2} \right) \\
\leq &\ \lambda_{q+1}^{1/2} \left( \Big( \sum_{k=1}^{n} r_{ki}^2 \Big)^{1/2} + \Big( \sum_{k=1}^{n} r_{kj}^2 \Big)^{1/2} \right) = 2\lambda_{q+1}^{1/2}, \tag{7.7}
\end{aligned}
$$

since by $RR^T = I_n$ the row vectors of the matrix $R$ have Euclidean length 1 as well. In the above Figures 7.3.1–7.3.3 we put $q = 2$. The errors in these plots can be estimated by the upper bound in (7.7), (Exercise 16).

## Nonmetric *MDS*

In *nonmetric MDS* a $n \times n$–matrix $(\delta_{ij})$ of distances or, better, *dissimilarities* is given, which is typically on an ordinal scaling. That means we can rank the $n$ objects, but we do not know their actual values. The dissimilarities $\delta_{ij}$ are, consequently, between integers 1 and $n^2$. The ranking of the similarities of teaching habitudes in Example 7.3.4 is a typical example.

Kruskal (1964) defined the following fit measure for a spatial configuration set $x_1, \ldots, x_n \in \mathbb{R}^p$:

$$S(x_1, \ldots, x_n, f) := \left( \frac{\sum_{1 \leq i,j \leq n} \left( d_{ij} - f(\delta_{ij}) \right)^2}{\sum_{1 \leq i,j \leq n} d_{ij}^2} \right)^{1/2},$$

where $f : [0, \infty) \to [0, \infty)$ is a monotone increasing function that relates the Euclidean distances $d_{ij} := \| x_i - x_j \|$ to $\delta_{ij}$, i.e.,

$$d_{ij} \approx f(\delta_{ij}) \quad \text{and} \quad \delta_{ij} < \delta_{kl} \iff f(\delta_{ij}) < f(\delta_{kl}).$$

One might interpret this approach as a *GLIM* with dependent variable $\delta_{ij}$, regressor $d_{ij}$ and monotone link function $f$, see Section 4.2. Since $S$ is some kind of standardized residual sum of squares, it is called *STRESS* function. The nonmetric *MDS* problem is the minimization of $S$ with respect to the $x_i$ *and* $f$.

Different to the metric *MDS* approach, the nonmetric *MDS* problem is solved in an iterative way as follows:

In the *initial phase* a dimension $p$ and an initial configuration $x_1^{(0)}, \ldots, x_n^{(0)} \in \mathbb{R}^p$ is selected. This can be done by means of metric *MDS* methods, as described above.

The *nonmetric phase* now minimizes $S(x_1^{(0)}, \ldots, x_n^{(0)}, f)$ with respect to monotone functions $f$, i.e., a function $f_{\min}^0$ is computed which satisfies

$$S(x_1^{(0)}, \ldots, x_n^{(0)}, f_{\min}^{(0)})$$

$$= \min \left\{ S(x_1^{(0)}, \ldots, x_n^{(0)}, f) : f : [0, \infty) \longrightarrow [0, \infty) \text{ is monotone increasing} \right\}.$$

The *metric phase* now minimizes $S(x_1, \ldots, x_n, f_{\min}^{(0)})$ with respect to $x_1, \ldots, x_n \in \mathbb{R}^p$, i.e.,

$$S(x_1^{(1)}, \ldots, x_n^{(1)}, f_{\min}^{(0)}) = \min \left\{ S(x_1, \ldots, x_n, f_{\min}^{(0)}) : x_1, \ldots, x_n \in \mathbb{R}^p \right\}.$$

If STRESS $S(x_1^{(1)}, \ldots, x_n^{(1)}, f_{\min}^{(0)})$ is small enough, typically less than 0.1 or 0.05, the process is stopped. Otherwise, the nonmetric phase and the metric phase are iterated.

This *Sheppard–Kruskal algorithm* is usually computed for various dimensions $p$. It ends with the *final selection* of the dimension $p$ and the corresponding spatial configuration and the *interpretation of the results*. The typical choice is, nevertheless, $p = 2$, in which case the objects can readily be displayed.

An alternative fit measure is

$$S_2(\boldsymbol{x}_1,\ldots,\boldsymbol{x}_n,f) := \left( \frac{\sum_{1 \le i,j \le n} \left(d_{ij}^2 - f(\delta_{ij})^2\right)^2}{\sum_{1 \le i,j \le n} f(\delta_{ij})^4} \right)^{1/2},$$

defined in terms of squared distances $d_{ij}^2$ and squared *disparities* $f(\delta_{ij})^2$. The algorithm typically used for finding minima of $S_2$ is the *ALSCAL algorithm*, created by Takane, Young and de Leeuw (1977). For a description of this algorithm we refer to Schiffman et al. (1981).

**7.3.6 Example.** We apply the nonmetric *MDS* approach to the economy data underlying Figure 7.3.2. The final values of the STRESS functions for dimensions 2, 3 and 4 are 0.1603, 0.0720 and 0.0254, so that a nonmetric *MDS* in $p = 2$ dimensions seems to be not quite adequate. The following figure displays the corresponding spatial configuration in $I\!\!R^2$ and the SAS output.

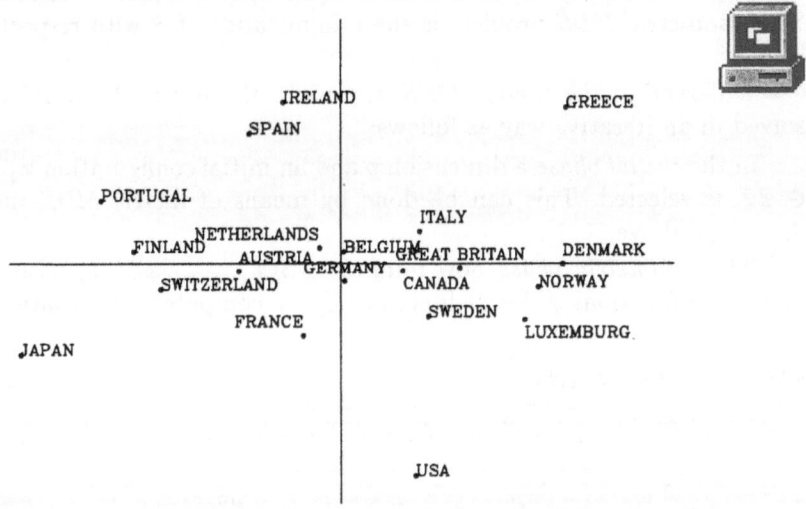

Figure 7.3.4. Nonmetric *MDS* alternative of Figure 7.3.2 in $I\!\!R^2$.

```
Multidimensional Scaling:  Data=WORK.DIST2
Shape=TRIANGLE Condition=MATRIX Level=ORDINAL
Coef=IDENTITY Dimension=2 Formula=1 Fit=1

Mconverge=0.01 Gconverge=0.01 Maxiter=100 Over=2 Ridge=0.0001
```

| | | Badness- | | Convergence Measures | |
|---|---|---|---|---|---|
| | | of-Fit | Change in | --------------------- | |
| Iteration | Type | Criterion | Criterion | Monotone | Gradient |
| 0 | Initial | 0.2863 | . | . | . |
| 1 | Monotone | 0.2241 | 0.0623 | 0.1531 | 0.5937 |

| | | | | | |
|---|---|---|---|---|---|
| 2 | Gau-New | 0.1776 | 0.0465 | . | . |
| 3 | Monotone | 0.1682 | 0.009446 | 0.0532 | 0.2254 |
| 4 | Gau-New | 0.1671 | 0.001071 | . | . |
| 5 | Monotone | 0.1620 | 0.005037 | 0.0340 | 0.1231 |
| 6 | Gau-New | 0.1617 | 0.000299 | . | . |
| 7 | Monotone | 0.1609 | 0.000815 | 0.0163 | 0.0835 |
| 8 | Gau-New | 0.1609 | 6.1855E-6 | . | . |
| 9 | Monotone | 0.1608 | 0.000157 | 0.006783 | 0.0739 |
| 10 | Gau-New | 0.1603 | 0.000437 | . | 0.005944 |

Convergence criteria are satisfied.
                    Multidimensional Scaling:  Data=WORK.DIST2
                Shape=TRIANGLE Condition=MATRIX Level=ORDINAL
                Coef=IDENTITY Dimension=3 Formula=1 Fit=1


Mconverge=0.01 Gconverge=0.01 Maxiter=100 Over=2 Ridge=0.0001


| | | Badness- | | Convergence Measures | |
|---|---|---|---|---|---|
| | | of-Fit | Change in | ---------------------- | |
| Iteration | Type | Criterion | Criterion | Monotone | Gradient |
| 0 | Initial | 0.2832 | . | . | . |
| 1 | Monotone | 0.1131 | 0.1701 | 0.1801 | 0.6360 |
| 2 | Gau-New | 0.0894 | 0.0236 | . | . |
| 3 | Monotone | 0.0818 | 0.007643 | 0.0291 | 0.3235 |
| 4 | Gau-New | 0.0808 | 0.001029 | . | . |
| 5 | Monotone | 0.0746 | 0.006199 | 0.0284 | 0.2093 |
| 6 | Gau-New | 0.0743 | 0.000255 | . | . |
| 7 | Monotone | 0.0734 | 0.000883 | 0.009319 | 0.1861 |
| 8 | Gau-New | 0.0722 | 0.001266 | . | 0.0444 |
| 9 | Gau-New | 0.0721 | 0.0000961 | . | 0.0229 |
| 10 | Gau-New | 0.0720 | 0.0000325 | . | 0.0164 |
| 11 | Gau-New | 0.0720 | 0.0000166 | . | 0.0115 |
| 12 | Gau-New | 0.0720 | 8.1025E-6 | . | 0.007929 |

Convergence criteria are satisfied.
                    Multidimensional Scaling:  Data=WORK.DIST2
                Shape=TRIANGLE Condition=MATRIX Level=ORDINAL
                Coef=IDENTITY Dimension=4 Formula=1 Fit=1


Mconverge=0.01 Gconverge=0.01 Maxiter=100 Over=2 Ridge=0.0001

|            |           | Badness-of-Fit | Change in  | Convergence Measures | |
| ---------- | --------- | --------------- | ---------- | --------- | --------- |
| Iteration  | Type      | Criterion       | Criterion  | Monotone  | Gradient  |
| 0          | Initial   | 0.3362          | .          | .         | .         |
| 1          | Monotone  | 0.0794          | 0.2568     | 0.1989    | 0.7821    |
| 2          | Gau-New   | 0.0531          | 0.0263     | .         | .         |
| 3          | Monotone  | 0.0371          | 0.0160     | 0.0329    | 0.5530    |
| 4          | Gau-New   | 0.0343          | 0.002760   | .         | .         |
| 5          | Monotone  | 0.0277          | 0.006601   | 0.0188    | 0.3462    |
| 6          | Gau-New   | 0.0271          | 0.000610   | .         | .         |
| 7          | Monotone  | 0.0261          | 0.001010   | 0.006937  | 0.2309    |
| 8          | Gau-New   | 0.0254          | 0.000699   | .         | 0.0125    |
| 9          | Gau-New   | 0.0254          | 2.064E-6   | .         | 0.001725  |

Convergence criteria are satisfied.

**Figure 7.3.5.** Nonmetric *MDS* solution for economy data in dimensions 2, 3 and 4.

```
***    Program 7_3_4    ***;
TITLE1 'Nonmetric MDS Solution';
TITLE2 'Economy Data';
LIBNAME datalib 'c:\data';

PROC STANDARD DATA=datlib.economy OUT=economy STD=1;
   VAR invest inflatn gnp nukes unempld;

PROC IML;
***    import dataset economy;
   USE economy;
   READ ALL VAR{invest inflatn gpn nukes unempld} INTO x;
***    computation of distance matrix;
   dist=J(20,20,.);
   DO k=1 TO 20;
     DO l=1 TO 20;
       d1=(x[k,]-x[l,]);
       dist[k,l]=d1*d1';
     END;
   END;
CREATE dist1 FROM dist;
APPEND FROM dist;
QUIT;            *   quit IML;
                          ↓
```

```
                                    ↑
DATA dist2 (KEEP=country col1-col20);
   MERGE economy dist1;
PROC MDS DATA=dist2 DIM=2 TO 4 OUT=mdsout;
   ID country;
DATA annomds;
   SET mdsout(WHERE=(_DIMENS_=2));
   FUNCTION='LABEL'; TEXT=country; XSYS='2'; YSYS='2'; HSYS='4';
   POSITION='C'; SIZE=1.5; X=DIM2; Y=DIM1;
      IF country IN('DENMARK' 'ITALY' 'AUSTRIA') THEN POSITION='3';
      IF country IN('GREAT BRITAIN' 'GERMANY') THEN POSITION='2';
      IF country IN('LUXEMBURG' 'CANADA') THEN POSITION='F';
      IF country IN('NETHERLANDS' 'FRANCE') THEN POSITION='1';
SYMBOL1 V=DOT C=GREEN H=0.6;
PROC GPLOT DATA=mdsout(WHERE=(_TYPE_='CONFIG' AND _DIMENS_=2));
   PLOT DIM1*DIM2 / ANNOTATE=annomds NOAXIS HREF=0 VREF=0;
RUN; QUIT;
```

The variables are standardized by the use of PROC STANDARD.

PROC MDS now calculates by default a nonmetric MDS as described above, minimizing the stress function S. Because of the used options this is done for dimensions 2, 3 and 4 and written into the dataset mdsout. The WHERE dataset option in the DATA step creating the annotate dataset and in the DATA option of PROC GPLOT selects the cases of mdsout providing the two–dimensional representation of the data.

# 7.4  Hierarchical Clustering Methods

Here and in the following section we will consider the set $I = \{1, \ldots, n\}$ of objects, equipped with a distance measure. Objects, which are close to each other, will be gathered into a subset, i.e., into a cluster. The iterative clustering process will be done by *hierarchical clustering methods*.

Hierarchical clustering methods sequentially define arrangements of the set of objects $I$ in disjoint groups. In the following we consider *agglomerative processes*, where at each step two groups are joined. We assume that a distance measure $D$ is defined on the set of all nonempty subsets of $\{1, \ldots, n\}$. The divisions of $I$ into clusters is sequentially defined as follows:

1. Put $C^{(n)} = \{\{1\}, \ldots, \{n\}\}$. This is the initial division of the set of objects $I = \{1, \ldots, n\}$. Each object is viewed as a group of its own.

2. Suppose $C^{(\nu)}$ is a partition containing $\nu$ clusters. Two clusters, whose distance measure $D$ is the smallest one among all pairs of clusters, are then joined, yielding the partition $C^{(\nu-1)}$ containing $\nu - 1$ clusters.

3. Step 2 is iterated until $C^{(1)} = \{I\}$.

This procedure supposes that the pair of clusters with smallest distance is unique. If this is not true, i.e., if there are ties among the distances, then one can use an arbitrary rule of joining. It can, for example, depend on the number of objects that are joined or on the smallest or largest object identification number. In the sequel we introduce various specific agglomerative procedures.

## Single Linkage Method

The *single linkage method*, also called *minimum distance* or *nearest neighbor method*, defines the distance of two nonempty sets $C_1, C_2$ as the minimum distance between an object and an object from $C_2$: Let $d$ be a distance measure on $I = \{1, \ldots, n\}$. Then

$$D(C_1, C_2) := \min_{i \in C_1, j \in C_2} d(i, j)$$

defines a distance measure on the set of all nonempty subsets of $I$. If the distance $d$ is based on a density estimator, this procedure is called *density linkage method*. It will be investigated in the next section.

We now join those two groups $C_{k\bullet}^{(\nu)}, C_{l\bullet}^{(\nu)}$ of partition $\mathcal{C}^{(\nu)}$, which have minimum distance $D$:

$$D(C_{k\bullet}^{(\nu)}, C_{l\bullet}^{(\nu)}) = \min_{k \neq l} D(C_k^{(\nu)}, C_l^{(\nu)}) = \min_{k \neq l} \min_{i \in C_k^{(\nu)}, j \in C_l^{(\nu)}} d(i, j).$$

The following simple example illustrates this procedure. We start with the $5 \times 5$–distance matrix of 5 objects:

$$(d(i,j)) = \begin{pmatrix} 0 & 7 & 1 & 9 & 8 \\ 7 & 0 & 6 & 3 & 5 \\ 1 & 6 & 0 & 8 & 7 \\ 9 & 3 & 8 & 0 & 4 \\ 8 & 5 & 7 & 4 & 0 \end{pmatrix}. \tag{7.8}$$

The objects 1 and 3 have minimum distance $d(1,3) = 1$. By joining them we obtain the 4 groups $C_1^{(4)} = \{1, 3\}, C_2^{(4)} = \{2\}, C_3^{(4)} = \{4\}$ and $C_4^{(4)} = \{5\}$. Since

$$D(C_1^{(4)}, C_2^{(4)}) = \min\{d(1,2), d(3,2)\} = 6$$
$$D(C_1^{(4)}, C_3^{(4)}) = \min\{d(1,4), d(3,4)\} = 8$$
$$D(C_1^{(4)}, C_4^{(4)}) = \min\{d(1,5), d(3,5)\} = 7,$$

the distance matrix of these groups is the $4 \times 4$–matrix

$$\left(D(C_i^{(4)}, C_j^{(4)})\right) = \begin{pmatrix} 0 & 6 & 8 & 7 \\ 6 & 0 & 3 & 5 \\ 8 & 3 & 0 & 4 \\ 7 & 5 & 4 & 0 \end{pmatrix}.$$

This matrix shows that the groups $C_2^{(4)} = \{2\}$ and $C_3^{(4)} = \{4\}$ have minimum distance 3. The next iteration, therefore, yields the 3 groups

$$C_1^{(3)} = \{1, 3\}, C_2^{(3)} = \{2, 4\} \text{ and } C_3^{(3)} = \{5\}.$$

This yields the $3 \times 3$–distance matrix

$$\left( D(C_i^{(3)}, C_j^{(3)}) \right) = \begin{pmatrix} 0 & 6 & 7 \\ 6 & 0 & 4 \\ 7 & 4 & 0 \end{pmatrix}.$$

The groups $C_2^{(3)}$ and $C_3^{(3)}$ have minimum distance $D = 4$. The next step yields the 2 groups $C_1^{(2)} = \{1, 3\}$ and $C_2^{(2)} = \{2, 4, 5\}$, whose distance is 6. These are joined to the final group $\{1, 2, 3, 4, 5\}$.

## Dendrograms

The above sequential clustering process can be visualized by means of a tree diagram, usually called *dendrogram*. The classes with minimum distance are joined with a bar, which starts at the value of this minimum distance.

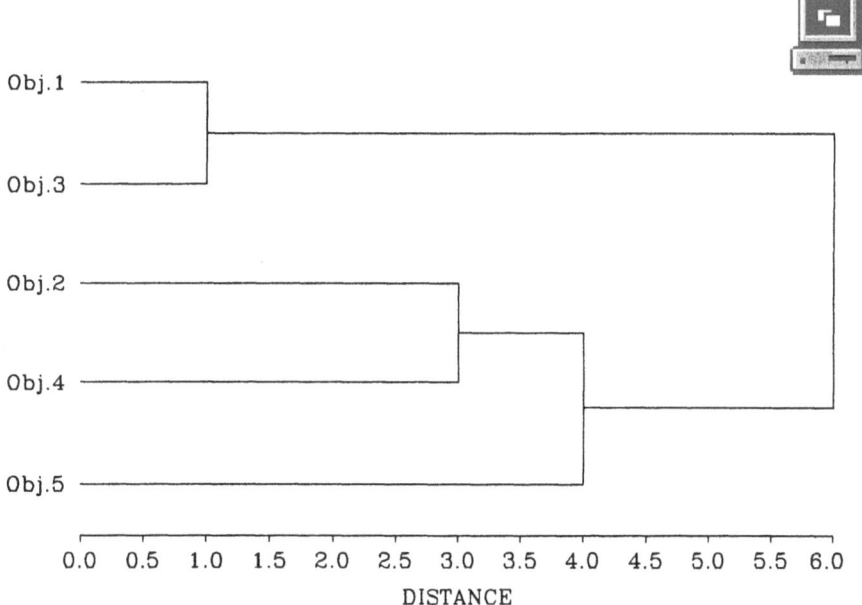

**Figure 7.4.1.** Dendrogram of the single linkage method for the distance matrix in (7.8).

```
***    Program 7_4_1    ***;
TITLE1 'Dendrogram of Single Linkage';
TITLE2 'Distance Matrix in (7.8)';
LIBNAME datalib 'c:\data';

DATA dist1(TYPE=DISTANCE);
    INPUT id $ (i1-i5) (2.);
    CARDS;
Obj.1 0 7 1 9 8
Obj.2 7 0 6 3 5
Obj.3 1 6 0 8 7
Obj.4 9 3 8 0 4
Obj.5 8 5 7 4 0
;
PROC CLUSTER METHOD=SINGLE DATA=dist1 OUTTREE=datalib.dendro NONORM;
    VAR i1-i5;
    ID id;
AXIS1 LABEL=NONE;
AXIS2 LABEL=('DISTANCE');
PROC TREE DATA=datalib.dendro HORIZONTAL HAXIS=AXIS2 VAXIS=AXIS1;
RUN; QUIT;
```

The DATA step of this program generates a SAS data set of 'TYPE=DISTANCE', which is used as the input data set in PROC CLUSTER, see the comments on Program 7_2_2.

PROC CLUSTER requires the specification of the clustering method used by the procedure. Here we specify METHOD=SINGLE, which determines the single linkage method. The option OUTTREE= *data set* creates an output data set, which contains the history of the clustering process and which can be used by the TREE procedure to draw a tree diagram (dendro-

gram). See also the comments on Programs 7_4_2 and 7_5_5.

The option NONORM prevents the distances from being normalized to unit mean or unit mean square. The values of the ID variable identify observations in the printed cluster history and in the OUTTREE= *data set*.

After the CLUSTER procedure, the procedure TREE is invoked, which uses the OUTTREE= datalib.dendro file to plot the high quality dendrogram in the graphics output. PROC TREE requires version 8 of SAS.

A vertical cut of the dendrogram at distance $h$ displays the clusters of those objects whose distance is at most $h$. Take in the above example $h = 3$. Objects 2 and 4 as well as objects 1 and 3 have distances less than or equal to 3. If we take $h = 1$, only objects 1 and 3 have a distance not larger than 1.

## How Many Clusters?

The agglomerative hierarchical clustering process does not provide a single cluster solution. Instead it provides a *sequence* of cluster solutions starting with $n$ clusters and ending with 1 cluster. The choice of an appropriate number of clusters $\nu^*$ comes with a trade–off situation just like the choice of a bandwidth in kernel density estimation: If $\nu^*$ is chosen too large, adjacent clusters are not joined. If $\nu^*$ is too small, wrong clusters are joined.

Since the determination of an appropriate number of clusters $\nu^*$ depends on the step at which the hierarchical process is terminated, the following fusion function can provide a *stopping rule* or *optimality criterion*. Put

$$\varphi(\nu) := D(C_{k\bullet}^{(\nu+1)}, C_{l\bullet}^{(\nu+1)}), \qquad \nu = 1, 2, \ldots, n = 1 \qquad (7.9)$$

This strictly monotone function $\varphi$ provides the minimum distance between two clusters in the partition $\mathcal{C}^{(\nu+1)}$. These two clusters will be joined in the next step, thus yielding $\mathcal{C}^{(\nu)}$. If the difference $\varphi(\nu^* - 1) - \varphi(\nu^*)$ for the step from $\nu^*$ to $\nu^* - 1$ clusters is very large, then the process should be terminated with $\nu^*$ clusters. In the above example we have $\varphi(4) = 1$, $\varphi(3) = 3$, $\varphi(2) = 4$, $\varphi(1) = 6$. Here the stopping rule suggests $\nu^* = 2$ yielding the two clusters

$$C_1^{(2)} = \{1, 3\}, \; C_2^{(2)} = \{2, 4, 5\}.$$

The dendrogram in Figure 7.4.1 shows that this suggestion for a cluster configuration seems to be quite reasonable.

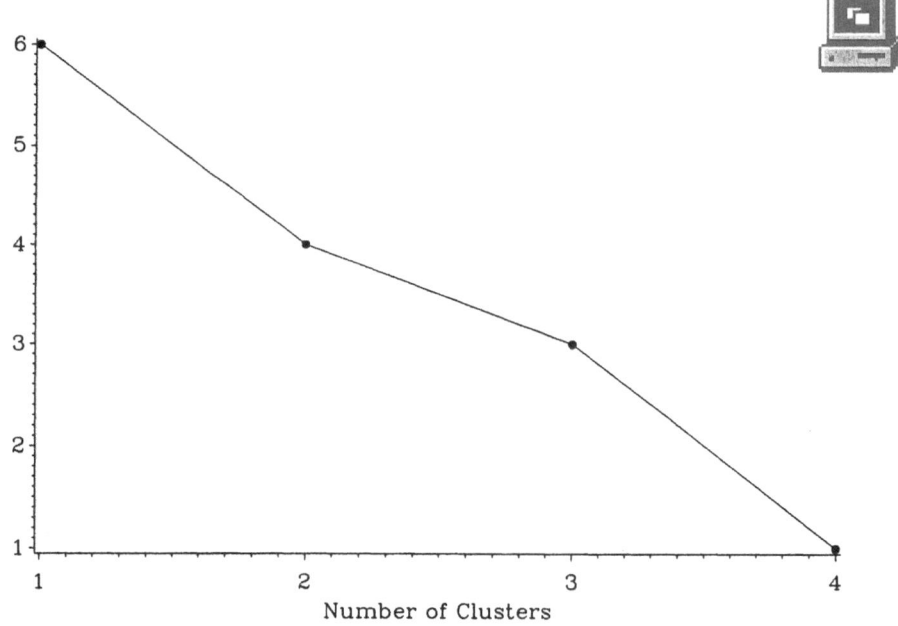

**Figure 7.4.2.** Function $\varphi$ corresponding to the dendrogram in Figure 7.4.1.

```
***    Program 7_4_2   ***;
TITLE1 'Fusion Function phi';
TITLE2 'Data from Figure 7.4.1';
LIBNAME datalib 'c:\data';

PROC SORT DATA=datalib.dendro;
    BY _NCL_;
SYMBOL1 V=DOT I=JOIN;
PROC GPLOT DATA=datalib.dendro(WHERE=(_NCL_<5));
    PLOT _HEIGHT_*_NCL_;
RUN; QUIT;
```

This program uses the OUTTREE= datalib.dendro file, which was created in Program 7_4_1. It contains in particular the automatic variables _HEIGHT_ and _NCL_, which denote the values of the function $\varphi$ and the corresponding arguments or number of clusters _NCL_. These variables are displayed in the plot. The DATA set option '(WHERE=(_NCL_<5))' restricts the plot of $\varphi$ to the set of arguments $\{1, 2, 3, 4\}$.

## Complete Linkage Method

The *complete linkage method*, maximum distance or farthest neighbor method defines the distance between two classes $C_1$ and $C_2$ as the *largest* distance between an object from $C_1$ and one from $C_2$:

$$D(C_1, C_2) = \max_{i \in C_1, j \in C_2} d(i, j).$$

## Average Linkage Method

The *average linkage method* defines the distance between the classes $C_1$ and $C_2$ as the arithmetic mean of all distances between objects from $C_1$ and $C_2$:

$$D(C_1, C_2) = \frac{1}{n_1 n_2} \sum_{i \in C_1} \sum_{j \in C_2} d(i, j),$$

where $n_k$ denotes the number of objects in $C_k$. Two classes are joined if the objects in the two classes are *on the average* quite similar.

In the sequel we will assume that our objects $1, \ldots, n$ are vectors $x_1, \ldots, x_n$ in $I\!R^p$. This will provide the initial distance $d$ between objects. The distance $D$ between clusters will then be defined on nonempty subsets of $\{1, \ldots, n\}$.

## Centroid Method

The following idea underlies the *centroid method*: Each cluster $C_k$ is represented by its *centroid*, i.e., by the vector of the averages of the corresponding coordinates of the points in $C_k$, which corresponds to the center of its mass:

$$\bar{x}_k := \frac{1}{n_k} \sum_{i \in C_k} x_i.$$

The distance between two clusters is then defined as the squared Euclidean distance between the corresponding centroids:

$$D(C_1, C_2) := d_2^2(\bar{x}_1, \bar{x}_2) = ||\bar{x}_1 - \bar{x}_2||^2.$$

The centroid $\bar{x}$ of the union $C_1 \cup C_2$ satisfies

$$\bar{x} = \frac{n_1 \bar{x}_1 + n_2 \bar{x}_2}{n_1 + n_2}. \tag{7.10}$$

The centroid method is in this sense similar to the average linkage method as two clusters are joined if the objects in both classes are similar on the average. In particular, if we use the squared Euclidean distance $|| \ ||^2$ as the distance between objects in the average linkage method, then we can directly compare both methods, see Exercise 18:

$$\frac{1}{n_1 n_2} \sum_{i \in C_1} \sum_{j \in C_2} ||x_i - x_j||^2$$

$$= ||\bar{x}_1 - \bar{x}_2||^2 + \frac{1}{n_1} \sum_{i \in C_1} ||x_i - \bar{x}_1||^2 + \frac{1}{n_2} \sum_{j \in C_2} ||x_j - \bar{x}_2||^2$$

$$= ||\bar{x}_1 - \bar{x}_2||^2 + s_1^2 + s_2^2. \tag{7.11}$$

By $s_j^2$ we denote the empirical variance of the Euclidean distances of the vectors within $C_j$. Unlike the centroid method the average linkage method requires for the union of two clusters, that the squared Euclidean distance between their centroids is small *and* that the variances within the classes are small as well.

## Ward's Method

*Ward's method* joins those two classes $C_{k \bullet}$ and $C_{l \bullet}$, such that the increase $D(C_{k \bullet}, C_{l \bullet})$ of the sum of squared Euclidean distances from the centroid (7.10) is minimized:

$$D(C_1, C_2) := \sum_{i \in C_1 \cup C_2} ||x_i - \bar{x}||^2 - \left\{ \sum_{i \in C_1} ||x_i - \bar{x}_1||^2 + \sum_{j \in C_2} ||x_j - \bar{x}_2||^2 \right\}$$

$$= n_1 ||\bar{x}_1 - \bar{x}||^2 + n_2 ||\bar{x}_2 - \bar{x}||^2$$

$$= \frac{n_1 n_2}{n_1 + n_2} ||\bar{x}_1 - \bar{x}_2||^2, \tag{7.12}$$

see Exercise 18. For single objects $i$ and $j$ we obtain in particular

$$D(\{i\}, \{j\}) = \frac{1}{2} ||x_i - x_j||^2.$$

## Median Method

The *median method* is quite similar to the centroid method, which compares Euclidean distances between centroids. But now the centroid (7.10) of the union of two classes $C_k \cup C_l$ is sequentially replaced by

$$\bar{x} = \frac{1}{2}(\bar{x}_k + \bar{x}_l).$$

This method aims at neutralizing the following tendency of the centroid method: If a cluster with only a few elements and one with a large number of elements are joined, the centroid (7.10) of the union will commonly be close to the centroid of the large cluster.

## An Update Algorithm

For each of the above methods, the following update algorithm computes the distance matrix $(D(C_k^{(\nu-1)}, C_l^{(\nu-1)}))$ of the partition $\mathcal{C}^{(\nu-1)}$ with $\nu-1$ clusters from the distance matrix $(D(C_i^{(\nu)}, C_j^{(\nu)}))$. The distance of the two classes $C_m$ and $C_k \cup C_l$ of a partition satisfies

$$D(C_m, C_k \cup C_l) = \alpha_k \, D(C_m, C_k) + \alpha_l \, D(C_m, C_l) + \beta \, D(C_k, C_l)$$
$$+\gamma|D(C_m, C_k) - D(C_m, C_l)|.$$

The following table lists the parameters for the above methods. By $n_i$ we denote the number of objects in the cluster $C_i$, $i \in \{k, l, m\}$ (Exercise 20).

| Method | Parameter | | |
|---|:---:|:---:|:---:|
| | $\alpha_i$ | $\beta$ | $\gamma$ |
| Single Linkage | $\frac{1}{2}$ | $0$ | $\frac{1}{2}$ |
| Complete Linkage | $\frac{1}{2}$ | $0$ | $\frac{1}{2}$ |
| Average Linkage | $\frac{n_i}{n_k+n_l}$ | $0$ | $0$ |
| Centroid | $\frac{n_i}{n_k+n_l}$ | $\frac{-n_k n_l}{(n_k+n_l)^2}$ | $0$ |
| Ward | $\frac{n_i+n_m}{n_k+n_l+n_m}$ | $\frac{-n_m}{n_k+n_l+n_m}$ | $0$ |
| Median | $\frac{1}{2}$ | $-\frac{1}{4}$ | $0$ |

(7.13)

The following two figures show dendrograms corresponding to the Mahalanobis distance matrix of 15 U.S. cities from Example 7.1.1, based on the 6 variables *smean*, *pmean*, *pm2*, *perwh*, *nonpoor* and *ge65*, see Figure 7.2.2.

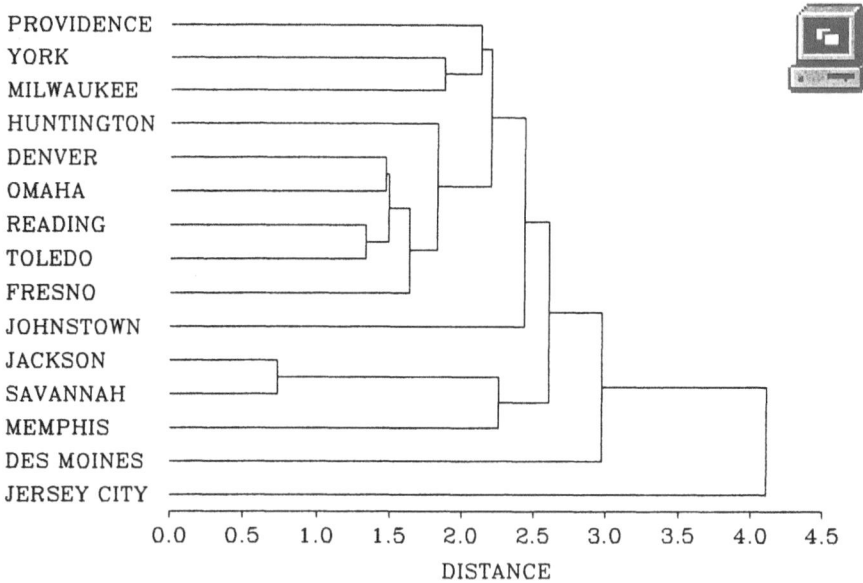

**Figure 7.4.3.** Single linkage dendrogram corresponding to distance matrix in Figure 7.2.2.

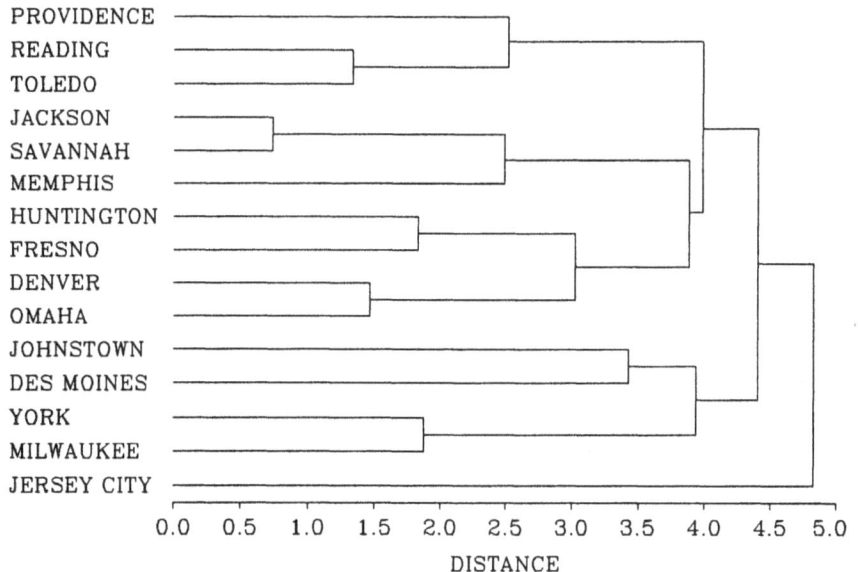

**Figure 7.4.4.** Complete linkage dendrogram corresponding to the distance matrix in Figure 7.2.2.

Program 7_4_3 generates the dendrogram in Figure 7.4.3 based on single linkage. It differs from the program for the above complete linkage dendrogram only in the specification METHOD=COMPLETE, see Program 7_4_1. Other possible choices of METHOD=*specification* are AVERAGE, CENTROID, MEDIAN and WARD, among others.

Due to the involved definition of the distance of two objects, the single linkage method has the tendency to provide chainlike clusters, whereas complete linkage leads to compact clusters. Equally, an isolated point becomes relatively close to an existing cluster with the complete linkage method rather than with the single linkage method, where outliers remain isolated until very late in the hierarchical process. For an extensive comparison of the different clustering methods we refer to Section 10.2.1 of Jobson (1992). It is often reasonable first to reduce the dimension of highdimensional observations, before a clustering process is started. A possible data reduction technique is factor analysis, which will be introduced in Section 8.3.

## 7.5   Density Linkage

*Density linkage methods* use density estimators for clustering. Suppose that the given objects in $I\!R^p$ are realizations of independent random vectors $\boldsymbol{X}_1, \ldots, \boldsymbol{X}_n$, which have a common density $f$. Agglomerations of observations can then be viewed as areas in $I\!R^p$, where $f$ has large values so that observations tend to cluster.

A popular model is the assumption that the $\boldsymbol{X}_i$ are independent copies of some random vector $\boldsymbol{X}$, which has a *mixture distribution*, i.e., its density $f$ has the representation

$$f(x) = p_1 f_1(x) + p_2 f_2(x) + \cdots + p_K f_K(x). \tag{7.14}$$

We have $p_i > 0$ with $p_1 + \cdots + p_K = 1$, and the $f_i$ are unimodal densities with different modes, see Section 1.2. Equation (7.14) can then be interpreted as a two–step random procedure as follows: In a first step some random mechanism determines the class index, where $j \in \{1, \ldots, K\}$ is chosen with probability $p_j$. Given that class index, the random vector $\boldsymbol{X}$ is generated in a second step according to the density $f_j$. The density $f$ is now the density of the complete two–step generation of $\boldsymbol{X}$. Think of the size of a human being, which is stratified according to the gender. Model (7.14) with a *known* number of classes $K$ was investigated in the framework of discriminant analysis in the preceding section. Here we consider the case, where $K$ is unknown and, thus, also $p_i$ and $f_i$ are unknown.

Since we require that the $f_i$ have different modes, the realizations of the random vectors $\boldsymbol{X}_1, \ldots, \boldsymbol{X}_n$ will agglomerate about these $K$ modes. An estimator of the number of classes is then given by the number of modes of a density estimator of $f$, see e.g. Müller and Sawitzki (1991). For one– or two–dimensional data this can be done visually, higher dimensions require the use of some search algorithm. This can be based on the methods introduced in Section 7.4.

**7.5.1 Example** (Random Data). The following scatterplot displays 140 realizations of independent random vectors in $I\!\!R^2$, generated by different bivariate normal distributions.

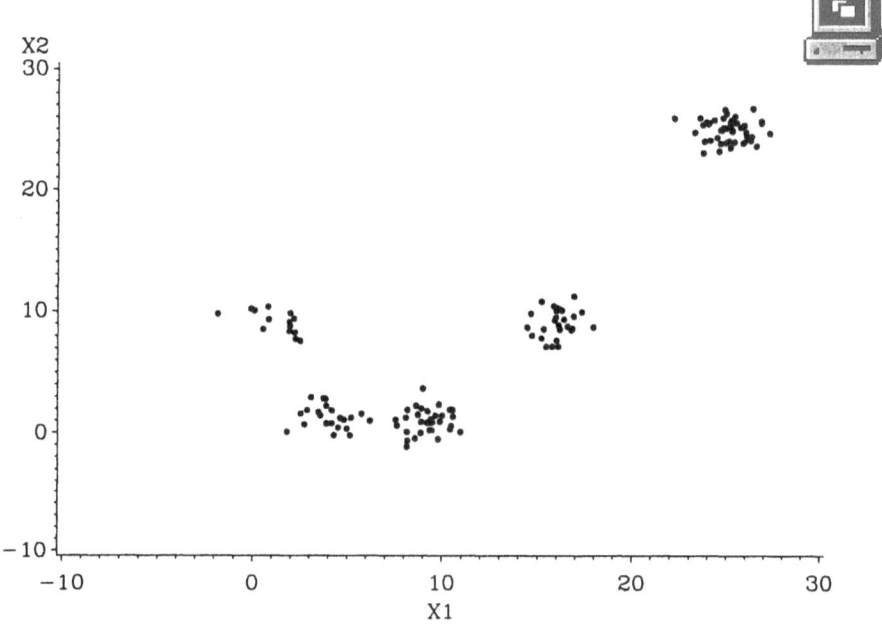

**Figure 7.5.1.** Scatterplot of 140 realizations of bivariate normal random vectors.

An estimator of the number of different underlying distributions is now given by the number of clusters of the data, for example by the number of modes of a bivariate density estimator. The following figure is based on a bivariate Epanechnikov kernel density estimator, with bandwidth $h = 1.5$.

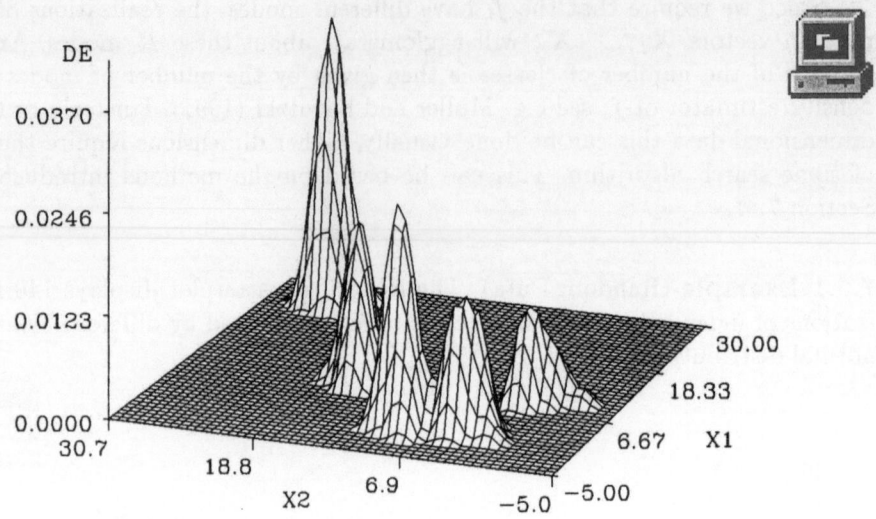

**Figure 7.5.2.** Kernel density estimation of random data; bivariate Epanechnikov kernel with bandwidth $h = 1.5$.

```
***    Program 7_5_2    ***;
TITLE1 'Kernel Density Estimator';
TITLE2 'Random Data';
LIBNAME datalib 'c:\data';

PROC MEANS DATA=datalib.random MIN MAX NOPRINT; VAR x1 x2;
    OUTPUT OUT=margin MIN=x1min x2min MAX=x1max x2max;
DATA datalib.density(KEEP=x y de i); SET margin;
    h=1.5;                              * bandwidth;
    DO x=INT(x1min)-4 TO INT(x1max)+4 BY .7;
        DO y=INT(x2min)-4 TO INT(x2max)+4 BY .7; d=0;
        DO i=1 TO 140;
            SET datalib.random POINT=i;
            z=((x-x1)**2+(y-x2)**2)/h**2;
            f=(2/3.14159)*(1-z);        * Epanechnikov kernel;
            f=max(0,f); d=d+f;
        END;
    de=d/(140*(h**2));
    OUTPUT; END; END; STOP;

SYMBOL1 C=GREEN I=JOIN V=NONE WIDTH=2;
PROC G3D DATA=datalib.density(RENAME=(y=x1 x=x2));
    PLOT x2*x1=de;
RUN; QUIT;
```

For the generation of this bivariate kernel density estimator plot we refer to Programs 1_1_4 and 6_4_2 with the accompanying comments. Here we use the Epanechnikov kernel with bandwidth h=1.5.

## The Nearest Neighbor Method

Let $x_1, \ldots, x_n \in \mathbb{R}^p$ be realizations of independent random vectors with common density $f$. Denote by $h_k(x)$ the (random) Euclidean distance of an arbitrary point $x \in \mathbb{R}^p$ from its $k$–th nearest neighbor among $\{x_1, \ldots, x_n\}$ and denote by $v_p(h_k(x)) = h_k^p v_p$ the volume of a ball in $\mathbb{R}^p$ of radius $h_k(x)$, see Example 6.4.1. The $k$–th nearest neighbor estimator of $f(x)$ is

$$\hat{f}_n(x) = \frac{k/n}{v_p(h_k(x))},$$

see Section 6.5. Two points $x_i$ and $x_j$ are called *adjacent* if they are mutually at most $k$–th nearest neighbors, i.e., if

$$\|x_i - x_j\| \leq \max\{h_k(x_i), h_k(x_j)\}.$$

The distance measure $d$ of two sample points $x_i$ and $x_j$ is now defined as the average of the reciprocals of the kernel density estimator at $x_i$ and $x_j$:

$$d(x_i, x_j) = \begin{cases} 0, & x_i = x_j, \\ 2^{-1}(1/\hat{f}_n(x_i) + 1/\hat{f}_n(x_j)) & \text{if } x_i, x_j \text{ are adjacent, } x_i \neq x_j, \\ \infty & \text{elsewhere,} \end{cases}$$

$$= \begin{cases} 0, & x_i = x_j, \\ \frac{n}{2k}\left(v_p(h_k(x_i)) + v_p(h_k(x_j))\right) & \text{if } x_i, x_j \text{ are adjacent, } x_i \neq x_j, \\ \infty & \text{elsewhere.} \end{cases}$$

The clustering process now commonly uses the single linkage method. The process terminates with more than 1 cluster if there are no more adjacent sample points. The determination of a proper choice of $k$ is analogous to the problem of choosing a bandwidth for a kernel density estimator, see Remark 1.1.10 and Sections 6.4 and 6.5.

## Uniform Kernel Method

This method is similar to the $k$–th nearest neighbor method. But now we apply the kernel density estimator based on the uniform kernel, see Example 6.4.1 (i):

$$\hat{f}_n(x) = \frac{R/n}{v_p(h)}.$$

This estimator of $f(x)$ uses a *nonrandom* bandwidth $h > 0$ and the *random* number $R = R(x) \in \{0, 1, \ldots, n\}$ of observations, which fall into the ball with center $x \in \mathbb{R}^p$ and radius $h$. Two sample points $x_i, x_j \in \mathbb{R}^p$ are now called adjacent if

$$\|x_i - x_j\| \leq h.$$

The distance of two sample points is again defined by

$$d(x_i, x_j) = \begin{cases} 0, & x_i = x_j, \\ 2^{-1}(1/\hat{f}_n(x_i) + 1/\hat{f}_n(x_j)) & \text{if } x_i, x_j \text{ are adjacent, } x_i \neq x_j, \\ \infty & \text{elsewhere,} \end{cases}$$

$$= \begin{cases} 0, & x_i = x_j, \\ 2^{-1} n v_p(h)(1/R(x_i) + 1/R(x_j)) & \text{if } x_i, x_j \text{ are adjacent, } x_i \neq x_j, \\ \infty & \text{elsewhere.} \end{cases}$$

The clustering process again commonly uses the single linkage method. This process can terminate with more than 1 cluster as well.

A useful tool for the determination of the number of clusters generated by this single linkage method is the *normalized fusion density* $\tilde{\varphi}$, which is similar to the function $\varphi$ in (7.9): The above density estimators $\hat{f}_n(x_i)$ are normalized by putting $\tilde{f}_n(x_i) := 100 \hat{f}_n(x_i) / \max_{1 \leq j \leq n} \hat{f}_n(x_j) \in [0, 100]$, thus yielding percentages of the maximum value. Then $d$ is computed as above but with the normalized values, and the distance $D$ of two clusters is derived by applying the single linkage method. The monotone increasing function

$$\tilde{\varphi}(\nu) := \frac{1}{D(C_{k\bullet}^{(\nu+1)}, C_{l\bullet}^{(\nu+1)})} = \frac{1}{\frac{1}{2}\left(\frac{1}{\tilde{f}_n(x_{(k\bullet)})} + \frac{1}{\tilde{f}_n(x_{(l\bullet)})}\right)}$$

$$= \frac{\tilde{f}_n(x_{(k\bullet)}) \tilde{f}_n(x_{(l\bullet)})}{(\tilde{f}_n(x_{(k\bullet)}) + \tilde{f}_n(x_{(l\bullet)}))/2} \in [0, 100]$$

is the normalized fusion density. It can be interpreted as a density estimator computed at the middle of the two minimum sample points $x_{(k\bullet)} \in C_{k\bullet}^{(\nu+1)}$, $x_{(l\bullet)} \in C_{l\bullet}^{(\nu+1)}$ with $d(x_{(k\bullet)}, x_{(l\bullet)}) = D(C_{k\bullet}^{(\nu+1)}, C_{l\bullet}^{(\nu+1)})$. If $\tilde{\varphi}(\nu^*)/\tilde{\varphi}(\nu^* - 1)$ is considerably greater than 1, then the density at the middle of the joined clusters is too small, thus indicating that they are actually separated. The process should, therefore, be terminated with $\nu^*$ clusters. The plot of the function $\nu^*$ in Figure 7.5.4 suggests $\nu^* = 5$ clusters for the random data, compare with Figure 7.5.5.

In the following example we apply the nearest neighbor method with $k = 8$ neighbors to the 140 random data in Figure 7.5.1. The fusion of two *modal* or *multimodal* clusters is indicated by SAS if each of the two clusters contains at least $k$ sample points. In addition, the number of modal clusters in the final cluster configuration is reported. The clustering process terminates with $\nu = 4$ clusters, since there are no more adjacent sample points at this stage.

```
                  Density Linkage Cluster Analysis
                                                    Normalized
                                                  Maximum Density
                                       Normalized  in Each Cluster  T
                                         Fusion                     i
 NCL Clusters Joined       FREQ         Density    Lesser  Greater  e

 139 OB1      OB4            2          95.8220    91.9791 100.0000
 138 CL139    OB139          3          81.9833    69.4676 100.0000
 137 CL138    OB20           4          77.6361    63.4470 100.0000
 136 CL137    OB5            5          76.5016    61.9455 100.0000
  .
  .           116 further rows of cluster hierarchy         .
  .                                                         .

  19 CL20     OB90          26           9.3141     5.6286  29.9178
  18 CL25     OB30           5           9.1497     7.3719  12.0574 T
  17 CL18     OB71           6           9.1497     7.3719  12.0574
  16 CL31     OB56          31           8.9439     5.4722  45.5536
  15 CL17     OB118          7           8.8307     6.9664  12.0574
  14 CL16     CL21          52           8.4397 *  23.0121  45.5536
  13 CL15     OB76           8           7.9976     5.9831  12.0574
  12 CL13     OB140          9           7.4827     5.5791  12.0574
  11 CL12     OB59          10           7.1214     5.0528  12.0574
  10 CL41     OB108         46           7.0129     3.9852 100.0000
   9 CL14     OB70          53           6.6762     3.7307  45.5536
   8 CL11     OB122         11           6.2295     4.2917  12.0574
   7 CL8      OB22          12           5.7414     3.7677  12.0574
   6 CL7      OB55          13           5.3366     3.4876  12.0574
   5 CL9      OB24          54           5.2205     3.0227  45.5536
   4 CL6      OB134         14           2.2535     1.2508  12.0574
```

* indicates fusion of two modal or multimodal clusters

5 modal clusters have been formed.

**Figure 7.5.3.** Nearest neighbor cluster analysis of the random data in Figure 7.5.1.

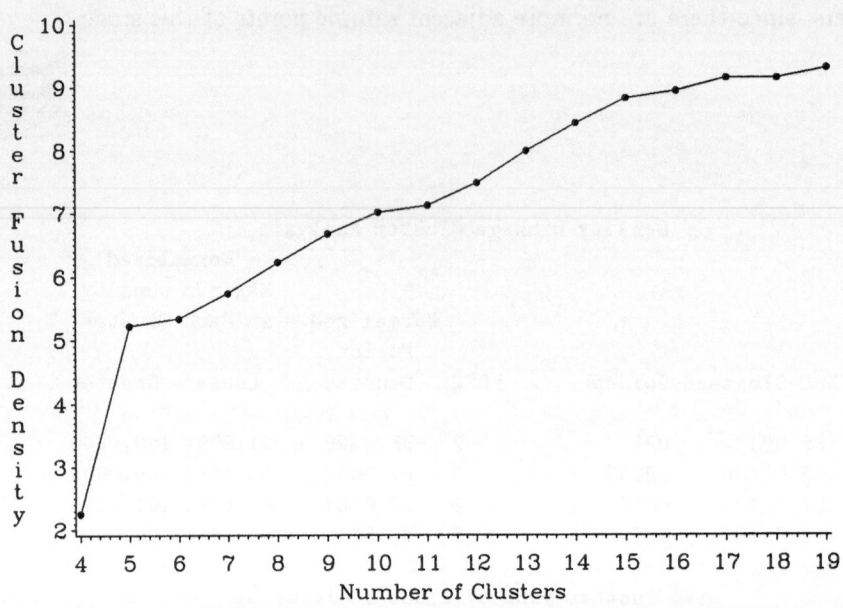

**Figure 7.5.4.** Normalized fusion density $\bar{\varphi}$ of the cluster process in Figure 7.5.3.

```
***   Program 7_5_3   ***;
TITLE1 'Nearest Neighbor Cluster Analysis';
TITLE2 'Random Data';
LIBNAME datalib 'c:\daten';

PROC CLUSTER DATA=datalib.random METHOD=DENSITY K=8
    OUTTREE=datalib.tree; VAR x1 x2;
RUN; QUIT;
```

The option METHOD=DENSITY requests density linkage for clustering with the uniform kernel. The other option K=8 specifies the nearest neighbor approach with 8 nearest neighbors.

The function $\bar{\varphi}$ is plotted the same way as in Program 7_4_2.

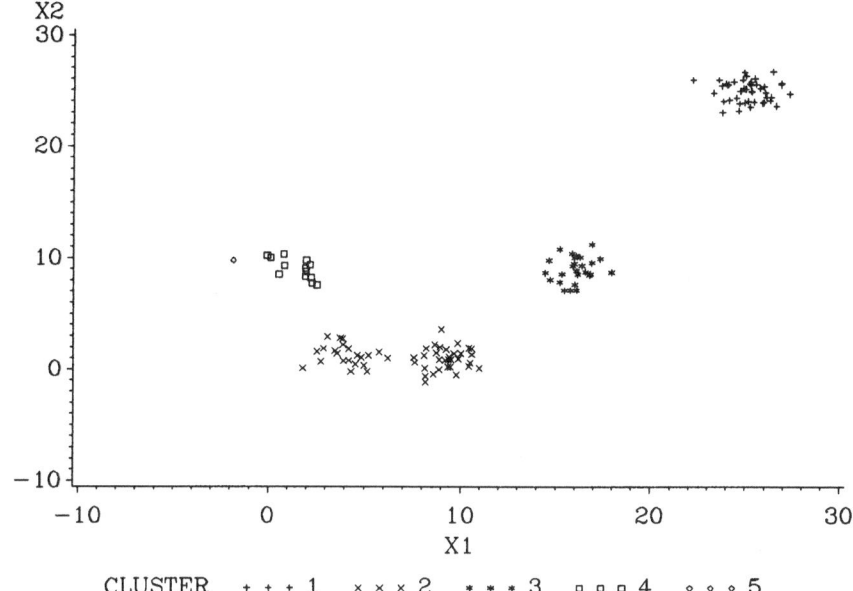

**Figure 7.5.5.** Random data with 5 clusters.

| _NAME_ | X1 | X2 | CLUSTER | CLUSNAME |
|--------|------|------|---------|----------|
| . | | | | |
| . | | | | |
| . | | | | |
| OB132 | 15.1818 | 7.7966 | 3 | CL19 |
| OB131 | 14.6093 | 9.7738 | 3 | CL19 |
| OB94 | 15.7492 | 7.0936 | 3 | CL19 |
| OB67 | 14.6712 | 8.0115 | 3 | CL19 |
| OB128 | 15.4291 | 7.0844 | 3 | CL19 |
| OB53 | 16.0690 | 7.0922 | 3 | CL19 |
| OB133 | 14.4146 | 8.6497 | 3 | CL19 |
| OB75 | 15.1748 | 10.7701 | 3 | CL19 |
| OB92 | 16.8950 | 11.1987 | 3 | CL19 |
| OB90 | 17.9290 | 8.6918 | 3 | CL19 |
| OB27 | 0.8211 | 9.2814 | 4 | CL6 |
| OB65 | 1.9414 | 8.7570 | 4 | CL6 |
| OB93 | 1.9192 | 9.0572 | 4 | CL6 |
| OB98 | 1.9092 | 8.3479 | 4 | CL6 |
| OB30 | 2.2017 | 8.2233 | 4 | CL6 |
| OB71 | 1.9591 | 9.7865 | 4 | CL6 |
| OB118 | 2.1568 | 9.3392 | 4 | CL6 |
| OB76 | 0.5262 | 8.5250 | 4 | CL6 |
| OB140 | 0.7897. | 10.3243 | 4 | CL6 |
| OB59 | 2.2568 | 7.7149 | 4 | CL6 |

| OB122 | 0.0860  | 10.0256 | 4 | CL6   |
| OB22  | 2.5055  | 7.5355  | 4 | CL6   |
| OB55  | -0.0950 | 10.1674 | 4 | CL6   |
| OB134 | -1.8559 | 9.7625  | 5 | OB134 |

**Figure 7.5.6.** Excerpt of the clustering of the random data into 5 clusters.

```
GOPTIONS RESET=GLOBAL COLORS=(BLUE);
***    Program 7_5_5    ***;
TITLE1 'Scatterplot after Nearest Neighbor Cluster Analysis';
TITLE2 'Random Data';
LIBNAME datalib 'c:\data';

PROC TREE DATA=datalib.tree NOPRINT OUT=clusdat NCL=5;
    COPY x1 x2;
PROC SORT;
    BY CLUSTER;
PROC PRINT;
    ID _NAME_;
PROC GPLOT DATA=clusdat;
    PLOT x2*x1=CLUSTER;
RUN; QUIT;
```

The TREE procedure uses the output OUT-TREE=*data set* created by PROC CLUSTER, here datalib.tree, see Program 7_5_3. It prints a tree diagram and it can also create an output data set identifying disjoint clusters at a specified level in the tree. The option 'NCL=k', here 'NCL=5', specifies the number of clusters desired in the OUT= *data set*, here 'clusdat'. The OUT= *data set* contains among others the automatic numeric variable CLUSTER taking values from 1 to $k$. The cluster belonging to the first observation is given the number 1, the cluster belonging to the next observation, that does not belong to cluster 1, is given the number 2, etc.

The COPY statement lists one or more variables to be copied to the OUT= *data set*. PROC SORT together with the statement BY CLUSTER arranges the observations according to their cluster membership and PROC PRINT prints them.

Figure 7.5.5 reveals that the number of clusters suggested by the function $\tilde{\varphi}$ in Figure 7.5.4, $k = 5$, is correct. But the 5 clusters are not due to the 5 different bivariate normal distribution, which actually underly the random data. The sample points corresponding to two bridging normal distributions are fused to 1 cluster, whereas one isolated point is taken as a cluster of its own. The fusion of the two bridging clusters is marked in the SAS output in Figure 7.5.3 by an asterisk '*', indicating the fusion of two modal clusters.

A cluster analysis ends with the interpretation of the derived cluster configuration. This is often by no means a simple task. One approach for an interpretation in the case of observations in $I\!\!R^p$ is the comparison of the means of the variables in each cluster. Sigificant differences can provide descriptions of the clusters if they have large means with respect to several variables and small means with respect to others.

## Exercises

**1.** Let $C = (c_{ij})$ be an arbitrary positive semidefinite $n \times n$–matrix. Show that $d$ defined by $d^2(i,j) := c_{ii} + c_{jj} - 2c_{ij}$ is a metric distance measure on $I = \{1, \ldots, n\}$.

**2.** A distance measure $u$ on $I = \{1, \ldots, n\}$ is called *ultrametric* if

$$u(i,j) \leq \max\{u(i,k), u(j,k)\}, \quad i,j,k \in I.$$

An example is given in Exercise 17. Prove that an ultrametric distance $u$ satisfies $u(i,k) = u(j,k)$ if $u(i,j) \leq u(i,k) \leq u(j,k)$. Conclude that $u$ is a metric distance measure. Is the reverse implication true that each metric distance measure is ultrametric as well?

**3.** Let $d$ be a metric on $I\!\!R^p$, see Section 7.2. Show that $d$ is nonnegative and symmetric, i.e., $d(x,y) \geq 0$ and $d(x,y) = d(y,x)$, $x,y \in I\!\!R^p$.

**4.** Show that

$$d(x,y) := \frac{|x-y|}{x+y}$$

defines a metric on $(0, \infty)$.

**5.** (i) Let $d$ be a metric on $I\!\!R^p$ and $w > 0$ an arbitrary number. Prove that

$$\tilde{d}(x,y) = \frac{d(x,y)}{w + d(x,y)},$$

is a metric on $I\!\!R^p$.

(ii) Let $d^{(1)}$ and $d^{(2)}$ be metrics on $I\!\!R^p$. Prove that

$$d((x_1, y_1), (x_2, y_2)) := d^{(1)}(x_1, x_2) + d^{(2)}(y_1, y_2), \quad x_i, y_i \in I\!\!R^p, \ i = 1, 2,$$

defines a metric on $I\!\!R^p \times I\!\!R^p$.

**6.** Show that any metric $d : I\!\!R^p \times I\!\!R^p \to [0, \infty)$ is a uniformly continuous function, where $I\!\!R^p \times I\!\!R^p$ is equipped with the metric defined in Exercise 5 (ii) with $d^{(1)} = d^{(2)} = d$ and the interval $[0, \infty)$ is equipped with the Euclidean distance.

**7.** Show that any metric $d$ on $I\!\!R^p$ which satisfies in addition

$$d(x+z, y+z) = d(x,y), \quad x,y,z \in I\!\!R^p, \qquad \text{(translation invariance)}$$

and

$$d(\alpha x, \alpha y) = |\alpha| d(x,y), \quad \alpha \in I\!\!R, \ x,y \in I\!\!R^p, \qquad \text{(homogeneity)}$$

is a *norm*: The function $||x|| := d(x, 0)$, $x \in \mathbb{R}^p$ satisfies

(N1)     $||x|| \geq 0$ and $||x|| = 0 \Leftrightarrow x = 0$,

(N2)     $||\alpha x|| = |\alpha| \, ||x||$,

(N3)     $||x + y|| \leq ||x|| + ||y||$.

8. Prove, for $1 \leq q_1 \leq q_2 < \infty$, the inequality (see Example 7.2.2)

$$d_{q_1}(x, y) \leq p^{1/q_1 - 1/q_2} d_{q_2}(x, y), \quad x, y \in \mathbb{R}^p.$$

Hint: Use Hölder's inequality (cf Section 3.1 in Berberian (1999)).

9. Show that $\lim_{q \to \infty} d_q(x, y) = \max_i |x_i - y_i|$.

10. Show that the squared Euclidean distance $d_2^2$ is not a metric.

11. Let $B$ be a positive definite matrix. Prove that the generalized squared distance

$$\Delta_B(x, y) = ((x - y)^T B^{-1}(x - y))^{1/2}$$

is a metric.

12. Prove that the Mahalanobis distance corresponding to the points $x_1, \ldots, x_n$ ($n \geq 2$) satisfies

$$\frac{1}{n(n-1)} \sum_{j=1}^{n} \sum_{i=1}^{n} \Delta^2(x_i, x_j) = 2p.$$

Hint: Check that $\sum_{1 \leq i, j \leq n} \Delta^2(x_i, x_j) = 2n \sum_{i=1}^{n} \Delta^2(x_i, \bar{x})$. Then use the equation $a^T a = tr(aa^T)$, $a = (a_1, \ldots, a_p)^T$.

13. Prove that the rank of a symmetric matrix coincides with the number of nonvanishing eigenvalues.

14. Generate the table of driving mileages in Figure 7.2.1 using SAS.

15. Compute the Mahalanobis distance matrix of the studentized economy data in Example 3.3.1 and generate a metric and a nonmetric *MDS* plot.

16. Use (7.7) to give an upper bound for the errors in the *MDS* plots in Figures 7.3.1 – 7.3.3. Generate corresponding *MDS* plots in 3 dimensions. Hint: Use the procedure G3D together with the PLOT statement.

17. Consider a distance matrix $D = (d(i, j))_{1 \leq i, j \leq n}$ and the single linkage method. Let $U = (u(i, j))_{1 \leq i, j \leq n}$, where $u_{i,j}$ is the smallest distance, such that the objects $i$ and $j$ are elements of the same class. Prove that

(i)  $U$ is ultrametric, i.e., $u(i, j) \leq \max\{u(i, k), u(j, k)\}$ (see Exercise 2).

(ii) $D = U \Leftrightarrow D$ is ultrametric.

(iii) Check that the distance matrix

$$\begin{pmatrix} 0 & 4 & 1 & 4 & 3 \\ & 0 & 4 & 2 & 4 \\ & & 0 & 4 & 3 \\ & & & 0 & 4 \\ & & & & 0 \end{pmatrix}$$

is ultrametric and plot the single linkage dendrogram of the pertaining clustering process.

**18.** Prove (7.11) and (7.12).

**19.** Compute cluster analyses of the economy data in Example 3.3.1 with different
methods and compare the corresponding dendrograms (see Exercise 15). Which cluster configuration do you determine? How do you interpret the clusters?

**20.** Prove the update algorithm (7.13).

**21.** Modify Program 7_5_2 such that the Epanechnikov kernel density estimator
with a sphering of the data is employed as in Section 6.5.

# Chapter 8

# Principal Components

Consider $p$ measurements, which are taken on each of $n$ objects. Principal component analysis aims at replacing these $p$ variables by $k$ new ones, where the number $k$ is small relative to $p$. These new variables, called principal components, are not generated by simply dropping several of the original variables, since this might result in the loss of important information. Instead, each of the principal components is a linear combination of the original variables designed to carry most of the information in the initial sample. Principal component analysis, therefore, is mainly a technique for reducing the dimension of the data. But it is also a tool for the detection of outliers in a high dimensional data set.

We will introduce principal components as an exploratory data analysis tool, i.e., we do not assume any model underlying the data (cf Exercises 6 – 11). Then we will introduce exploratory factor analysis by means of principal components and we will discuss its additional features in detail. A case study of the economy data concludes this chapter.

## 8.1 Principal Components in $I\!\!R^2$

To give a geometric interpretation of principal components, we start with observations in $I\!\!R^2$. The following scatterplot displays a sample of $n = 100$ highly correlated points $x_1, \ldots, x_n$ in $I\!\!R^2$.

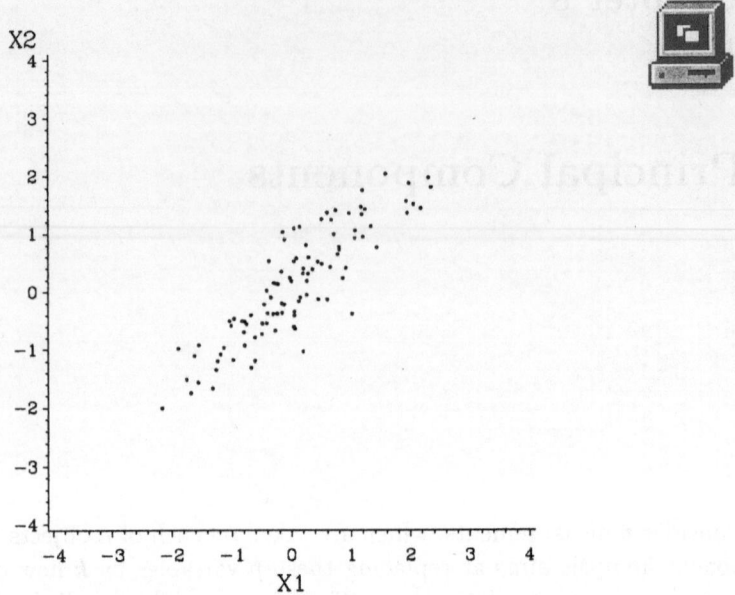

**Figure 8.1.1.** Scatterplot of 100 highly correlated points in $\mathbb{R}^2$.

```
***     Program 8_1_1     ***;
TITLE1 'Scatterplot';
TITLE2 '100 random vectors with correlation 0.9';
LIBNAME datalib 'c:\data';

DATA datalib.pca1;
    SET datalib.corr1(WHERE=(rho=0.9));
    RENAME x=x1 y=x2;

SYMBOL1 V=DOT C=RED H=.3 I=NONE;
AXIS1 ORDER=(-4 TO 4) LENGTH=6CM VALUE=(H=1.5);
PROC GPLOT DATA=datalib.pca1;
    PLOT x2*x1 / HAXIS=AXIS1 VAXIS=AXIS1;
RUN; QUIT;
```

The above program uses those 100 observations from the file datalib.corr1 generated in Program 3_1_1, for which the variable 'rho' attains the value 0.9. The variables 'x' and 'y' are written to the file datalib.pca1 under the names 'x1' and 'x2' by using the RENAME statement.

This data cloud obviously has one principal direction $r_1$ in $\mathbb{R}^2$, along which the variation of the data is maximized. There is another principal direction $r_2$, which is at right angles to $r_1$. Imagine to capture the sample by an ellipsoid with its two principal axes providing the directions. We will give below a precise definition of the principal directions by means of the projection pursuit technique in Section 6.3.

A direction in $\mathbb{R}^p$ can in general be defined by a vector $r \in \mathbb{R}^p$ of length 1, i.e., $||r|| = 1$. Think of a directional arrow that points for instance to the north. In the above example we are now looking for that vector $r_1$, whose direction maximizes the variation of the data: We project the data onto the line $\{sr_1 : s \in \mathbb{R}\}$, thus obtaining their coordinates along this line

$$s_i := r_1^T x_i, \qquad i = 1, \ldots, n,$$

with maximum variation

$$\sum_{i=1}^{n}(s_i - \bar{s})^2 = \sum_{i=1}^{n}(r_1^T x_i - r_1^T \bar{x})^2 = \max_{r \in \mathbb{R}^2, ||r||=1} \sum_{i=1}^{n}(r^T(x_i - \bar{x}))^2. \qquad (8.1)$$

By $\bar{s}$ we denote the arithmetic mean of the $s_i$ and by $\bar{x}$ the mean vector of $x_1, \ldots, x_n$. Note that the arithmetic mean of the projected data $\bar{s}$ equals the projection of the mean vector of the data $n^{-1}\sum_{i=1}^{n} r^T x_i = r^T \bar{x}$.

Except for its sign, a solution $r_1$ of (8.1) will commonly be unique; the vector $-r_1$ obviously solves (8.1) as well. Next we are looking for the second principal direction $r_2 \in \mathbb{R}^2$, $||r_2|| = 1$, which is orthogonal to $r_1$, i.e., $r_2^T r_1 = 0$. Except for its sign, this second principal direction $r_2$ obviously is uniquely determined.

We thus obtain the two functions or new variables

$$z_1(y) := r_1^T(y - \bar{x}) \text{ and } z_2(y) := r_2^T(y - \bar{x}), \qquad y \in \mathbb{R}^2.$$

These variables are called the first and second *principal components* of $x_1, \ldots, x_n$ and the function

$$z(y) := \begin{pmatrix} z_1(y) \\ z_2(y) \end{pmatrix} = \begin{pmatrix} r_1^T(y - \bar{x}) \\ r_2^T(y - \bar{x}) \end{pmatrix}, \qquad y \in \mathbb{R}^2,$$

is the *principal axes transformation*. The vector $z(y)$ provides the coordinates $s_1, s_2$ of the point $y$ with respect to the orthogonal system of coordinates $\{\bar{x} + s_1 r_1 + s_2 r_2 : s_1, s_2 \in \mathbb{R}\}$ in $\mathbb{R}^2$. The vectors

$$z(x_i) = \begin{pmatrix} z_1(x_i) \\ z_2(x_i) \end{pmatrix}, \qquad i = 1, \ldots, n$$

contain the *principal component scores* $z_1(x_i), z_2(x_i)$ of the data.

Suppose now that the variation of $x_1, \ldots, x_n$ in the direction of the second principal component is small, i.e., suppose that the number

$$\frac{1}{n-1} \sum_{i=1}^{n} z_2^2(x_i) = \frac{1}{n-1} \sum_{i=1}^{n} (r_2^T(x_i - \bar{x}))^2$$

is close to 0. Then all the points $x_1, \ldots, x_n \in \mathbb{R}^2$ lie approximately on the line $\bar{x} + s r_1$, since their second coordinates $z_2(x_i)$ almost vanish. In this case we could reduce the two variables $z_1, z_2$ to the single variable $z_1$ without losing essential information about the sample points. This is what the reduction to principal components aims at. We will introduce this technique for arbitrary dimensions in the next section.

Those points $x_i$ in the sample, whose first or second principal component is significantly large in absolute value, are isolated points in the data set. In particular, a significantly large *second* principal component indicates an outlier, since this point is far away from the principal direction of the data cloud. The following two figures display again the scatterplot in Figure 8.1.1. Figure 8.1.2 contains, in addition, the two principal axes $s_1 r_1 + \bar{x}$, $s_2 r_2 + \bar{x}$, $s_1, s_2 \in \mathbb{R}$, and Figure 8.1.3 displays the principal axis transformation. Figure 8.1.2 suggests the idea that the first principal component is the regression line from Section 3.2 (see Figure 3.2.1). But the regression line is defined as the minimizer of the residual sum of squares using *vertical* projection of the data onto the line, whereas the first principal component is derived using *orthogonal* projection (see Exercise 4).

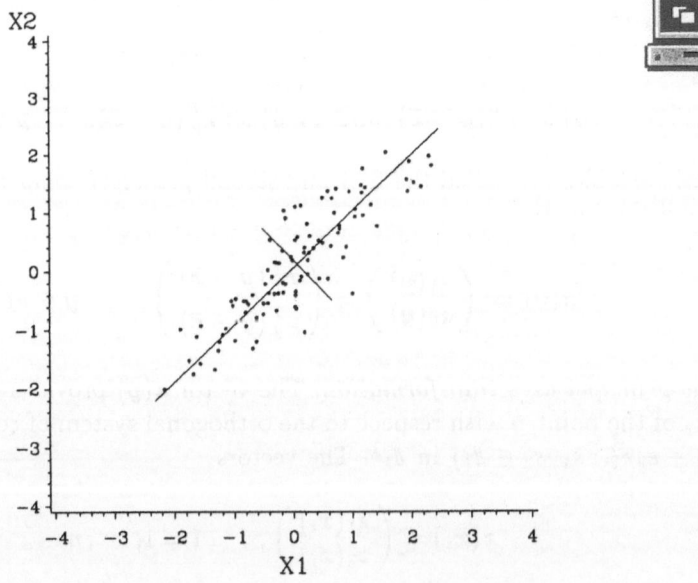

**Figure 8.1.2.** Scatterplot from Figure 8.1.1 with principal axes.

```
***    Program 8_1_2    ***;
TITLE1 'Principal Components'; TITLE2 'Data from Figure 8.1.1';
LIBNAME datalib 'c:\data';

PROC PRINCOMP DATA=datalib.pca1 COV OUTSTAT=stats NOPRINT;
    VAR x1 x2;

DATA anno;
    LENGTH FUNCTION $8.;
    SET stats;   RETAIN mx my l1 l2;
    IF _TYPE_='MEAN' THEN DO; mx=x1; my=x2; END;
    IF _TYPE_='EIGENVAL' THEN DO;
        l1=SQRT(x1); l2=SQRT(x2); END;
    IF _TYPE_='SCORE' THEN DO;
            IF _NAME_='Prin1' THEN l=2.5*l1;   ELSE l=2.5*l2;
            XSYS='2'; YSYS='2'; COLOR='RED';
            FUNCTION='MOVE'; X=mx-l*x1; Y=my-l*x2; OUTPUT;
            FUNCTION='DRAW'; X=mx+l*x1; Y=my+l*x2; OUTPUT;   END;
SYMBOL1 V=DOT H=0.3 I=NONE C=G;
AXIS1 ORDER=(-4 TO 4) LENGTH=6CM VALUE=(H=1.5);
PROC GPLOT DATA=datalib.pca1;
    PLOT x2*x1 / VAXIS=AXIS1 HAXIS=AXIS1 ANNOTATE=anno;   RUN; QUIT;
```

The principal axes in the data cloud are drawn by means of an ANNOTATE data set, see Program 7_1_1. The information about the location of the axes is contained in the OUTSTAT=*data set*, here 'stats', generated in the PRINCOMP procedure.

Without the option COV, SAS computes a principal component analysis of the correlation matrix.

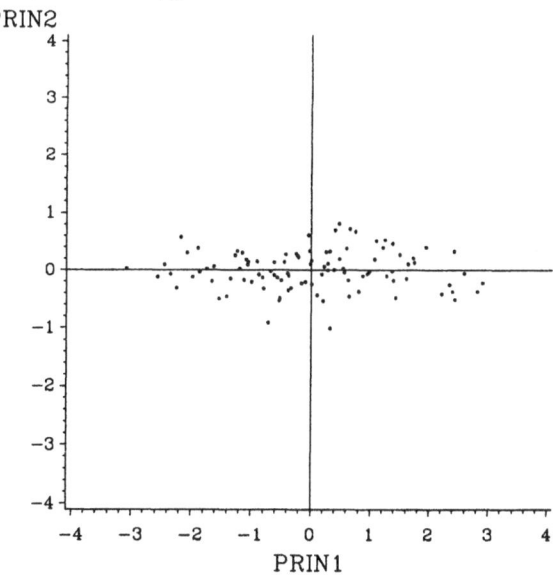

Figure 8.1.3. Principal axes transformation of the scatterplot in Figure 8.1.2.

```
***    Program 8_1_3   ***;
TITLE1 'Principal Components';
TITLE2 'Data from Figure 8.1.2';
LIBNAME datalib 'c:\data';

PROC PRINCOMP DATA=datalib.pca1 OUT=pca2 COV NOPRINT;
    VAR x1 x2;
SYMBOL1 V=DOT H=0.3 I=NONE C=G;
AXIS1 ORDER=(-4 TO 4) LENGTH=6CM VALUE=(H=1.5);
PROC GPLOT DATA=pca2;
    PLOT PRIN2*PRIN1 / VREF=0 HREF=0 HAXIS=AXIS1 VAXIS=AXIS1;
RUN; QUIT;
```

This figure displays the 100 observations from the data set datalib.pca1 with respect to the new system of coordinates, which is given by the principal components. The coordinates are contained in the variables 'PRIN1' and 'PRIN2' in the OUT= *data set*, generated in the PRINCOMP procedure.

## 8.2   Principal Components in $I\!R^p$

We consider in the following $n$ objects together with $p$ variables that were measured for each of the objects, i.e., our sample can be represented by $n$ vectors $x_i$ in $I\!R^p$. Denote by $\bar{x} = n^{-1} \sum_{i=1}^{n} x_i$ the mean vector and by

$$S := \frac{1}{n-1} \sum_{i=1}^{n} (x_i - \bar{x})(x_i - \bar{x})^T$$

the corresponding empirical $p \times p$–covariance matrix. Since the diagonalization of the matrix $S$ is the mathematical key to principal components analysis, we give its detailed derivation below, although we already introduced the diagonalization of a matrix in Section 7.3 in the framework of multidimensional scaling.

### Diagonalization of $S$

Since the $p \times p$–matrix $S$ is symmetric and positive semidefinite (see Exercise 20 in Chapter 3), it has $p$ nonnegative eigenvalues $\lambda_i \geq 0$, which we arrange according to their size: $\lambda_1 \geq \lambda_2 \geq \cdots \geq \lambda_p \geq 0$. Denote by $r_1, \ldots, r_p$ corresponding orthogonal eigenvectors of length 1, i.e., we have

$$r_j^T r_k = \begin{cases} 1 & \text{if } j = k, \\ 0 & \text{if } j \neq k, \end{cases} \tag{8.2}$$

and

$$Sr_j = \lambda_j r_j, \qquad j = 1, \ldots, p, \tag{8.3}$$

see, for example, Chapter VIII in Lang (1987). Denote, as in Section 7.3, by $R$ that $p \times p$–matrix, whose columns are the $p$ eigenvectors $r_1, \ldots, r_p$ of $S$. Then we can write (8.3) in closed form as

$$SR = R\Lambda, \tag{8.4}$$

where $\Lambda$ is the $p \times p$–diagonal matrix of the eigenvalues:

$$\Lambda = \begin{pmatrix} \lambda_1 & & & 0 \\ & \lambda_2 & & \\ & & \ddots & \\ 0 & & & \lambda_p \end{pmatrix}.$$

$R$ is, by (8.2), an orthogonal matrix, i.e., it satisfies $R^T R = I_p$. This implies $R^{-1} = R^T$ and, thus, $RR^T = I_p$. If we multiply both sides of equation (8.4) with $R^T$ from the left, we obtain the diagonalization of $S$:

$$R^T SR = \Lambda \quad \text{and} \quad S = R\Lambda R^T. \tag{8.5}$$

Thus,

$$S = \lambda_1 r_1 r_1^T + \lambda_2 r_2 r_2^T + \ldots + \lambda_p r_p r_p^T,$$

which is called the *spectral decomposition* of $S$.

## Principal Components

We will show in the sequel that the eigenvectors $r_1, \ldots, r_p$ are the $p$ principal directions of the data cloud $\{x_i : i = 1, \ldots, n\}$ in $\mathbb{R}^p$.

**8.2.1 Definition.** The variable

$$z_j(y) := r_j^T (y - \bar{x}), \qquad y \in \mathbb{R}^p,$$

is the $j$–th *principal component* of $x_1, \ldots, x_n$ with $z_j(x_i)$ being the $j$–th principal component score, and the function

$$z(y) := \begin{pmatrix} z_1(y) \\ \vdots \\ z_p(y) \end{pmatrix} = R^T (y - \bar{x}), \qquad y \in \mathbb{R}^p$$

is the *principal component* or *principal axes transformation*.

The transformation to principal components $z(y) = (s_1, \ldots, s_p)^T$ of a vector $y \in \mathbb{R}^p$ yields the coordinates $(s_1, \ldots, s_p)^T$ of $y$ with respect to the system of coordinates $T := \{\bar{x} + s_1 r_1 + \cdots + s_p r_p : s_j \in \mathbb{R}, j = 1, \ldots, p\}$.

**8.2.2 Theorem.** *The principal component scores of* $x_1, \ldots, x_n$ *satisfy*

(i) $\dfrac{1}{n} \displaystyle\sum_{i=1}^{n} z_j(x_i) = 0, \qquad j = 1, \ldots, p,$

(ii) $\dfrac{1}{n-1} \displaystyle\sum_{i=1}^{n} z_j^2(x_i) = \lambda_j, \qquad j = 1, \ldots, p,$

(iii) $\dfrac{1}{n-1} \displaystyle\sum_{i=1}^{n} z_j(x_i) z_k(x_i) = 0, \qquad j, k = 1, \ldots, p, \quad j \neq k,$

*i.e., the principal component scores have mean 0, variance* $\lambda_j, j = 1, \ldots, p$, *and are uncorrelated. Finally, the covariance matrix* $S$ *satisfies*

(iv) $tr(S) = \displaystyle\sum_{j=1}^{p} \lambda_j,$

(v) $\det(S) = \displaystyle\prod_{j=1}^{p} \lambda_j.$

**Proof:** (i) is obvious:

$$\frac{1}{n} \sum_{i=1}^{n} z_j(x_i) = \frac{1}{n} \sum_{i=1}^{n} r_j^T(x_i - \bar{x}) = r_j^T(\bar{x} - \bar{x}) = 0.$$

Parts (ii) and (iii) follow from:

$$\frac{1}{n-1} \sum_{i=1}^{n} z(x_i) z(x_i)^T = \frac{1}{n-1} \sum_{i=1}^{n} R^T(x_i - \bar{x})(R^T(x_i - \bar{x}))^T$$

$$= \frac{1}{n-1} \sum_{i=1}^{n} R^T(x_i - \bar{x})(x_i - \bar{x})^T R$$

$$= R^T S R = \Lambda.$$

(iv) is a consequence of (8.5) and the equation $tr(AB) = tr(BA)$ for arbitrary $p \times p$–matrices $A, B$ (see Exercise 26, Chapter 3):

$$tr(S) = tr(SI_p) = tr(SRR^T)$$

$$= tr(R^T SR) = tr(\Lambda) = \sum_{j=1}^{p} \lambda_j.$$

Finally, part (v) follows from (8.5) and the fact that the determinant of the product of $p \times p$-matrices $\boldsymbol{A}, \boldsymbol{B}$ is the product of their determinants, i.e., $\det(\boldsymbol{AB}) = \det(\boldsymbol{A})\det(\boldsymbol{B})$. This implies that $\det(\boldsymbol{A}^{-1}) = 1/\det(\boldsymbol{A})$ if the inverse matrix $\boldsymbol{A}^{-1}$ exists, i.e., if $\det(\boldsymbol{A}) \neq 0$ (see, for instance, Section 2.1 in Kwak and Hong (1997)). And hence,

$$\det(\boldsymbol{S}) = \det(\boldsymbol{R\Lambda R}^T) = \det(\boldsymbol{R})\det(\boldsymbol{\Lambda})\det(\boldsymbol{R}^T)$$

$$= \det(\boldsymbol{\Lambda}) = \prod_{j=1}^{p} \lambda_j,$$

since $\boldsymbol{R}^T = \boldsymbol{R}^{-1}$.                                                                                □

## Maximum Variation

The consequence of part (ii) in Theorem 8.2.2 is that, among their $p$ principal components, the points $\boldsymbol{x}_1, \ldots, \boldsymbol{x}_n$ have maximum variation $\lambda_1$ in the direction of their first principal component, second to maximum variation $\lambda_2$ along their second principal component and so on. Actually $\boldsymbol{x}_1, \ldots, \boldsymbol{x}_n$ have maximum variation in the direction of their first principal component among *any* direction $\boldsymbol{r} \in \mathbb{R}^p$. This is the content of the next theorem.

**8.2.3 Theorem.** *Let $\boldsymbol{r} \in \mathbb{R}^p$ be an arbitrary vector of length 1. Then we have*

$$\frac{1}{n-1} \sum_{i=1}^{n} (\boldsymbol{r}^T(\boldsymbol{x}_i - \bar{\boldsymbol{x}}))^2 \leq \lambda_1.$$

**Proof:** Any vector $\boldsymbol{r} \in \mathbb{R}^p$ with $||\boldsymbol{r}|| = 1$ can be written as a linear combination $\boldsymbol{r} = \sum_{j=1}^{p} c_j \boldsymbol{r}_j$ of the principal directions $\boldsymbol{r}_j$. Due to the orthonormality (8.2) of the $\boldsymbol{r}_j$, the coefficients $c_j$ are given by $c_j = \boldsymbol{r}^T \boldsymbol{r}_j$ and they satisfy

$$1 = \boldsymbol{r}^T \boldsymbol{r} = \sum_{j=1}^{p} c_j^2.$$

From equation (8.5) we obtain

$$\frac{1}{n-1} \sum_{i=1}^{n} (\boldsymbol{r}^T(\boldsymbol{x}_i - \bar{\boldsymbol{x}}))^2 = \frac{1}{n-1} \sum_{i=1}^{n} (\boldsymbol{r}^T(\boldsymbol{x}_i - \bar{\boldsymbol{x}}))((\boldsymbol{x}_i - \bar{\boldsymbol{x}})^T \boldsymbol{r}) = \boldsymbol{r}^T \boldsymbol{S} \boldsymbol{r}$$

$$= \boldsymbol{r}^T \boldsymbol{R\Lambda R}^T \boldsymbol{r} = (\boldsymbol{r}^T \boldsymbol{R})\boldsymbol{\Lambda}(\boldsymbol{r}^T \boldsymbol{R})^T$$

$$= (c_1, \ldots, c_p)\boldsymbol{\Lambda}(c_1, \ldots, c_p)^T = \sum_{j=1}^{p} c_j^2 \lambda_j \leq \lambda_1.$$                      □

The next result states that $x_1, \ldots, x_n$ have, in the direction of their $k$-th principal components, the maximum variation among all directions $r$, which are orthogonal to the first $k - 1$ principal directions $r_1, \ldots, r_{k-1}$, $k = 2, \ldots, p$.

**8.2.4 Theorem.** *Any vector $r \in \mathbb{R}^p$ of length 1 with $r^T r_j = 0$ for $j = 1, \ldots, k - 1$, $k \geq 2$, satisfies the inequality*

$$\frac{1}{n-1} \sum_{i=1}^{n} \left( r^T(x_i - \bar{x}) \right)^2 \leq \lambda_k.$$

**Proof:** From the proof of Theorem 8.2.3 we obtain the representation $r = \sum_{j=1}^{p} c_j r_j$, where $c_j = r^T r_j = 0$, $j = 1, \ldots, k - 1$, and thus,

$$\frac{1}{n-1} \sum_{i=1}^{n} (r^T(x_i - \bar{x}))^2 = \sum_{j=1}^{p} c_j^2 \lambda_j = \sum_{j=k}^{p} c_j^2 \lambda_j \leq \lambda_k. \qquad \square$$

## Dimension Reduction

So far we have shown that the problem of determining principal components consists of the problem of computing the eigenvalues and corresponding orthonormal eigenvectors of the matrix $S$.

If an eigenvalue $\lambda_{k+1}$ is close to 0, the variation about 0 of the principal component scores $z_j(x_i)$, $i = 1, \ldots, n$, is small for $j \geq k + 1$. Taken with respect to the system of coordinates $T$, the last $p - k$ coordinates of the points $x_i \in \mathbb{R}^p$ are, in this case, close to 0 and can, therefore, be dropped without losing important information:

$$x_i \mapsto \begin{pmatrix} z_1(x_i) \\ \vdots \\ z_k(x_i) \end{pmatrix} = \begin{pmatrix} r_1^T(x_i - \bar{x}) \\ \vdots \\ r_k^T(x_i - \bar{x}) \end{pmatrix} \in \mathbb{R}^k, \qquad i = 1, \ldots, n.$$

If $k$ is small relative to $p$, we have solved the initial dimension reduction problem by replacing the high dimensional data $x_i \in \mathbb{R}^p$ by the lower dimensional observations $(z_j(x_i))_{j=1}^{k}$.

## How Many Principal Components?

A measure for the total variation of the observations $x_i = (x_{i1}, \ldots, x_{ip})^T$, $i = 1, \ldots, n$, is the sum of the variation of each of the $p$ components, see Theorem 8.2.2, (iv):

$$\sum_{j=1}^{p} \frac{1}{n-1} \sum_{i=1}^{n} (x_{ij} - \bar{x}_j)^2 = tr(S) = \sum_{j=1}^{p} \lambda_j.$$

By $\bar{x}_j$ we denote the $j$-th component of the mean vector $\bar{x} = (\bar{x}_1, \ldots, \bar{x}_p)^T$. The trace $tr(S)$ is usually called *total variation*, and the first $k$ principal components of the $x_i$ explain

$$100 \frac{\lambda_1 + \cdots + \lambda_k}{\lambda_1 + \cdots + \lambda_p}$$

percent of the total variation, see Theorem 8.2.2. This percentage can be used to determine the number $k$ of principal components, which are then used for a further analysis of the data. We can take e.g. the minimum number $k$ such that 90% of the total variation is explained. If the rank of $S$ is $k$, the first $k$ principal components explain 100% of the total variation, since in this case $\lambda_{k+1} = \cdots = \lambda_p = 0$ (see Exercise 13 in Chapter 7).

Another popular selection rule is to use only those principal components for a further analysis, whose eigenvalues $\lambda_k$ are larger than the average of all eigenvalues, i.e., $\lambda_k > \sum_{j=1}^{p} \lambda_j / p$. Suppose that the $x_i$ have variations equal to 1 in each component, i.e.,

$$\frac{1}{n-1} \sum_{i=1}^{n} (x_{ij} - \bar{x}_j)^2 = 1, \qquad j = 1, \ldots, p,$$

which holds in the case of standardized observations. Then we have $\sum_{j=1}^{p} \lambda_j = tr(S) = p$ and, thus, $\sum_{j=1}^{p} \lambda_j / p = 1$. The selection rule $\lambda_k > \sum_{j=1}^{p} \lambda_j / p$ is, thus, simply the rule: Use those principal components, whose eigenvalues are larger than 1; see also the standardized principle components analysis below.

A graphical selection rule can be based on a plot of the points $(i, \lambda_i)$, $i = 1, \ldots, p$. The interpolation of these points by a polygon often generates a plot, which is shaped like the slope of a hill with an accumulation of loose stones at its base. Those principal components, whose eigenvalues visually belong to this *scree*, are excluded from further analysis. The selection rule based on this *scree plot* is, therefore, to take those $k$ principal components whose eigenvalues lie on the slope, see Figure 8.2.2.

In the sequel, we compute a principal component analysis of the economy data in Example 3.3.1 with 20 countries and 9 variables, see also Figure 7.3.2. The observations are standardized by dividing each variable by its empirical standard deviation. The matrix $S$ is then the correlation matrix of the observations and, thus, $tr(S) = 9$. This standardization is computed by SAS by default; its advantages are explained below. The scree plot in Figure 8.2.2 suggests the reduction to the first 4 principal components. Equally, we have $\lambda_4 > tr(S)/p = 9/9 = 1$, whereas $\lambda_5$ is essentially smaller than 1; see also the factor analysis of these data in Example 8.3.3 and Exercise 17.

The PRINCOMP Procedure

Observations          20
Variables              9

Eigenvalues of the Correlation Matrix

|   | Eigenvalue | Difference | Proportion | Cumulative |
|---|---|---|---|---|
| 1 | 2.53617448 | 0.13806738 | 0.2818 | 0.2818 |
| 2 | 2.39810710 | 0.72679188 | 0.2665 | 0.5483 |
| 3 | 1.67131522 | 0.48448479 | 0.1857 | 0.7340 |
| 4 | 1.18683043 | 0.77302228 | 0.1319 | 0.8658 |
| 5 | 0.41380815 | 0.07276066 | 0.0460 | 0.9118 |
| 6 | 0.34104749 | 0.06722330 | 0.0379 | 0.9497 |
| 7 | 0.27382419 | 0.16003965 | 0.0304 | 0.9801 |
| 8 | 0.11378453 | 0.04867612 | 0.0126 | 0.9928 |
| 9 | 0.06510841 |            | 0.0072 | 1.0000 |

Eigenvectors

|          | Prin1 | Prin2 | Prin3 | Prin4 | Prin5 |
|----------|-------|-------|-------|-------|-------|
| INVEST   | -.025532 | 0.282593 | -.575634 | -.294937 | -.026467 |
| INFLATN  | 0.489223 | 0.169237 | 0.209113 | -.390947 | -.038211 |
| GNP      | 0.160140 | 0.372307 | -.482488 | 0.245135 | -.164870 |
| TAX      | 0.151289 | -.522763 | 0.163120 | -.001939 | -.513589 |
| NUKES    | -.353728 | 0.390644 | 0.376754 | -.051161 | -.183212 |
| UNEMPLD  | 0.190305 | 0.099087 | 0.103374 | 0.818604 | 0.160141 |
| LABCOST  | -.415273 | -.368600 | -.156634 | -.073072 | 0.579590 |
| POPULATN | -.350793 | 0.411834 | 0.370448 | -.037118 | 0.038609 |
| STRIKE   | 0.504607 | 0.108225 | 0.226093 | -.143635 | 0.557007 |

Eigenvectors

|         | Prin6 | Prin7 | Prin8 | Prin9 |
|---------|-------|-------|-------|-------|
| INVEST  | 0.444773 | -.545445 | -.070864 | -.009595 |
| INFLATN | -.159347 | -.162905 | 0.670958 | 0.178543 |
| GNP     | 0.141172 | 0.650499 | 0.268944 | -.016288 |
| TAX     | 0.615048 | 0.062253 | 0.132649 | -.117005 |
| NUKES   | 0.298872 | 0.100243 | -.079174 | 0.662197 |
| UNEMPLD | 0.164251 | -.411167 | 0.201590 | 0.104151 |

| LABCOST | 0.231342 | 0.129005 | 0.475416 | 0.172345 |
|---------|----------|----------|----------|----------|
| POPULATN | 0.187988 | -.009076 | 0.242474 | -.687765 |
| STRIKE | 0.416426 | 0.230428 | -.350922 | -.044706 |

**Figure 8.2.1.** Principal component analysis of the economy data in Example 3.3.1.

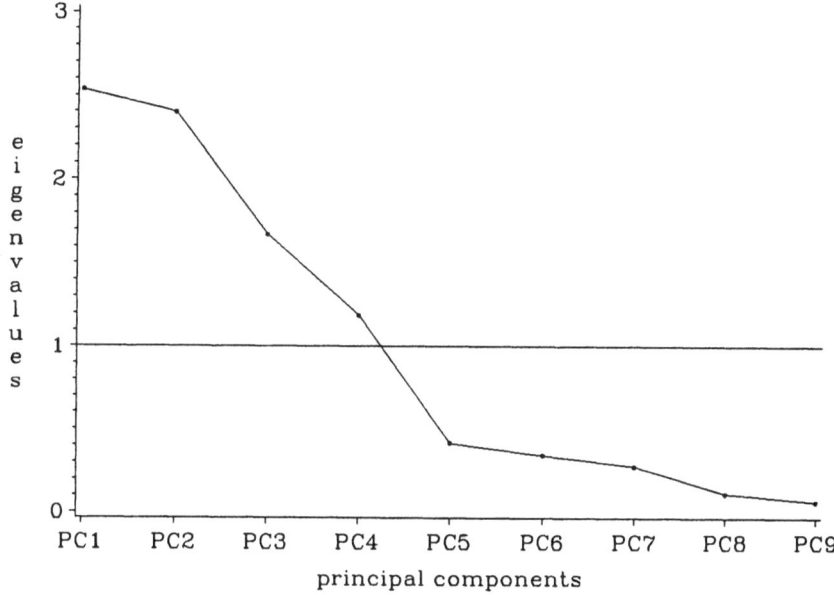

**Figure 8.2.2.** Scree plot corresponding to the principal component analysis in Figure 8.2.1.

```
***    Program 8_2_2    ***;
TITLE1 'Scree Plot'; TITLE2 'Economy Data';
LIBNAME datalib 'c:\data';
PROC PRINCOMP DATA=datalib.economy OUTSTAT=stat1 NOPRINT;
DATA stat2(KEEP=ev pc);
    SET stat1(WHERE=(_TYPE_='EIGENVAL'));
    LABEL pc='principal components' ev='eigenvalues';
    ARRAY pr {9} invest--strike;
    DO i=1 TO 9;
        pc='PC'||LEFT(i);
        ev=pr{i};
OUTPUT; END;
```
↓

```
                               ↑
   SYMBOL1 V=DOT H=.5 I=JOIN;
   PROC GPLOT DATA=stat2;
      PLOT ev*pc / VREF=1;
   RUN; QUIT;
```

The scree plot can be derived from the OUT-STAT= *data set*, here 'stat1', in the PRIN-COMP procedure. The data set stat1 contains a row, where the eigenvalues of the principal components are stored. This row is assigned the value 'EIGENVAL' of the automatic SAS variable _TYPE_. It can be selected by the corresponding DATA SET option, see Program 8_1_1.

From this row the DATA step generates the new data set 'stat2', which contains the 9 eigenvalues and the variables 'ev' and 'pc'. Note the function LEFT and the operator '||'. This operator joins two strings. Here the values of the variable 'i' are added to the string 'PC'. Since the values of 'i' are integers, which are right aligned by default, we use LEFT to left align them yielding PC1, PC2,..., PC9.

Arrays are set up in the ARRAY statement. It can appear anywhere in the DATA step prior to its first use. Its syntax is 'AR-RAY *name {dimension}* [$] *variables*', where $ would indicate that the elements of the array are character variables and *dimension* is the number of variables in the array.

The option COV in the PRINCOMP procedure would compute a principal components analysis of the covariance matrix in place of the correlation matrix, which is the default, see Programs 8_1_2 and 8_1_3. Without the option NOPRINT the results of PROC PRINCOMP are printed as in Figure 8.2.1.

## Identification of Outliers

Principal components are useful tools for the detection of outliers among the $x_i$. We consider the first $k$ principal component scores of the $x_i$, thus obtaining $k$ sets of one dimensional data $z_j(x_i)$, $i = 1, \ldots, n$. These can be checked for outliers using the methods in Section 1.5, for example by boxplots. A further useful tool is a scatterplot matrix, where each entry is the scatterplot of two principal components $(z_j(x_i), z_l(x_i))$, $i = 1, \ldots, n$. Those points $(z_j(x_m), z_l(x_m))$, which are isolated in this plot, indicate that $x_m$ is an outlier.

The following figure is a scatterplot matrix of all 5 principal components of the cns data in Example 1.1.3. There are, for instance, three obviously isolated points in the final column; two of these have relatively large 5th principal component scores and one is very small, see also the subsequent Figures 8.2.4 and 8.2.6. There are several scatterplots such as the plot of the second against the fourth component which could be mixtures of two linear data clouds, thus indicating a stratification of the data.

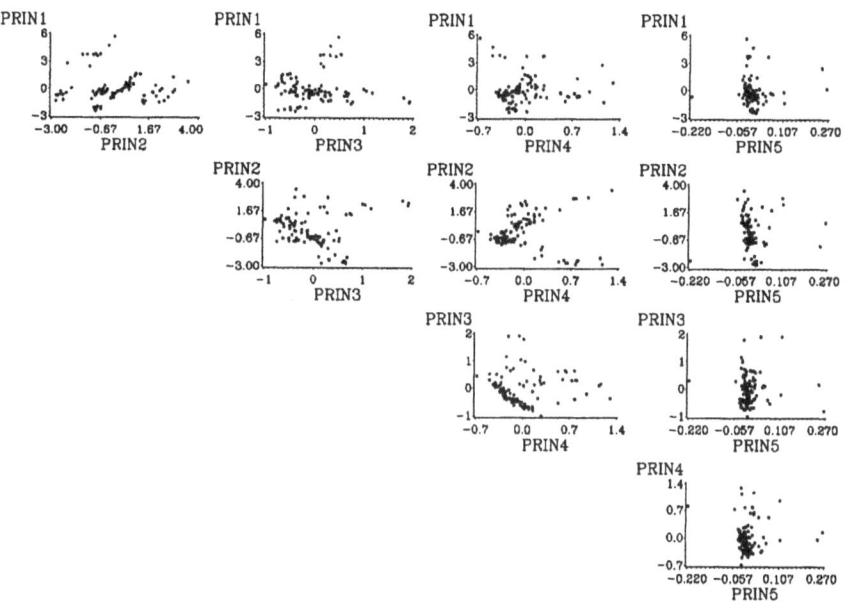

**Figure 8.2.3.** Scatterplot matrix of all 5 principal component scores of the cns data.

```
***    Program 8_2_3    ***;
TITLE1 'Scatterplot Matrix Principal Component Scores';
TITLE2 'CNS Data'
LIBNAME datalib 'c:\data';

PROC PRINCOMP DATA=datalib.cns OUT=datalib.princ NOPRINT;
    VAR an on mn gn ao;

GOPTIONS NODISPLAY HTEXT=4; TITLE1;
AXIS1 MAJOR=(N=4) LABEL=(H=5.5) VALUE=(H=4);
SYMBOL1 V=DOT C=G H=1 I=NONE;
PROC GPLOT DATA=datalib.princ GOUT=fig8_2_3;
    PLOT PRIN1*(PRIN2 PRIN3 PRIN4 PRIN5)/HAXIS=AXIS1 VAXIS=AXIS1;
    PLOT PRIN2*(      PRIN3 PRIN4 PRIN5)/HAXIS=AXIS1 VAXIS=AXIS1;
    PLOT PRIN3*(            PRIN4 PRIN5)/HAXIS=AXIS1 VAXIS=AXIS1;
    PLOT PRIN4*                  PRIN5 /HAXIS=AXIS1
VAXIS=AXIS1;
                            ↓
```

↑

```
RUN; QUIT;
GOPTIONS DISPLAY HTEXT=2;
%mkfields(4,4)
PROC GREPLAY IGOUT=fig8_2_3 TC=TEMPCAT TEMPLATE=NEWTEMP NOFS;
     TREPLAY 1:1   2:2    3:3     4:4
                   6:5    7:6     8:7
                         11:8    12:9
                                 16:10;
RUN; DELETE _ALL_; QUIT;
```

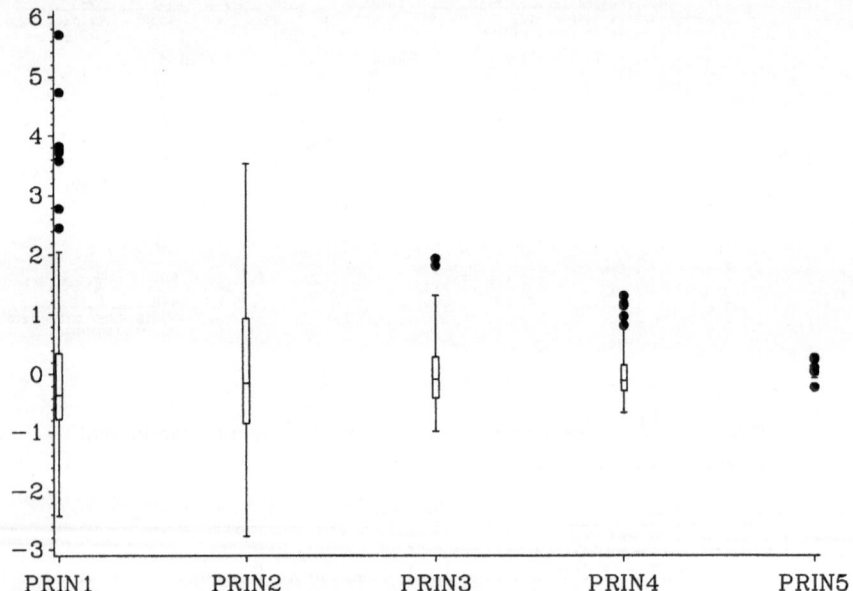

Figure 8.2.4. Boxplots of all 5 principal components of the cns data.

```
***    Program 8_2_4    ***;
TITLE1 'Boxplots of Principal Component Scores';
TITLE2 'CNS Data';
LIBNAME datalib 'c:\data';

PROC TRANSPOSE DATA=datalib.princ OUT=data1;
    VAR PRIN1-PRIN5;
    BY no;
AXIS1 LABEL=NONE;
SYMBOL1 C=GREEN V=DOT H=1 I=BOXT;
PROC GPLOT DATA=data1;
    PLOT COL1*_NAME_=1 / VAXIS=AXIS1 HAXIS=AXIS1;
RUN; QUIT;
```

The scatterplot matrix is computed in analogy to the code of Program 3_3_1. The plot of the principal component scores in Figure 8.2.4 requires an appropriate structure of the data, which were generated by the OUT option of PROC PRINCOMP. This structure is provided by the TRANSPOSE procedure, see also Program 5_1_1.

The option I=BOXT of the SYMBOL statement of PROC GPLOT selects boxplots for the display of the data. Those data corresponding to outliers in the plots can be identified using PROC UNIVARIATE together with the statement ID NR. An easy alternative approach would be the use of the interactive module SAS/INSIGHT.

## Principal Components of Standardized Data

A disadvantage of the ordinary principal components is the fact that they are quite sensitive to the scale on which the measurements are put. If distances are, for instance, given in kilometers instead of miles, the principal components can change drastically: The second scatterplot in the following figure is derived from the first one by multiplying the second coordinate $x_{i2}$ of each point $\boldsymbol{x}_i = (x_{i1}, x_{i2})^T \in I\!\!R^2$ by 1.6093. The points $\boldsymbol{y}_i = (y_{i1}, y_{i2})^T$ in the second scatterplot consequently have the coordinates

$$\boldsymbol{y}_i = \begin{pmatrix} y_{i1} \\ y_{i2} \end{pmatrix} = \begin{pmatrix} x_{i1} \\ 1.6093x_{i2} \end{pmatrix}, \qquad i = 1, \dots, 100.$$

This corresponds to distances given in kilometers in the second component instead of miles. The two principal axes of the data clouds obviously are heavily influenced.

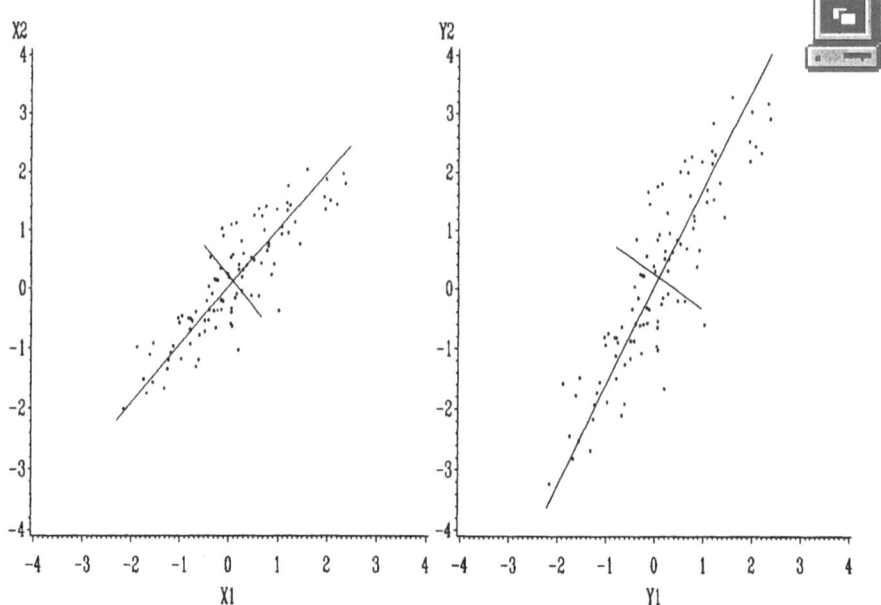

**Figure 8.2.5.** Scatterplots of the points in Figure 8.1.1; $y_1 = x_1$, $y_2 = 1.6093x_2$.

```
***    Program 8_2_5    ***;
TITLE1 'Principal Components';
TITLE2 'Data from Figure 8.1.2';
LIBNAME datalib 'c:\data';

DATA pca2; SET datalib.pca1; y1=x1; y2=1.6093*x2;

SYMBOL1 V=DOT C=RED H=.3 I=NONE;
PROC PRINCOMP DATA=pca2 OUT=pca3 OUTSTAT=stats COV NOPRINT;
    VAR x1 x2;
DATA anno;
    LENGTH FUNCTION $8.; SET stats; RETAIN mx my l1 l2;
    IF _TYPE_='MEAN' THEN DO; mx=x1; my=x2; END;
    IF _TYPE_='EIGENVAL' THEN DO; l1=SQRT(x1); l2=SQRT(x2); END;
    IF _TYPE_='SCORE' THEN DO;
        IF _NAME_='Prin1' THEN l=2.5*l1; ELSE l=2.5*l2;
        XSYS='2'; YSYS='2'; COLOR='RED';
        FUNCTION='MOVE'; X=mx-l*x1; Y=my-l*x2; OUTPUT;
        FUNCTION='DRAW'; X=mx+l*x1; Y=my+l*x2; OUTPUT;
    END; GOPTIONS NODISPLAY; AXIS1 ORDER=(-4 TO 4);
PROC GPLOT DATA=pca2 GOUT=fig8_2_5;
    PLOT x2*x1 / VAXIS=AXIS1 HAXIS=AXIS1 ANNOTATE=anno NAME='miles';
RUN; QUIT;

PROC PRINCOMP DATA=pca2 OUT=pca3 OUTSTAT=stats COV NOPRINT;
    VAR y1 y2;
DATA anno;
    LENGTH FUNCTION $8.; SET stats; RETAIN mx my l1 l2;
    IF _TYPE_='MEAN' THEN DO; mx=y1; my=y2; END;
    IF _TYPE_='EIGENVAL' THEN DO; l1=SQRT(y1); l2=SQRT(y2); END;
    IF _TYPE_='SCORE' THEN DO;
        IF _NAME_='Prin1' THEN l=2.5*l1; ELSE l=2.5*l2;
        XSYS='2'; YSYS='2'; COLOR='RED';
        FUNCTION='MOVE'; X=mx-l*y1; Y=my-l*y2; OUTPUT;
        FUNCTION='DRAW'; X=mx+l*y1; Y=my+l*y2; OUTPUT; END;

PROC GPLOT DATA=pca2 GOUT=fig8_2_5;
   PLOT y2*y1 / VAXIS=AXIS1 HAXIS=AXIS1 ANNOTATE=anno NAME='km';
TITLE2 'transformed data'; RUN; QUIT;

GOPTIONS DISPLAY;
PROC GREPLAY IGOUT=fig8_2_5 NOFS
    TC=SASHELP.TEMPLT TEMPLATE=H2;
    TREPLAY 1:miles   2:km;
RUN; DELETE _ALL_; RUN; QUIT;
```

The first DATA step in this program computes the transformation $y_2 = 1.6093x_2$, the    rest is analogous to Program 8_1_2.

We want to remove this disadvantage by considering principal components, which are invariant under a linear scale shift of the data. Precisely, they are invariant under any transformation of the observations $\boldsymbol{x}_i = (x_{i1}, \ldots, x_{ip})^T \in \mathbb{R}^p$, which is of the form

$$\boldsymbol{x}_i \mapsto \begin{pmatrix} c_1 & & 0 \\ & \ddots & \\ 0 & & c_p \end{pmatrix} \boldsymbol{x}_i + \begin{pmatrix} d_1 \\ \vdots \\ d_p \end{pmatrix} = \begin{pmatrix} c_1 x_{i1} + d_1 \\ \vdots \\ c_p x_{ip} + d_p \end{pmatrix}, \qquad i = 1, \ldots, n, \quad (8.6)$$

with $c_j > 0, d_j \in \mathbb{R}, j = 1, \ldots, p$. This is achieved by replacing the $\boldsymbol{x}_i$ by the standardized vectors $\boldsymbol{y}_i = (y_{i1}, \ldots, y_{ip})^T$, where

$$y_{ij} = \frac{x_{ij} - \bar{x}_j}{s_j}, \qquad i = 1, \ldots, n, \quad j = 1, \ldots, p.$$

By

$$\bar{x}_j = \frac{1}{n} \sum_{i=1}^{n} x_{ij}$$

we denote the mean and by

$$s_j^2 := \frac{1}{n-1} \sum_{i=1}^{n} (x_{ij} - \bar{x}_j)^2$$

the empirical variance of the $j$-th component of the $\boldsymbol{x}_i$, $j = 1, \ldots, p$. The vectors $\boldsymbol{y}_i$ then have the mean vector $\boldsymbol{0}$ and the $p \times p$–covariance matrix

$$\boldsymbol{W} = (w_{jl}) = \frac{1}{n-1} \sum_{i=1}^{n} \boldsymbol{y}_i \boldsymbol{y}_i^T$$

with

$$w_{jl} = \frac{\sum_{i=1}^{n} (x_{ij} - \bar{x}_j)(x_{il} - \bar{x}_l)}{\sqrt{\sum_{i=1}^{n} (x_{ij} - \bar{x}_j)^2} \sqrt{\sum_{i=1}^{n} (x_{il} - \bar{x}_l)^2}}.$$

The matrix $\boldsymbol{W}$ is the correlation matrix of the $\boldsymbol{x}_i$ and, thus, its entries satisfy $|w_{jl}| \leq 1$ and $w_{jj} = 1$, see Section 3.1. These values obviously are not touched by a linear shift of the $\boldsymbol{x}_i$ as in (8.6) and, thus, the principal components derived from the matrix $\boldsymbol{W}$ are not touched either. For the principal component scores of the $\boldsymbol{y}_i$ or, equivalently, for the orthogonal eigenvectors $\boldsymbol{r}_j$ and the eigenvalues $\lambda_j$ of the matrix $\boldsymbol{W}$, we obtain from Theorem 8.2.2 that $p = tr(\boldsymbol{W}) = \sum_{j=1}^{p} \lambda_j$ and

$$z_j(\boldsymbol{y}_i) = \boldsymbol{r}_j^T \boldsymbol{y}_i, \qquad i = 1, \ldots, n, \quad j = 1, \ldots, p.$$

The selection rule, by which one considers only those principal components in a further analysis, whose eigenvalues $\lambda_k$ satisfy the inequality $\lambda_k > \sum_{j=1}^{p} \lambda_j / p$, now selects those with eigenvalue $\lambda_k > 1$, see Figure 8.2.2.

Finally, the principal component scores are often standardized themselves, i.e., one considers $z_j(y_i)/\sqrt{\lambda_j}$, $i = 1,\ldots,n$, $j = 1,\ldots,p$. These have by Theorem 8.2.2 (i)–(iii) mean vector $0$ and covariance matrix $I_p$. They will be investigated in the factor analysis in the next section.

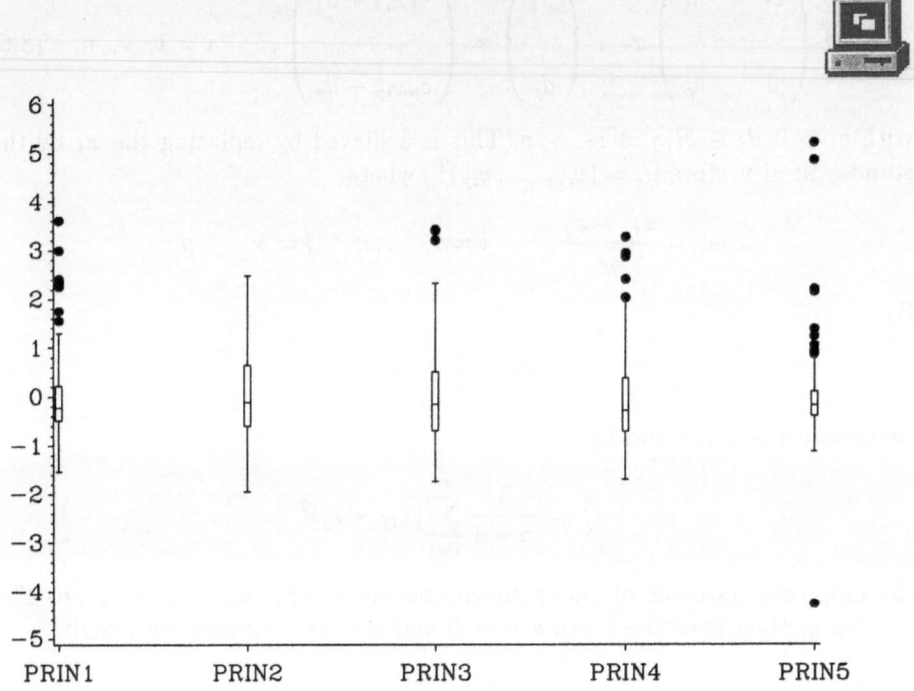

**Figure 8.2.6.** Boxplots of the standardized principal component scores in Figure 8.2.4.

Program 8_2_6, which generates this figure, is not listed here, since it coincides with Program 8_2_3 up to the additional option STD in the PRINCOMP procedure. This option computes standardized principal components, i.e., the principal component scores in the OUT= *data set* have unit variance.

The boxplots are generated with Program 8_2_4. For the identification of the outliers in the boxplots see the comments on Program 8_2_4.

The boxplots in Figure 8.2.4 visualize the decrease of the variance of the principal components. The above Figure 8.2.6 marks in particular those observations in the scatterplot matrix in Figure 8.2.3 as outliers, whose 5-th principal components are the two largest and the smallest ones.

## 8.3 Factor Analysis based on Principal Components

Factor analysis aims at the reduction of high dimensional data $x_i \in \mathbb{R}^p$, $i = 1, \ldots, n$ to only a few *factors* $f_1, \ldots, f_k$, according to the representation

$$x_i = \bar{x} + L \begin{pmatrix} f_1(x_i) \\ \vdots \\ f_k(x_i) \end{pmatrix} + \text{error}, \qquad i = 1, \ldots, n,$$

where $L$ is a $p \times k$–matrix. Instead of the initial vectors $x_i \in \mathbb{R}^p$, one analyzes the vectors of the *factor scores* $(f_1(x_i), \ldots, f_k(x_i))^T \in \mathbb{R}^k$; a cluster analysis of these is a typical example.

The particular feature of the factor analysis is the *factor rotation* using any orthogonal $k \times k$–matrix $A$, i.e., $A^T = A^{-1}$. For such a matrix we have

$$L \begin{pmatrix} f_1(x_i) \\ \vdots \\ f_k(x_i) \end{pmatrix} = (LA)A^T \begin{pmatrix} f_1(x_i) \\ \vdots \\ f_k(x_i) \end{pmatrix}$$

$$=: (LA) \begin{pmatrix} \tilde{f}_1(x_i) \\ \vdots \\ \tilde{f}_k(x_i) \end{pmatrix} =: \tilde{L} \begin{pmatrix} \tilde{f}_1(x_i) \\ \vdots \\ \tilde{f}_k(x_i) \end{pmatrix}$$

and hence, we obtain, with *one* solution $(f_1, \ldots, f_k)$, an *infinite* number of solutions $\tilde{f}_1, \ldots, \tilde{f}_k$ with the same error in the above representation. The selection of a factor solution, is, therefore, not only based on the size of the error, but also on a different reasoning. This will lead us to the varimax or quartimax rotation.

### Derivation of Factors Using Principal Components

Suppose that we have computed for $x_i \in \mathbb{R}^p$, $i = 1, \ldots, n$, the $p$ principal components. With the notations in Definition 8.2.1 we obtain

$$\begin{pmatrix} z_1(x_i) \\ \vdots \\ z_p(x_i) \end{pmatrix} = R^T(x_i - \bar{x}), \qquad i = 1, \ldots, n,$$

or, equivalently,

$$x_i - \bar{x} = R \begin{pmatrix} z_1(x_i) \\ \vdots \\ z_p(x_i) \end{pmatrix}, \qquad i = 1, \ldots, n.$$

If the eigenvalue $\lambda_{k+1}$ of the matrix $S$ is sufficiently small, the principal component scores $z_j(x_i)$ are close to 0 for $j = k+1,\ldots,p$ and $i = 1,\ldots,n$, see Theorem 8.2.2 (i) and (ii). The eigenvalue could, for example, be considered to be sufficiently small if it satisfies $\lambda_{k+1} \leq \sum_{j=1}^{p} \lambda_j/p$, see the previous section. We divide the right–hand side of the above representation of $x_i - \bar{x}$ into a leading term and an error:

$$
x_i - \bar{x} = R \begin{pmatrix} z_1(x_i) \\ \vdots \\ z_k(x_i) \\ 0 \\ \vdots \\ 0 \end{pmatrix} + R \begin{pmatrix} 0 \\ \vdots \\ 0 \\ z_{k+1}(x_i) \\ \vdots \\ z_p(x_i) \end{pmatrix}
$$

$$
= R_1 \begin{pmatrix} z_1(x_i) \\ \vdots \\ z_k(x_i) \end{pmatrix} + R_2 \begin{pmatrix} z_{k+1}(x_i) \\ \vdots \\ z_p(x_i) \end{pmatrix}. \tag{8.7}
$$

$R_1$ and $R_2$ are $p\times k-$ and $p\times(p-k)$-matrices, respectively, which are generated from the matrix $R$ by dropping the last $p - k$ or the first $k$ columns of $R$, respectively. The vector

$$
\varepsilon(x_i) := \begin{pmatrix} \varepsilon_1(x_i) \\ \vdots \\ \varepsilon_p(x_i) \end{pmatrix} := R_2 \begin{pmatrix} z_{k+1}(x_i) \\ \vdots \\ z_p(x_i) \end{pmatrix}, \qquad i = 1,\ldots,n,
$$

denotes the error. Its coordinates $\varepsilon_j(x_i)$ are by the following result close to 0.

**8.3.1 Theorem.** *With the above notations, we have,*

(i) $\frac{1}{n} \sum_{i=1}^{n} \varepsilon_j(x_i) = 0, \qquad j = 1,\ldots,p,$

(ii) $v_j^2 := \frac{1}{n-1} \sum_{i=1}^{n} \varepsilon_j^2(x_i) \leq \lambda_{k+1}, \qquad j = 1,\ldots,p,$

(iii) $\sum_{j=1}^{p} v_j^2 = \frac{1}{n-1} \sum_{i=1}^{n} \varepsilon(x_i)^T \varepsilon(x_i) = \lambda_{k+1} + \cdots + \lambda_p.$

**Proof:** Put $\bar{\varepsilon} := n^{-1} \sum_{i=1}^{n} \varepsilon(x_i)$ and $\bar{z}_j := n^{-1} \sum_{i=1}^{n} z_j(x_i)$. Part (i) is immediate from Theorem 8.2.2 (i):

$$
\bar{\varepsilon} = \frac{1}{n} \sum_{i=1}^{n} R_2 \begin{pmatrix} z_{k+1}(x_i) \\ \vdots \\ z_p(x_i) \end{pmatrix} = R_2 \begin{pmatrix} \bar{z}_{k+1} \\ \vdots \\ \bar{z}_p \end{pmatrix} = R_2 \begin{pmatrix} 0 \\ \vdots \\ 0 \end{pmatrix} = 0.
$$

Denote by $t_j \in \mathbb{R}^{p-k}$ that *column vector*, which is given by the $j$-th *row* of the matrix $R_2, j = 1, \ldots, p$. Part (ii) now follows from Theorem 8.2.2 (ii), (iii):

$$\frac{1}{n-1} \sum_{i=1}^n \varepsilon_j^2(x_i) = \frac{1}{n-1} \sum_{i=1}^n t_j^T \begin{pmatrix} z_{k+1}(x_i) \\ \vdots \\ z_p(x_i) \end{pmatrix} \begin{pmatrix} z_{k+1}(x_i) \\ \vdots \\ z_p(x_i) \end{pmatrix}^T t_j$$

$$= t_j^T \begin{pmatrix} \lambda_{k+1} & & 0 \\ & \ddots & \\ 0 & & \lambda_p \end{pmatrix} t_j \leq \lambda_{k+1} t_j^T t_j$$

$$\leq \lambda_{k+1} s_j^T s_j = \lambda_{k+1},$$

where $s_j^T$ denotes the $j$-th row of the matrix $R$. Observe that $RR^T = I_p$ implies, in particular, that $s_j^T s_j = 1$, $j = 1, \ldots, p$. Part (iii) follows from similar arguments:

$$\frac{1}{n-1} \sum_{i=1}^n \varepsilon(x_i)^T \varepsilon(x_i)$$

$$= \frac{1}{n-1} \sum_{i=1}^n \left( R \begin{pmatrix} 0 \\ \vdots \\ 0 \\ z_{k+1}(x_i) \\ \vdots \\ z_p(x_i) \end{pmatrix} \right)^T R \begin{pmatrix} 0 \\ \vdots \\ 0 \\ z_{k+1}(x_i) \\ \vdots \\ z_p(x_i) \end{pmatrix}$$

$$= \frac{1}{n-1} \sum_{i=1}^n \begin{pmatrix} 0 \\ \vdots \\ 0 \\ z_{k+1}(x_i) \\ \vdots \\ z_p(x_i) \end{pmatrix}^T R^T R \begin{pmatrix} 0 \\ \vdots \\ 0 \\ z_{k+1}(x_i) \\ \vdots \\ z_p(x_i) \end{pmatrix}$$

$$= \frac{1}{n-1} \sum_{i=1}^n (z_{k+1}^2(x_i) + \cdots + z_p^2(x_i)) = \lambda_{k+1} + \cdots + \lambda_p.$$

$\square$

## Common and Unique Factors

Denote by

$$f_j(y) := z_j(y)/\sqrt{\lambda_j}, \qquad y \in \mathbb{R}^p,$$

the $j$-th principal component, divided by its standard deviation. Equation (8.7)
then becomes

$$
x_i - \bar{x} = R_1 \begin{pmatrix} \sqrt{\lambda_1} & & 0 \\ & \ddots & \\ 0 & & \sqrt{\lambda_k} \end{pmatrix} \begin{pmatrix} f_1(x_i) \\ \vdots \\ f_k(x_i) \end{pmatrix} + \varepsilon(x_i)
$$

$$
=: Lf(x_i) + \varepsilon(x_i), \qquad i = 1, \ldots, n, \tag{8.8}
$$

where $f(y) = (f_1(y), \ldots, f_k(y))^T$ and

$$
L = (l_{jm}) = R_1 \begin{pmatrix} \sqrt{\lambda_1} & & 0 \\ & \ddots & \\ 0 & & \sqrt{\lambda_k} \end{pmatrix}.
$$

We assume that $\lambda_k > 0$.

The function $f_j(y), y \in \mathbb{R}^p$, which takes values in $\mathbb{R}$, is the $j$-th *common factor* of $x_1, \ldots, x_n$ and the $p \times k$-matrix $L$ is the *factor pattern matrix*
consisting of the *factor loadings* $l_{jm}$. These numbers determine the influence,
which each of the factors $f_1, \ldots, f_k$ has on the coordinates of $x_i, i = 1, \ldots, n$.
The remaining influence is due to the *unique factor* $\varepsilon_j(x_i)$. Since we have, by
Theorem 8.2.2 (i)–(iii),

$$
\frac{1}{n} \sum_{i=1}^{n} f(x_i) = 0, \qquad \frac{1}{n-1} \sum_{i=1}^{n} f(x_i) f(x_i)^T = I_k, \tag{8.9}
$$

the mean vector and the covariance matrix of the approximation

$$
\hat{x}_i := Lf(x_i) \in \mathbb{R}^p, \qquad i = 1, \ldots, n
$$

of $x_i - \bar{x}$ satisfy:

$$
\frac{1}{n} \sum_{i=1}^{n} \hat{x}_i = 0, \qquad \frac{1}{n-1} \sum_{i=1}^{n} \hat{x}_i \hat{x}_i^T = LL^T. \tag{8.10}
$$

## Communalities

The variance of the $j$-th component of the $\hat{x}_i = (\hat{x}_{i1}, \ldots, \hat{x}_{ip})^T$, $i = 1, \ldots, n$,
is by (8.10) the scalar product of the $j$-th row of $L$ and the $j$-th column of
$L^T$. But note that these two coincide and, thus, the variance is

$$
d_j^2 := \frac{1}{n-1} \sum_{i=1}^{n} \hat{x}_{ij}^2 = \sum_{m=1}^{k} l_{jm}^2, \qquad j = 1, \ldots, p.
$$

The number $d_j^2$ is called *communality* of the $j$-th coordinate of the $x_i$, $i = 1, \ldots, n$. The communality $d_j^2$ is that part of the variance of the $j$-th coordinate, which is explained by the common factors $f_1, \ldots, f_k$, see (8.12) below.

Consequently, the part of the total variation $tr(S)$ of the variables, which is explained by the common factors, is

$$\sum_{j=1}^{p} \left( \frac{1}{n-1} \sum_{i=1}^{n} \hat{x}_{ij}^2 \right) = \sum_{j=1}^{p} d_j^2 = \sum_{j=1}^{p} \sum_{m=1}^{k} l_{jm}^2 = \sum_{m=1}^{k} c_m^2.$$

By

$$c_m^2 := \sum_{j=1}^{p} l_{jm}^2, \qquad m = 1, \ldots, k,$$

we denote that part of the total variance $\sum_{j=1}^{p} d_j^2$ explained by the common factors which is explained by the single factor $f_m$:

$$\sum_{j=1}^{p} d_j^2 = \sum_{j=1}^{p} \frac{1}{n-1} \sum_{i=1}^{n} \hat{x}_{ij}^2 = \sum_{j=1}^{p} \frac{1}{n-1} \sum_{i=1}^{n} \left( \sum_{m=1}^{k} l_{jm} f_m(x_i) \right)^2$$

$$= \sum_{j=1}^{p} \sum_{m=1}^{k} l_{jm}^2 = \sum_{m=1}^{k} c_m^2.$$

Recall that the factors are uncorrelated and have variance 1 by (8.9).

**8.3.2 Theorem.** *The common and the unique factors are uncorrelated:*

$$\frac{1}{n-1} \sum_{i=1}^{n} f(x_i) \varepsilon^T(x_i) = 0 = \frac{1}{n-1} \sum_{i=1}^{n} \varepsilon(x_i) f^T(x_i).$$

**Proof:** Theorem 8.2.2 (iii) implies that

$$\frac{1}{n-1} \sum_{i=1}^{n} f(x_i) \varepsilon^T(x_i) = \frac{1}{n-1} \sum_{i=1}^{n} \begin{pmatrix} z_1(x_i)/\sqrt{\lambda_1} \\ \vdots \\ z_k(x_i)/\sqrt{\lambda_k} \end{pmatrix} (z_{k+1}(x_i), \ldots, z_p(x_i)) R_2^T$$

$$= \left( \frac{1}{\sqrt{\lambda_l}(n-1)} \sum_{i=1}^{n} z_l(x_i) z_m(x_i) \right)_{\substack{1 \leq l \leq k \\ k+1 \leq m \leq p}} R_2^T = 0.$$

$\square$

By Theorem 8.3.2, the covariance matrix $S$ of $x_1, \ldots, x_n$ has the representation

$$S = \frac{1}{n-1} \sum_{i=1}^{n} (x_i - \bar{x})(x_i - \bar{x})^T$$

$$= \frac{1}{n-1} \sum_{i=1}^{n} \left( Lf(x_i) + \varepsilon(x_i) \right) \left( f^T(x_i)L^T + \varepsilon^T(x_i) \right)$$

$$= \frac{1}{n-1} \sum_{i=1}^{n} \left( Lf(x_i)f^T(x_i)L^T \right) + \frac{1}{n-1} \sum_{i=1}^{n} \varepsilon(x_i)\varepsilon^T(x_i)$$

$$= LL^T + \frac{1}{n-1} \sum_{i=1}^{n} \varepsilon(x_i)\varepsilon^T(x_i). \tag{8.11}$$

The variance $s_j^2$ of the $j$-th coordinate of the $x_i$ can, therefore, be written as

$$s_j^2 = \sum_{m=1}^{k} l_{jm}^2 + v_j^2, \qquad j = 1, \ldots, p. \tag{8.12}$$

If the unique factors are, in addition, uncorrelated, i.e., if

$$\frac{1}{n-1} \sum_{i=1}^{n} \varepsilon(x_i)\varepsilon^T(x_i) = \begin{pmatrix} v_1^2 & & 0 \\ & \ddots & \\ 0 & & v_p^2 \end{pmatrix} =: V,$$

then (8.11) implies that the covariance matrix $S$ of the $x_i$ can be written as

$$S = LL^T + V. \tag{8.13}$$

## Factor Rotation

Let $A$ be an arbitrary orthogonal $k \times k$-matrix, i.e., $A^T = A^{-1}$. Then we have

$$Lf(y) = (LA)(A^T f(y)) =: \tilde{L}\tilde{f}(y), \qquad y \in \mathbb{R}^p,$$

where

$$\tilde{f}(y) = \begin{pmatrix} \tilde{f}_1(y) \\ \vdots \\ \tilde{f}_k(y) \end{pmatrix} := A^T f(y) = A^T \begin{pmatrix} 1/\sqrt{\lambda_1} & & 0 \\ & \ddots & \\ 0 & & 1/\sqrt{\lambda_k} \end{pmatrix} \begin{pmatrix} z_1(y) \\ \vdots \\ z_k(y) \end{pmatrix}$$

$$= A^T \begin{pmatrix} 1/\sqrt{\lambda_1} & & 0 \\ & \ddots & \\ 0 & & 1/\sqrt{\lambda_k} \end{pmatrix} R_1^T(y - \bar{x}), \qquad y \in \mathbb{R}^p, \tag{8.14}$$

see Definition 8.2.1 and (8.7). The orthogonal transformation is a rotation of the $k$ coordinate axes in $\mathbb{R}^k$. The multiplication of the factors with the matrix $A^T$ yields a new set of common factors $\tilde{f}_j$, $j = 1, \ldots, k$, which also satisfy equations (8.8)–(8.10), and for which Theorem 8.3.2 is valid as well: With $\tilde{L} = LA$ and $\tilde{f} = A^T f$, equation (8.8) is not affected, i.e.,

$$x_i - \bar{x} = Lf(x_i) + \varepsilon(x_i) = \tilde{L}\tilde{f}(x_i) + \varepsilon(x_i), \qquad i = 1, \ldots, n.$$

The mean vectors and the covariance matrices in (8.9) and (8.10) are unaffected as well,

$$\frac{1}{n}\sum_{i=1}^{n}\tilde{f}(x_i) = 0, \quad \frac{1}{n}\sum_{i=1}^{n}\tilde{f}(x_i)\tilde{f}^T(x_i) = A^T\left(\frac{1}{n-1}\sum_{i=1}^{n}f(x_i)f^T(x_i)\right)A = I_k$$

and

$$\frac{1}{n}\sum_{i=1}^{n}\tilde{L}\tilde{f}(x_i) = 0, \qquad \tilde{L}\tilde{L}^T = LAA^TL^T = LL^T,$$

and, thus, also the communalities as

$$\sum_{m=1}^{k}\tilde{l}_{jm}^2 = \sum_{m=1}^{k}l_{jm}^2 = d_j^2. \tag{8.15}$$

The part $\sum_{j=1}^{p}\tilde{l}_{jm}^2$ of the total explained variation $\sum_{j=1}^{p}d_j^2 = \sum_{m=1}^{k}\sum_{j=1}^{p}\tilde{l}_{jm}^2$, which is explained by the single new factor $\tilde{f}_m$ will, however, be different: Denote by $a_m \in \mathbb{R}^k$ the $m$-th column of the matrix $A$ and by $e_m \in \mathbb{R}^k$ the $m$-th column of the $k \times k$-unity matrix $I_k$. Then we obtain

$$c_m^2 = \sum_{j=1}^{p}l_{jm}^2 = (Le_m)^T(Le_m) = e_m^T L^T Le_m$$

while, on the other hand,

$$\tilde{c}_m^2 := \sum_{j=1}^{p}\tilde{l}_{jm}^2 = (LAe_m)^T(LAe_m) = a_m^T L^T La_m.$$

The new common factors $\tilde{f}_j$ and the unique factors $\varepsilon_m$ are by Theorem 8.3.2 uncorrelated as well, i.e.,

$$\frac{1}{n-1}\sum_{i=1}^{n}\tilde{f}(x_i)\varepsilon^T(x_i) = 0 = \frac{1}{n-1}\sum_{i=1}^{n}\varepsilon(x_i)\tilde{f}^T(x_i).$$

The additional rotation $A^T f$ of the factors $f$ by an orthogonal $k \times k$-matrix $A^T$ typically aims at the generation of such factors $\tilde{f}_1, \ldots, \tilde{f}_k$, which have high loadings in some of the coordinates of the $x_i$, moderate to low in a few ones and negligible in the remaining ones. Suppose the factor loadings $|\tilde{l}_{jm}|$ of the factor pattern matrix $\tilde{L}$ are either large in absolute value or close to 0. Then the observations $x_{ij}, i = 1, \ldots, n$, are approximately linear combinations of a few factors with high loadings:

$$x_{ij} \approx \bar{x}_j + \tilde{l}_{jm_1}\tilde{f}_{m_1}(x_i) + \cdots + \tilde{l}_{jm_s}\tilde{f}_{m_s}(x_i), \qquad i = 1, \ldots, n.$$

## Varimax Rotation

*Normalized varimax rotation* (Kaiser (1958)) aims at the use of an orthogonal $k \times k$–matrix $A$, such that the corresponding factor pattern matrix $\tilde{L} = LA$ has in each *column* only a few high loadings in absolute values, whereas the others are close to 0. To this end, the variances of the squared factor loadings, weighted by the communalities, are computed for each column of $\tilde{L} = LA$ and summed up. The variance rotation now uses that matrix $A$, which maximizes this sum among all orthogonal matrices:

$$\sum_{m=1}^{k} \sum_{j=1}^{p} \left( \frac{\tilde{l}_{jm}^2}{d_j^2} - \frac{1}{p} \sum_{i=1}^{p} \frac{\tilde{l}_{im}^2}{d_i^2} \right)^2$$

$$= \sum_{m=1}^{k} \left( \sum_{j=1}^{p} \frac{\tilde{l}_{jm}^4}{d_j^4} - \frac{1}{p} \Big( \sum_{i=1}^{p} \frac{\tilde{l}_{im}^2}{d_i^2} \Big)^2 \right)$$

$$=: S_V(A) \longrightarrow \text{maximum with respect to } A. \tag{8.16}$$

Note that the variance of a set of numbers in $[0, 1]$ is heuristically maximized if their distances are large, i.e., if they are either close to 0 or 1. Observe that $\tilde{l}_{jm}^2/d_j^2 \in [0, 1]$.

## Quartimax Rotation

*Normalized quartimax rotation* aims at a factor pattern matrix, which has in each *row* only a few high loadings. To this end, the sum of the variances of the weighted squared loadings in each *row* of $\tilde{L} = LA$ is maximized with respect to the matrix $A$:

$$\sum_{j=1}^{p} \sum_{m=1}^{k} \left( \frac{\tilde{l}_{jm}^2}{d_j^2} - \frac{1}{k} \sum_{i=1}^{k} \frac{\tilde{l}_{ji}^2}{d_j^2} \right)^2 =: S_Q(A) \longrightarrow \text{maximum with respect to } A.$$

While varimax looks for a rotation such that each factor loads on only a few coordinates of the $x_i$, quartimax aims at a rotation such that each coordinate is explained by only a few factors.

Since $\sum_{i=1}^{k} \tilde{l}_{ji}^{2}$ is the communality $d_{j}^{2}$ of the $j$–th coordinates of $\boldsymbol{x}_{i}$, which is, by (8.15), not affected by a rotation, we obtain

$$\sum_{j=1}^{p}\sum_{m=1}^{k}\Big(\frac{\tilde{l}_{jm}^{2}}{d_{j}^{2}}-\frac{1}{k}\sum_{i=1}^{k}\frac{\tilde{l}_{ji}^{2}}{d_{j}^{2}}\Big)^{2}=\sum_{j=1}^{p}\sum_{m=1}^{k}\Big(\frac{\tilde{l}_{jm}^{2}}{d_{j}^{2}}-\frac{1}{k}\Big)^{2}=\sum_{j=1}^{p}\sum_{m=1}^{k}\frac{\tilde{l}_{jm}^{4}}{d_{j}^{4}}-\frac{p}{k}.$$

The quartimax maximization is, therefore, equivalent to the maximization of

$$\sum_{j=1}^{p}\sum_{m=1}^{k}\frac{\tilde{l}_{jm}^{4}}{d_{j}^{4}}\longrightarrow \text{maximum}$$

with respect to $\boldsymbol{A}$.

## Orthomax Rotation

Varimax and quartimax rotation are special cases of the *normalized orthomax rotation*, where a number $\gamma \in [0,1]$ is fixed and

$$\sum_{m=1}^{k}\Big(\sum_{j=1}^{p}\frac{\tilde{l}_{jm}^{4}}{d_{j}^{4}}-\frac{\gamma}{p}\Big(\sum_{i=1}^{p}\frac{\tilde{l}_{im}^{2}}{d_{i}^{2}}\Big)^{2}\Big)=:S_{O}(\boldsymbol{A})\longrightarrow \text{maximum}$$

is maximized with respect to $\boldsymbol{A}$. With $\gamma = 0$ this is the quartimax rotation and $\gamma = 1$ yields the varimax rotation.

## Interpretation of the Factors

Factor analysis ends with an interpretation of the factors by grouping the $p$ variables, or coordinates, of the $\boldsymbol{x}_{i} \in \mathbb{R}^{p}$ according to the loadings in the factor pattern matrix as follows. First, the $j$–th variable is identified with the $j$–th row vector $(\tilde{l}_{j1}, \ldots, \tilde{l}_{jk})$ of the matrix $\tilde{\boldsymbol{L}}$, which generates this variable, $j = 1, \ldots, p$. This yields $p$ points in $\mathbb{R}^{k}$. Those variables, whose corresponding vectors in $\mathbb{R}^{k}$ build a cluster, are now grouped. The factor rotations aim at generating clusters of points in $\mathbb{R}^{k}$, which are clearly separated. The kind of variables which are contained in a cluster can often be used to interpret the corresponding factors with high factor loadings, see e.g. Figure 8.3.7.

**8.3.3 Example.** We want to compute a factor analysis for the economy data in Example 3.3.1, see also the principal components analysis in Figure 8.2.1. We have $n = 20$ industrial nations (observations) with $p = 9$ measurements (variables) taken in 1990 for each of these countries. To make the results of the factor analysis independent of the scale of the initial variables, it is reasonable to substitute the $\boldsymbol{x}_{i} \in \mathbb{R}^{9}$ by their standardized counterparts $\boldsymbol{y}_{i} = ((x_{ij} - \bar{x}_{j})/s_{j})_{1 \le j \le 9}$, $i = 1, \ldots, 20$; see the end of section 8.2 for details. The covariance matrix of the $\boldsymbol{y}_{i}$ is the correlation matrix $\boldsymbol{W}$ of the initial observations. The total variation is, consequently, $tr(\boldsymbol{W}) = 9$ and the communalities are, by (8.12) and (8.15), in the interval $[0,1]$. The entries of the rotated factor pattern matrix are, therefore, between $-1$ and $1$.

| _NAME_ | INVEST | INFLATN | GNP | TAX | NUKES | UNEMPLD | LABCOST | POPULATN | STRIKE |
|---|---|---|---|---|---|---|---|---|---|
| INVEST | 1.000 | 0.014 | 0.544 | -.432 | -.024 | -.248 | -.041 | -.014 | -.101 |
| INFLATN | 0.014 | 1.000 | 0.054 | 0.014 | -.141 | -.044 | -.675 | -.121 | 0.747 |
| GNP | 0.544 | 0.054 | 1.000 | -.457 | -.072 | 0.250 | -.384 | -.071 | 0.090 |
| TAX | -.432 | 0.014 | -.457 | 1.000 | -.426 | -.029 | 0.194 | -.510 | 0.088 |
| NUKES | -.024 | -.141 | -.072 | -.426 | 1.000 | -.067 | -.081 | 0.920 | -.192 |
| UNEMPLD | -.248 | -.044 | 0.250 | -.029 | -.067 | 1.000 | -.337 | -.028 | 0.195 |
| LABCOST | -.041 | -.675 | -.384 | 0.194 | -.081 | -.337 | 1.000 | -.059 | -.519 |
| POPULTN | -.014 | -.121 | -.071 | -.510 | 0.920 | -.028 | -.059 | 1.000 | -.168 |
| STRIKE | -.101 | 0.747 | 0.090 | 0.088 | -.192 | 0.195 | -.519 | -.168 | 1.000 |

**Figure 8.3.1.** Correlation matrix of the economy data in Example 3.3.1.

```
***    Program 8_3_1    ***;
TITLE1 'Correlation Matrix';
TITLE2 'Economy Data';
LIBNAME datalib 'c:\data';

PROC FACTOR DATA=datalib.economy METHOD=PRINCIPAL
    CORR SCREE ROTATE=VARIMAX OUTSTAT=datalib.stats
    NFACTORS=3 OUT=datalib.economy2;
PROC PRINT DATA=datalib.stats(WHERE=(_TYPE_='CORR'));
    VAR invest--strike;
    FORMAT invest--strike 5.3;
    ID _NAME_;
RUN; QUIT;
```

Except for the graphics, the output of PROC FACTOR contains by default all the results given here and in the following figures. The formatting of the output which SAS uses by default is, however, not suitable for our purposes. Therefore, we use the OUTSTAT=*data set*, here 'datalib.stats', which contains, among others, means, standard deviations, number of observations, correlations or covariances, eigenvalues, and eigenvectors. The type of statistic, which one wants to select from the OUTSTAT=*data set*, is specified by the value of the SAS variable _TYPE_, here _TYPE_='CORR'. This selects the correlation matrix.

The FORMAT statement in the PRINT procedure controls how the data are displayed; the data themselves are not affected. The format 5.3 tells SAS to point the data using 5 digits including sign and decimal point and with a decimal fraction of length 3.

To select the number $k$ of the principal components of factors for a further analysis, we need the eigenvalues of $W$:

| principal components | eigenvalues |
|---|---|
| PC1 | 2.53617 |
| PC2 | 2.39811 |
| PC3 | 1.67132 |
| PC4 | 1.18683 |
| PC5 | 0.41381 |
| PC6 | 0.34105 |
| PC7 | 0.27382 |
| PC8 | 0.11378 |
| PC9 | 0.06511 |

**Figure 8.3.2.** Eigenvalues of the correlation matrix in Figure 8.3.1.

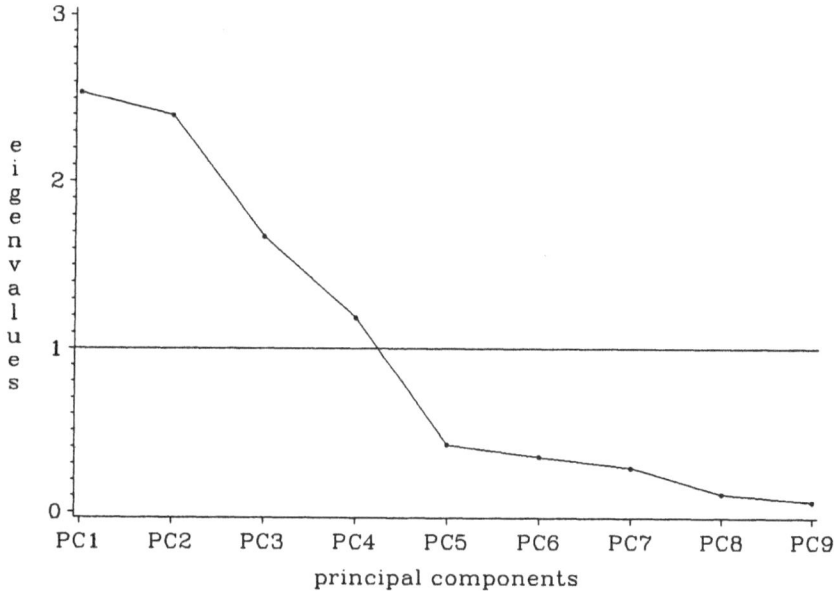

**Figure 8.3.3.** Scree plot of the eigenvalues in Figure 8.3.2.

The above figures were generated analogously to Program 8_2_2 using the OUTSTAT=datalib.stats file, which was created in Program 8_3_1. Here we use the value _TYPE_='EIGENVAL'.

Figures 8.3.4–8.3.6 below are generated in the same way. The values of _TYPE_ are as follows: 'UNROTATE' and 'COMMUNAL' is used for Figure 8.3.4, 'TRANSFOR' for Figure 8.3.5, and 'PATTERN' for Figure 8.3.6.

The three largest eigenvalues sum up to 6.6056, the sum of all is in the standardized case $p = 9$. The first 3 principal components explain, therefore,

$$100 \, \frac{6.6056}{9} \approx 73.40$$

percent of the total variation of the coordinates of the $y_i$.

The principal component approach with $k = 3$ factors yields the following factor pattern matrix $L^T$ and vector of communalities:

| _NAME_ | INVEST | INFLATN | GNP | TAX | NUKES | UNEMPLD | LABCOST | POPULATN | STRIKE |
|---|---|---|---|---|---|---|---|---|---|
| Factor1 | -.041 | 0.779 | 0.255 | 0.241 | -.563 | 0.303 | -.661 | -.559 | 0.804 |
| Factor2 | 0.438 | 0.262 | 0.577 | -.810 | 0.605 | 0.153 | -.571 | 0.638 | 0.168 |
| Factor3 | -.744 | 0.270 | -.624 | 0.211 | 0.487 | 0.134 | -.202 | 0.479 | 0.292 |

| INVEST | INFLATN | GNP | TAX | NUKES | UNEMPLD | LABCOST | POPULATN | STRIKE |
|---|---|---|---|---|---|---|---|---|
| 0.747 | 0.749 | 0.787 | 0.758 | 0.921 | 0.133 | 0.804 | 0.948 | 0.759 |

**Figure 8.3.4.** Transposed $L^T$ of the factor pattern matrix and vector of communalities pertaining to the first 3 principal components of the correlation matrix in Figure 8.3.1.

Up to the variance of the standardized rate of unemployed, which is explained by only 13%, the 3 factors explain more than 74% of the variance of each of the other 8 variables. The varimax rotation (8.16) yields the following rotation matrix $A^T$:

$$\begin{array}{ccc} 0.844 & 0.393 & 0.365 \\ -.534 & 0.682 & 0.500 \\ 0.052 & 0.617 & -.785 \end{array}$$

**Figure 8.3.5.** Transposed $A^T$ of the varimax rotation matrix.

The following matrix is the transposed factor pattern matrix after varimax rotation $\tilde{L} = LA$.

```
_NAME_  INVEST INFLATN GNP   TAX   NUKES UNEMPLD LABCOST POPULATN STRIKE

Factor1 -.134  0.859 0.214 -.038 -.060  0.365  -.856   -.046    0.851
Factor2 -.052  -.102 -.055 -.575 0.957  0.010  -.137   0.973    -.169
Factor3 0.852  -.010 0.859 -.652 -.039  0.006  -.228   -.012    -.084
```

**Figure 8.3.6.** Transpose of the factor pattern matrix $\tilde{L} = LA$ after varimax rotation.

The first factor has high positive loadings on the variables *inflation* and *days on strike* and a negative loading on *labor cost*. The other two factors have low loadings on these variables. The first factor, which summarizes these variables, might, consequently, be interpreted as the *strike potential*. The *number of running nuclear power stations* and the *population size* can obviously be explained by the second factor, which might, therefore, be interpreted as the *electrical power demand*. The third factor has high positive loadings on the *public investment rate* and the *increase of the gross national product* and a negative one on the *tax rate*. Hence, one might interpret it as the *economic power*:

| factor 1 | strike potential |
|----------|------------------|
| factor 2 | electrical power demand |
| factor 3 | economic power |

As can be seen from the vector of communalities in Figure 8.3.4, the unemployment is not explained by the three factors. This means that the unemployment rate in our data set is not essentially influenced by the factors *strike potential, electrical power demand* and *economic power*. This can also be observed in recent years in many industrialized nations.

A graphical analysis of the dependence of the variables on the factors can be based on plotting the rows of the matrix $L$ as points in $\mathbb{R}^3$.

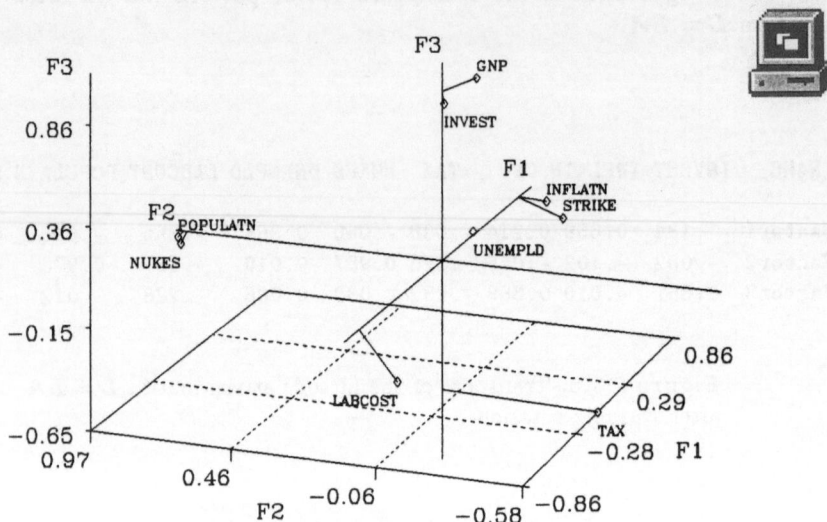

**Figure 8.3.7.** Plot of the rows of the factor pattern matrix in
Figure 8.3.6, causal connection between variables and factors.

```
***    Program 8_3_7    ***;
TITLE1 'Factor Space';
TITLE2 'Economy Data';
LIBNAME datalib 'c:\data';

PROC TRANSPOSE DATA=datalib.stats(WHERE=(_TYPE_='PATTERN')) OUT=rot;
DATA anno1;
    SET rot;
    YSYS='2'; XSYS='2'; ZSYS='2'; TEXT=_NAME_;
    Y=FACTOR2; X=FACTOR1; Z=FACTOR3; SIZE=1.3;
    IF _NAME_ IN('UNEMPLD' 'INVEST' 'TAX') THEN POSITION='9';
    ELSE IF _NAME_ IN('INFLATN' 'STRIKE' 'POPULATN' 'GNP')
       THEN POSITION='3';
       ELSE POSITION='7';
       FUNCTION='LABEL'; OUTPUT; SIZE=1;
    IF _NAME_ IN('INFLATN' 'LABCOST' 'STRIKE') THEN DO;
       FUNCTION='MOVE'; OUTPUT;
       FUNCTION='DRAW'; Y=0; Z=0; OUTPUT; END;
    IF _NAME_ IN('NUKES' 'POPULATN') THEN DO;
       FUNCTION='MOVE'; OUTPUT;
       FUNCTION='DRAW'; X=0; Z=0; OUTPUT; END;
    IF _NAME_ IN('INVEST' 'GNP' 'TAX') THEN DO;
       FUNCTION='MOVE'; OUTPUT;
       FUNCTION='DRAW'; X=0; Y=0; OUTPUT; END;
                              ↓
```

```
                                    ↑
DATA anno2;
    XSYS='2'; YSYS='2'; ZSYS='2'; HSYS='4';
    LINE=1; POSITION='1';
    INPUT FUNCTION $8. X Y Z SIZE TEXT $2.;
cards;
  LABEL     1    0    0   2 F1
  MOVE      1    0    0   1
  DRAW     -1    0    0   1
  LABEL     0    1    0   2 F2
  MOVE      0    1    0   1
  DRAW      0   -1    0   1
  LABEL     0    0    1   2 F3
  MOVE      0    0    1   1
  DRAW      0    0   -1   1
;
PROC G3D DATA=rot ANNOTATE=anno2;
    LABEL FACTOR1='F1' FACTOR2='F2' FACTOR3='F3';
    SCATTER FACTOR2*FACTOR1=FACTOR3 / ANNOTATE=anno1
              NONEEDLE SHAPE='DIAMOND' SIZE=.4;
RUN; QUIT;
```

This graphical representation of the factor pattern matrix from Figure 8.3.6 also uses the OUTSTAT=datalib.stats file, which was created by the FACTOR procedure in Program 8_3_1. The procedure G3D accepts, like almost all SAS/GRAPH procedures, AN-NOTATE=*data sets* for customizing graphics output.

The labels of the points and the connecting lines to the axes are defined in the ANNOTATE data set 'anno1'. The first IF/ELSE statement specifies the positions of the text strings with respect to the displayed points. This prevents the strings to overlap in the graphics output. The subsequent 3 IF/DO/END statements draw the connecting lines from the data points to the axes.

The corresponding system of axes is drawn in the ANNOTATE data set 'anno2'.

The 20 observations $y_i \in \mathbb{R}^9$ of the economy data are now analyzed by means of the corresponding vectors of *factor scores*

$$f(y_i) = \Big( f_1(y_i), \ f_2(y_i), \ f_3(y_i) \Big)^T, \qquad i = 1, \ldots, 20$$

in $\mathbb{R}^3$. See (8.14) for the computation of the factor scores.

| COUNTRY | FACTOR1 | FACTOR2 | FACTOR3 |
|---|---|---|---|
| BELGIUM | -0.50104 | -0.37876 | 0.05825 |
| DENMARK | -0.54368 | -0.84215 | -1.87061 |
| GERMANY | -1.00409 | 0.40196 | 0.00423 |

| FRANCE | -0.13300 | 0.92391 | 0.22949 |
| GREECE | 3.36661 | -0.38456 | -0.43012 |
| GREAT BRITAIN | 0.48964 | 0.49332 | -0.76798 |
| IRELAND | 0.59465 | -0.58785 | 0.56726 |
| ITALY | 0.14806 | -0.03278 | -0.28445 |
| NETHERLANDS | -0.60356 | -0.43342 | 0.24170 |
| PORTUGAL | 1.12433 | -0.36555 | 1.41183 |
| SPAIN | 0.71101 | 0.10324 | 1.10384 |
| FINLAND | 0.06419 | -0.88739 | 0.87759 |
| NORWAY | -0.76429 | -0.55573 | -1.32585 |
| AUSTRIA | -0.79000 | -0.57019 | 0.62036 |
| SWEDEN | -0.26147 | -0.74453 | -1.01306 |
| SWITZERLAND | -1.01790 | -0.43251 | 1.11982 |
| CANADA | 0.25272 | -0.05761 | -0.55457 |
| USA | 0.20083 | 3.49923 | -0.95557 |
| JAPAN | -0.79825 | 1.22848 | 1.91830 |
| LUXEMBURG | -0.53474 | -0.37712 | -0.95046 |

**Figure 8.3.8.** Factor scores of the 20 countries in Example 3.3.1 after varimax rotation.

```
***    Program 8_3_8    ***;
TITLE1 'Factor Scores'; TITLE2 'Economy Data';
LIBNAME datalib 'c:\data';

PROC PRINT DATA=datalib.economy2;
   VAR FACTOR1-FACTOR3;  ID country;
RUN; QUIT;
```

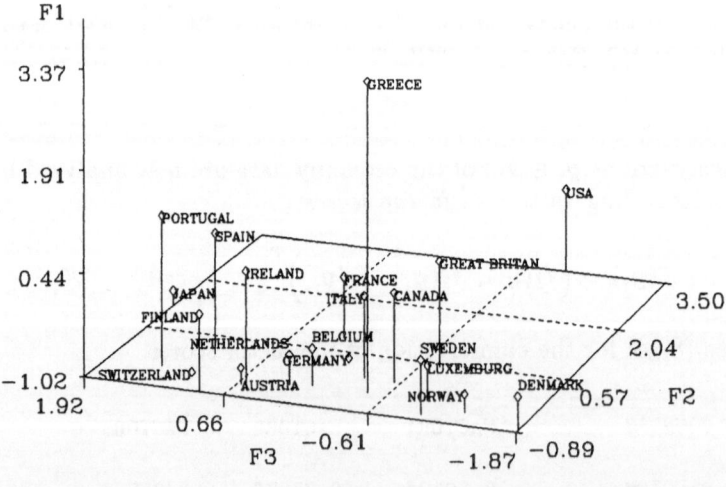

**Figure 8.3.9.** Plot of the vectors of factor scores in Figure 8.3.8.

```
***    Program 8_3_9    ***;
TITLE1 'Plot of Factor Scores';
TITLE2 'Economy Data';
LIBNAME datalib 'c:\data';

DATA anno1;
   SET datalib.economy2;
   YSYS='2'; XSYS='2'; ZSYS='2'; TEXT=country;
   Y=FACTOR3; X=FACTOR2; Z=FACTOR1; SIZE=1.2;
   FUNCTION='LABEL'; POSITION='6';
   IF country IN('FINLAND' 'ITALY' 'GERMANY'
                        'SWITZERLAND' 'NORWAY') THEN POSITION='4';
   IF country IN('SWEDEN' 'BELGIUM' 'DENMARK') THEN POSITION='3';
   IF country IN('AUSTRIA') THEN POSITION='9';
   IF country IN('NETHERLANDS') THEN POSITION='1';
PROC G3D DATA=datalib.economy2;
   LABEL FACTOR1='F1' FACTOR2='F2' FACTOR3='F3';
   SCATTER FACTOR3*FACTOR2=FACTOR1 / ANNOTATE=anno1
                  SHAPE='DIAMOND' SIZE=.4;
RUN; QUIT;
```

Programs 8_3_8 and 8_3_9 use the factor scores, which were written in Program 8_3_1 to the OUT=datalib.economy2 file as the values of the SAS variable FACTORi, i=1,2,3.

The above vectors of factor scores show, for example, that the strike potential in Japan in 1990 was below the average, whereas the demand for energy and the economical power were above the average. Recall that the factor scores have arithmetic mean 0 and variance 1, see (8.9). The strike potential in the U.S. was on the average, the demand for energy was extraordinarily high and the economical power considerably below the mean. Portugal's strike potential was in 1990 above the average, the demand for energy slightly less, its economical power, however, was considerably larger than the mean, see also the *MDS* plot in Figure 7.3.2 and the subsequent remarks.

The columns of the factor pattern matrix can be multiplied by -1 if the corresponding factors are simultaneously multiplied by -1 as well. By this, one can interpret positive and negative factor scores as values above and below the average.

### Exercises

1. Approximate the $n \times p$–data matrix $X = (x_{ij})$ by some matrix $\hat{X}$, where $\hat{X} = zr^T$ with $z = (z_1, \ldots, z_n)^T$ and a direction $r = (r_1, \ldots, r_n)^T$, $r^T r = 1$. Show that

(i) $\sum_{i=1}^{n} \sum_{j=1}^{p} (x_{ij} - \hat{x}_{ij})^2 = tr((X - \hat{X})^T (X - \hat{X})) = tr(X^T X) - 2z^T X r + z^T z$.
Hint to the second equation: Exercise 22, Chapter 3.

(ii) the function $f(z) = tr(X^T X) - 2z^T X r + z^T z$ has a minimum at $X r$.

(iii) minimizing the function $f(X r)$ with respect to $r$ under the constraint $r^T r = 1$
by using a Lagrange multiplier $\lambda$ yields the equation $(X^T X - \lambda I_p) r = 0$, i.e.,
$\lambda$ is an eigenvalue of $X^T X$ with eigenvector $r$.

2. Let $A$ be a symmetric $p \times p$–matrix with eigenvalues $\lambda_1 \geq \lambda_2 \geq \cdots \geq \lambda_p$ and
corresponding orthonormal eigenvectors $r_1, \ldots, r_p$. Show that

(i) $\sup\limits_{x \neq 0} \dfrac{x^T A x}{x^T x} = r_1^T A r_1 = \lambda_1$,

(ii) $\sup\limits_{x \neq 0 : R_k^T x = 0} \dfrac{x^T A x}{x^T x} = r_{k+1}^T A r_{k+1} = \lambda_{k+1}$,

where $R_k := (r_1, \ldots, r_k)$, $k = 1, \ldots, p - 1$.

3. Consider two positive definite (symmetric) $p \times p$–matrices $A$ and $B$, such that
$A - B$ is positive definite. Prove that the eigenvalues of $A B^{-1}$ are greater than or
equal to 1 and conclude that $\det A \geq \det B$.

Hint: There exists an invertible matrix $F$, such that $F^T A F = \mathrm{diag}(\lambda_1, \ldots, \lambda_p)$,
where $\lambda_1, \ldots, \lambda_p$ are the eigenvalues of $A B^{-1}$ and $F^T B F = I_p$. For the proof of this
auxiliary result, show first the existence of an invertible matrix $C$ with $C^T B C = I_p$.
The principal axes transformation (8.5) implies the existence of an orthogonal matrix
$R$ such that $R^T (C^T A C) R$ is a diagonal matrix. Put $F = C R$.

4. Consider $x_1, \ldots, x_n \in I\!\!R^p$. Denote by

$$S(a, r) = \sum_{i=1}^{n} \| x_i - (a + ((x_i - a)^T r) r) \|^2$$

the sum of the squared Euclidean distances of $x_i$ and their orthogonal projections
onto a line $a + \lambda r$ with $\lambda \in I\!\!R$, $a \in I\!\!R^p$ and directional vector $r \in I\!\!R^p$, i.e., $\|r\| = 1$.
Prove that minimizing $S(a, r)$ with respect to $a$ and $r$ is equivalent to maximizing
the variation of the orthogonal projections onto the line $a + \lambda r$. It turns out that
$a = \bar{x}$.

Hint: With $y_i := x_i - \bar{x}$ and $b := a - \bar{x}$ one obtains $S(a, r) = \sum_{i=1}^{n} \| y_i - b -$
$((y_i - b)^T) r) r \|^2$. Minimizing the residual sum of squares with respect to *orthogonal*
projections onto a line, therefore, leads to principal components, whereas the *vertical*
projections lead to the least squares estimator, see Section 3.2.

5. (Economy data) Compute a principal components analysis and generate scatterplots of the principal component scores.

Exercises 6 – 11 correspond to principal components in a *stochastic* model: Let $X =$
$(X_1, \ldots, X_p)^T$ be a random vector with mean vector $\mu$ and covariance matrix $\Sigma$.

Denote by $\lambda_1 \geq \cdots \geq \lambda_p \geq 0$ the eigenvalues of $\Sigma$ and $R = (r_1, \ldots, r_p)$ the matrix of corresponding orthonormal eigenvectors, i.e., $R^T \Sigma R = \Lambda = \text{diag}(\lambda_1, \ldots, \lambda_p)$. Then

$$Z_j = r_j^T(X - \mu), \quad j = 1, \ldots, p,$$

is called $j$–th principal component of $X$ and $\lambda_j^{-1/2} Z_j$ is the $j$–th standardized principal component of $X$.

**6.** Show that:

  (i) $Z_j, \ j = 1, \ldots, p$, are uncorrelated and that $Var(Z_j) = \lambda_j$,

  (ii) $Cov(\lambda_1^{-1/2} Z_1, \ldots, \lambda_p^{-1/2} Z_p) = I_p$,

  (iii) $\sup_{||r||=1} Var(r^T X) = Var(r_1^T X) = \lambda_1$,

  (iv) $\sup_{||r||=1, \ r^T r_i = 0, \ i=1,\ldots,j-1} Var(r^T X) = Var(r_j^T X) = \lambda_j$,

  (v) $\sum_{j=1}^{p} Var(Z_j) = \sum_{j=1}^{p} Var(X_j) = tr(\Sigma)$,

  (vi) $Cov(X, Z) = R\Lambda, \ Z = (Z_1, \ldots, Z_p)^T$.

Hint: Exercises 1 and 22 in Chapter 3.

**7.** Let $X = (X_1, X_2)^T$ be a bivariate random vector with $E(X) = 0$ and covariance matrix

$$Cov(X) = \begin{pmatrix} 1 & \varrho \\ \varrho & 1 \end{pmatrix}.$$

Compute the principal components of $X$.

**8.** Let $X_1, \ldots, X_p$ be independent random variables with $E(X) = 0$ und $Cov(X) = \Sigma = (\sigma_{ij})_{1 \leq i, j \leq p}$, $X = (X_1, \ldots, X_p)^T$. Denote by $Z_j = r_j^T X$ the $j$–th principal component of $X$ with variance $\lambda_j$. Show that:

$$\varrho(X_i, Z_j) = r_{ji}(\sigma_{ii}/\lambda_j)^{1/2} = r_{ji}(\lambda_j/\sigma_{ii})^{1/2},$$

where $r_{kl}$ denote the $l$–th component of the eigenvector $r_k$.

**9.** Let $N = (N_1, N_2)^T$ be a multinomial $B(1, p, q)$–distributed random variable, $p + q = 1$, see Lemma 4.1.6 for the definition of the multinomial distribution. Compute the principal components.

**10.** Let $X_i = Y_0 + Y_i, \ i = 1, \ldots, p$, where $Y_0, \ldots, Y_p$ are independent and identically distributed random variables with mean 0 and variance $\sigma^2$. Show that $X = (X_1, \ldots, X_p)^T$ has a principal component, which is proportional to the arithmetic mean $\bar{X}$. Compute its variance and conclude that this is the first principal component.

**11.** Consider the covariance matrix

$$\Sigma = \begin{pmatrix} 1+\alpha & 1 & 1 \\ 1 & 1+\alpha & 1 \\ 1 & 1 & 1+\alpha \end{pmatrix}.$$

Compute the first principal component of $\Sigma$.

**12.** (cns data) Compute a factor analysis with varimax rotation.

– How many factors do you select?
– What is the explained total variation?
– Which variables correspond to which factors?

Perform a cluster analysis of the vectors of the factor scores and generate a graphic representation of the results.

Exercises 13 – 16 correspond to the following *stochastic factor analysis model*: Let $X = (X_1, \ldots, X_p)^T$ be a random vector in $\mathbb{R}^p$ with mean vector $\mu = (\mu_1, \ldots, \mu_p)^T$ and covariance matrix $\Sigma = (\sigma_{ij})$. Suppose that $X$ can be represented as

$$X = Lf + e + \mu, \qquad (8.17)$$

where $L = (l_{ij})$ is a $p \times k$–matrix, $f = (f_1, \ldots, f_k)^T$ is a $k$–dimensional random vector and $e = (e_1, \ldots, e_p)^T$ is a $p$–dimensional random vector. We assume that $1 \leq k \leq p$. The variables $f_i$ are the *common factors* and the noisy variables $e_i$ are the *unique factors*. $L$ is the *factor pattern matrix* with $l_{ij}$ representing the influence or *loading* of the $j$–th common factor onto the $i$–th variable. We suppose that:

$$E(f) = 0, \; Cov(f) = E(ff^T) = I_k$$
$$E(e) = 0, \; Cov(e) = E(ee^T) = V = diag(v_{11}, \ldots, v_{pp})$$
$$Cov(f, e) = E(fe^T) = 0.$$

**13.** Show that $Var(X_j) = d_j^2 + v_{jj}$, $j = 1, \ldots, p$, where $d_j^2 = \sum_{m=1}^k l_{jm}^2$ denotes the *communality* of $X_j$.

**14.** Show that the factor model (8.17) implies the decomposition $\Sigma = LL^T + V$ (see Exercise 15 for the reverse implication).

**15.** (i) Let $X$ be a $p$–dimensional random vector with mean vector $\mu$ and positive definite covariance matrix $\Sigma$. We assume the representation $\Sigma = LL^T + V$ with a $(p \times k)$–matrix $L$. Prove that there exist factors $f$ and $e$, such that $X$ satisfies the factor model (8.17) with $k$ factors.
Hint: Consider

$$\begin{pmatrix} e \\ f \end{pmatrix} = \begin{pmatrix} I_p & L \\ -L^T V^{-1} & I_k \end{pmatrix}^{-1} \begin{pmatrix} X - \mu \\ Y \end{pmatrix},$$

where $Y$ is $N(0, I_k + L^T V^{-1} L)$ distributed and is independent of $X$.

(ii) Show that if $X$ is normal distributed, then $(f, e)$ can be chosen as normal distributed as well.

**16.** Prove that

$$\det(\Sigma) = \det(V) \det(L^T V^{-1} L + I_k).$$

Hint: Use Exercise 14 together with the equation $\det(AA^T + I_p) = \det(A^T A + I_k)$, which holds for an arbitrary $(p \times k)$–matrix $A$.

**17.** (Economy data) Perform a factor analysis of the economy data as in Example 8.3.3 with 4 factors. Does the variable *unemployment* become a factor of its own?

**18.** (Mathematics data; Mardia et al. (1979), page 3f) 88 students took 5 tests in mechanics, vectors, algebra, analysis and statistics. For a discussion of these data we refer to Section 7.2 in Efron and Tibshirani (1993). Perform a factor analysis with just 1 factor $f$ for the test scores $x_i \in \mathbb{R}^5$, $i = 1, \ldots, 88$. The factor score $f(x_i)$ might be interpreted as the mathematical *Intelligence Quotient* (IQ) of the $i$-th student.

# Appendix: A Brief Introduction to SAS

## A.1 Preface

Originally the commonly used SAS–system was a pure statistics package (SAS = $\underline{S}$tatistical $\underline{A}$nalysis $\underline{S}$ystem), today, however, the manufacturer, the SAS Institute Inc. located at Cary, North Carolina, emphasizes the data warehouse concept of its product. It is conceived as a businesswide instrument for information and decision, but of course the core is still a huge analysis package for all different sorts of statistical problems. It is available for almost all computer platforms (PC, workstation, mainframe) and many different operating systems with an almost identical desktop.

This brief introduction gives only basic information about SAS and its use; it is not a substitute of SAS manuals. It is an effort to make the reader who does not have any knowledge of SAS understand the programs presented here and thus to develop programs for his own particular analyses. In general, it is meant for all versions 6.xx and higher, some features, however, are available only in version 8. Printed SAS manuals are available up to version 7; from version 8 on, the manuals are accessible through SAS OnlineDoc, which is delivered in html format on CD by SAS Institute.

Please note that in the following, for easier understanding, the standing denominations of the SAS–system are written in capital letters and the freely chosen denominations for files or variables are written in small letters.

### A.1.1 SAS–Modules

The SAS program system consists of several parts (modules), where only the BASICS–module is a necessary condition for the use of another module. In the following list the modules, which are used here, are marked with an asterisk *:

* SAS/BASICS        (Data and file management, simple analyses)
* SAS/STATISTICS   (Complex statical analyses, multivariate procedures)
* SAS/GRAPH        (Complex highly dissolved graphs, maps)
* SAS/IML            (Interactive matrix algebra)
  SAS/ETS            (Procedures of econometrics and time series analysis)
  SAS/QC              (Methods for quality control)
  SAS/OR              (Procedures of operation research)
  SAS/FSP           (Data input and administration program)
  SAS/ASSIST       (Complete menu control commands)
  SAS/AF              (Production of individual menus)
  SAS/ACCESS      (Interface to data bank systems)
  SAS/INSIGHT     (Interactive data visualization)

## A.1.2 Different Modes with SAS

There are two different modes to work with the SAS system:

- Batch Mode

- Interactive Mode

The Batch Mode, used principally in mainframe environments, needs an arbitrary editor to create a program with a series of SAS commands. This program file, which must have the extension *.sas*, is then transmitted to the computer for execution with an executable command. When the computer executes all SAS commands of the program the results are typically provided in two new data files. They have the same names as the program but different extensions: *.log* and *.lis*.

- The LOG file contains a record of the program execution with comments and error reports. The comments give hints about the proper working of the program and the error reports show the reasons for possible malfunctioning along with simple solutions for possible correction. The report also gives the CPU time.

- The LIS file is only created when at least one SAS procedure to create a task was executed without an error. It contains the actual results of the program.

It is especially advantageous to use SAS in the batch mode for a program, which you have successfully tested before and which will need much computer time for its execution. You can turn off your terminal and log in later to get the results.

The interactive mode certainly is considered with most systems as the standard mode. It enables you to give single statements to SAS for execution and is especially suitable to test programs, but also when you do not want to wait too long for your results. This mode will be presented in detail in Section A.2.1.

A useful tool for an interactive data analysis session is the SAS module INSIGHT. Instead of the command line approach discussed in this book, INSIGHT provides a graphical user interface (gui) for invoking various graphical tools. Bar charts can, for example, be plotted easily by clicking some icons. Some additional interactive features are available under this gui such as continuously rotating a scatterplot of the data in three dimensions. But you will not acquire the fundamentals of the SAS programming language by using SAS INSIGHT.

# A.2 Fundamentals of SAS Programming

In this chapter we will introduce the programming environment prefered on PC and workstations to develop SAS programs, the D̲isplay M̲anager S̲ystem (DMS), and present the concept and the fundamentals of the SAS programming language.

## A.2.1 The Display Manager System

When you invoke the SAS system on a PC or a workstation without any additional option using the command **sas** or clicking the corresponding icon, you normally get a graphical interface to SAS consisting of a series of windows called the Display Manager System. Note that there exists an interactive line mode, too.

The most important windows, which are opened by default, are the log windows, the output windows and the program editor windows. At the start of an interactive SAS session the log window is typically filled with some information about your configuration or some messages, the other two windows are empty. The output window is often hidden by the other two windows.

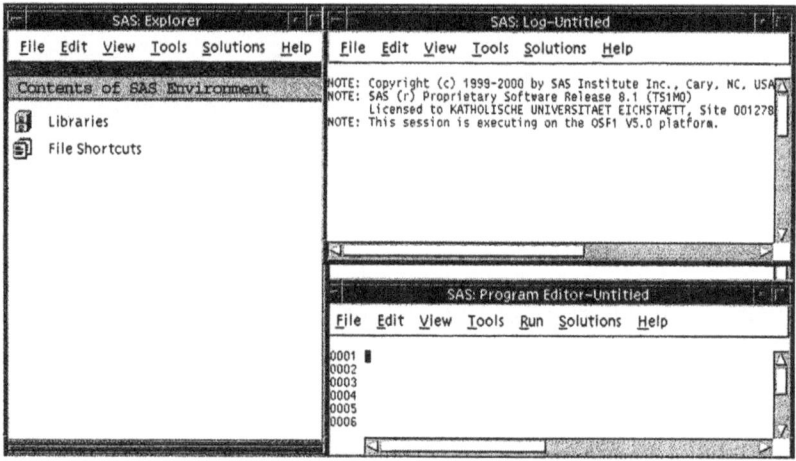

**Figure A.2.1.** SAS Window.

During a SAS session the log window will be filled with information about the execution of SAS commands like the *.log* file in batch mode, notes, warnings and perhaps error messages. The output of SAS procedures will be written in the output window (*.lis* file in batch mode).

There are two different ways to work in this environment: one can write commands in a command line or one can use the menu items in a menu line. In the above figure you see the menu line. In an environment like MS Windows or X Windows the menu will be the easier way, in a mainframe environment you probably prefer the command line. In most graphical user interface systems you have also a command line in an additional command bar when you are working with the menu line. (See Figure A.2.1 on the left side of the symbol line.)

In both modes you can make use of the function keys F1 to F12, perhaps in combination with the SHIFT, CTRL or ALT keys. The definition of these function keys can be quite different, so have a look at your keys window. You get there by choosing *Help* and then *Keys* from the menu.

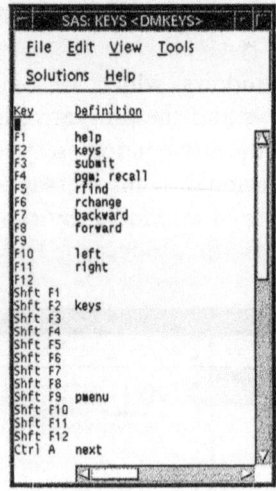

**Figure A.2.2.** Keys window.

The function keys correspond to commands of the command line, for example PGM, LOG or OUTPUT to get into one of these standard windows of DMS. With *Globals* from the menu line and the name of the window you can jump between these windows, too.

Typically you start in the program editor, edit your program statements and submit these statements for execution to SAS. The last step can be done by means of a function key, typing SUBMIT in the command line, clicking *Locals* and then *Submit* in the menu line or clicking the symbol with the running man icon in the symbol line. After that your program editor window will be empty again and you will get the log and (hopefully) the output of your statements. Note that the menu item *Locals* only appears in the menu line of the MS Windows SAS version if the program editor is the active window.

If you want to get the statements back into your program editor window, you can recall them with the command RECALL, a function key or clicking

*Locals* and then *Recall* from the menu line.

You can use editor functions like *Cut, Copy, Paste, Find* and *Change* from the *Edit* item in the menu line. You can mark text parts like in most other programs. The *File* submenu allows to save the statements as a program file on disk, usually with the *.sas* extension, and to load a program file into the program editor with the *Open* item. It is possible to save the contents of the output or log window, too.

The following Figure A.2.3 shows typical submenus of the menu items *File*, *Edit* and *Locals* in the MS Windows environment.

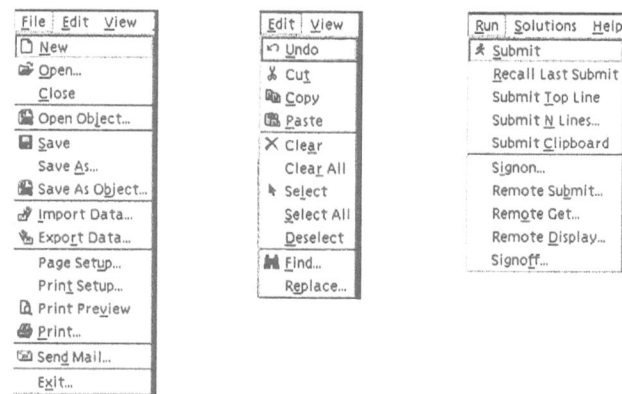

**Figure A.2.3.** Submenus.

## A.2.2 Data and PROC Steps

There are three different types of commands in the SAS programming language:

- global statements

- DATA steps

- PROC steps

By the global statements you can determine the layout, titles, footnotes etc. of the output. There exist defaults for most parameters you overwrite by these statements. They are of minor importance for correct results.

The main distinction in SAS programs is between DATA and PROC steps. In the DATA step you generate a SAS data set appropriate for your purpose, in the PROC step you analyze the data of the data set. DATA and PROC are the keywords to initialize these steps. A DATA or PROC step ends if a new DATA or PROC statement is used.

In a PROC step the data sets are normally kept unaltered, only the sorting procedure PROC SORT changes the order of the cases. But a lot of SAS procedures can create new data sets containing results of the statistical procedure. Typically the output of a procedure is written in the output window.

The data sets created, read or modified in a DATA step are stored in libraries. Each library is connected to a directory in the file system of your operating system. If you want to create a permanent SAS data set, the name specified in the DATA statement must consist of two parts separated by a point. The first part is the name of the library, the second one is the specific name of the data set in this library. Your operating system adds a suffix – MS Windows adds a *.sd2* – and stores it in a binary format in the directory of the library. If the library name is WORK or if the library name part is omitted (a name without a point), you just get a temporary data set. You can use this data set only in the current session. If you quit SAS, the file is deleted.

Some SAS libraries are created automatically when you start a session: WORK for temporary data sets and SASUSER for permanent data sets connected to special directories.

You can generate your own library by means of the LIBNAME statement explained in the following Section A.2.3. The libname window informs you about the current available libraries. In addition to the above mentioned WORK and SASUSER you normally find MAPS and SASHELP here, which you should not use for data sets. You get into this window by clicking the drawer symbol in the symbol line, the menu items *Globals, Access* and *Display Libraries* or the LIB command in the command line.

## A.2.3 Some Basic SAS Statements

A SAS program consists of a sequence of statements, each finished by a semicolon. Most of the statements can be used either in DATA steps or in PROC steps, others may be used in both or in a global context.

The first group of statements supports the creation of a data set: LIBNAME, DATA, INFILE, INPUT, CARDS and SET.

- LIBNAME *libref 'pathname'*;
  The first argument of the LIBNAME statement specifies the name of the library called the library reference and can be chosen by the user in consideration of some convention for SAS names. The second argument determines the directory where your files should be stored or should be read from. It depends on your choice and your operating system. The LIBNAME statement must appear anywhere in your program before you use this library.

- DATA *data set name;*
  The DATA statement opens a DATA step in the program and names the

SAS data set being created. This naming can be a one–level name for a temporary data set or a two–level name whose first part is the library. This library must have been created by a LIBNAME statement before the DATA statement is used. The two–level name creates a permanent data set. The DATA statement can be completed by a data set option.

- INFILE *file specification;*
  The INFILE statement is used to link the DATA step to a source outside the program by reading data from an external file. It identifies this source file by its full name including the path in the file system. The name can be the argument which then must be enclosed in quotes, or it is given by a fileref, which must have been defined earlier in the program by a FILENAME statement. Again additional options may be used.

  The following example generates a permanent data set of the mathematics data (Exercise 18, Chapter 8):

```
LIBNAME datalib 'c:\data';
DATA datalib.math;
    INFILE 'c:\rawdata\math.dat';
    INPUT mechanic vectors algebra analysis statist;
```

- INPUT *variable list;*
  The INPUT statement is used to tell SAS how to read data sets identified by an INFILE or a CARDS statement. The variables that are to be moved from the raw data set into the computer's memory by the INPUT statement must appear in the same order as they are stored in the external file, or they must be given in the data lines after a CARDS statement. The variable names are separated by a blank. The name of a character variable is followed by a $-sign.
  Example: `INPUT name $ age;`
  The variables can be followed by the numbers of the columns in which they are stored.
  Example: `INPUT name $ 1 - 8 age 10 - 11;`
  This is necessary if

  - a character variable value contains more than eight characters.

  - the data file contains missing values not defined by a point.

  - you don't want to read all variables from the external data file.

  - the data in the data file are not separated by a blank.

Normally the data for an observation are read consecutively by reading just one value from each line of the raw data file, additional values in a line are ignored.

But sometimes, especially if the data set contains just one variable, more than just one value of a variable is stored in one line of the data file. Then the option @@ at the end of the INPUT statement tells SAS to continue reading along the line of the raw data set until there are no more data.

- CARDS;

  The CARDS statement at the end of the DATA step informs SAS about the beginning of in–stream data. The CARDS statement is followed by the data lines, one line for each observation. The data in the data lines are separated by blanks. The lines are **not** ended by a semicolon, but the first line after the data lines must end with a semicolon. Typically, this line only contains a single semicolon. An example for the use of the CARDS statement is given in Program 4_1_2.

- SET *SAS data set*;

  To read data from a SAS data set, rather than from an external raw data set, the SET statement is used in place of the INFILE and INPUT statements. The lines

```
DATA data1;
    SET freqdata;
```

  in Program 1_1_3 create the new data set data1 and copy the data from the SAS data set freqdata to data1.

This group of statements is typical for a PROC step: PROC, VAR and BY.

- PROC *procedure* ⟨*options*⟩;

  This statement is used to begin a PROC step. The term directly behind PROC determines the procedure you want to use. If only these two terms are used, SAS works with the most recently created data set. Otherwise one has to specify the data set by a DATA option: DATA=*data set name*. For example, the content of the math data set is printed by:

  PROC PRINT DATA=datalib.math;

- VAR *variables*;

  The VAR statement is used by several procedures to determine the variables of the data set to be analyzed. Without this statement the procedures would select all possible variables. Two or more variables are separated by a space.

  The example of the PRINT procedure above could be continued by:

  VAR mechanic vectors;

- BY *variables*;
  If you want to process the data set in groups, you can specify in a BY statement a variable with different values for these groups. If there is more than one variable in the BY statement, each value combination forms a group of its own.
  Without the additional option NOTSORTED, SAS assumes that the data are sorted in ascending order by the BY variable. In PROC SORT the BY statement is necessary to define the order in which the data set should be sorted. Examples are given in Program 1_2_3 and Exercise 10, Chapter 1.

## A.2.4 Programming in a DATA Step

Reading data from an existing SAS data set, from an external file or explicitly listed data is not the only thing you can do in a DATA set. There are some statements and options you can use to manipulate the data set and to get the data in the form you need for your analysis with a SAS procedure.
In addition to special statements to select other data sets or variables, one can use loops and conditional instructions as in a normal programming language. SAS also provides the standard functions of programming languages and special statistical functions. Data set options are applicable not only in a DATA step but also in a PROC step, for instance in connection with a DATA option.

a) Statements Concerning Data Sets or Variables

- SET *data set(s)⟨(data set options)⟩*;
  If a SET statement is executed, SAS reads all variables and all observations from the data set into the new one unless you tell SAS to do otherwise, for example, with a data set option (see later in this section).
  If two or more data set names appear in the SET statement, the resulting data set is a concatenation of these data sets, i.e., they are stacked on top of each other. Every variable appearing in at least one of the listed data sets is part of the new one.
  Example: see Program 1_6_1.

- MERGE *data sets⟨(data set options)⟩*;
  The MERGE statement joins two or more data sets side by side. Without an additional BY statement, the first observation, the second observation and so on of each data set are joined. When a data set runs out of observations, the remaining observations from the others are joined with missing values.
  With an additional BY statement the observations with the same value of the variables specified are joined. The data sets must be

sorted by these variables.
Example: see Program 3_3_7.

- KEEP *variables*;
  DROP *variables*;
  These statements select some variables of the data set. If a KEEP
  statement is used, the variables named here are the only ones that
  stay in the data set. The DROP statement on the other hand determines the variables that are not included in the new data set.
  Example: See Program 8_2_2.

b) Loops and Conditional Instructions

- IF *expression* THEN *statement*;
  ELSE *statement*;
  For all observations where the expression after IF is true, the statement following THEN is executed. Optionally, you can use an additional ELSE statement. This statement is performed only in the
  case when the IF expression is false.
  Example: see Program 3_3_2.

- DO;
  DO *index variable=start* TO *stop* ⟨BY *increment*⟩; END;
  The DO statement with no extension will be executed exactly one
  time. It is often used in combination with an IF statement describing
  the condition for its execution. The other form of DO statements is
  used to define loops. In the iterative form with an index variable
  the group of statements will be performed until the index variable
  exceeds the stop value. If one does not specify an increment with
  the BY option, the default value 1 is used.
  Other forms of loops are also possible with a DO WHILE or DO
  UNTIL combination.
  Example: see Programs 6_4_2 and 8_3_7.

c) SAS Functions
   SAS provides a lot of functions which help to assign values to variables.
   A function returns a value computed in dependence of arguments which
   are enclosed in parentheses. There are different categories of functions:

- arithmetic functions like ABS (absolute value) and SQRT (square
  root),

- truncation functions like INT (integer value) and ROUND,

- mathematical functions like EVP, LOG (natural logarithm) and
  GAMMA,

- trigonometric functions like SIN, COS and ARCOS,

- probability functions like PROBIT (inverse standard distribution function) and PROBNORM (standard normal distribution function),

- sample statistic functions like MEAN and SUM,

- random number functions like RANNOR (observation of a standard normal distributed variable).

d) Data Set Options

Data set options appear after data set definitions, not only in a DATA step but also in a PROC step, for instance in a PROC statement with a DATA option. They are specified in parentheses. With data set options one can select variables to be included or dropped or one can determine the first or last observation of a data set to be used. In the following some of these options are listed.

| | |
|---|---|
| DROP=*variables* | specifies variables which are omitted from the data set. |
| KEEP=*variables* | specifies variables which are included. |
| FIRSTOBS=*variables* | determines the number of the first observation to be included. |
| OBS=*variables* | determines the number of the last observation to be included. |
| WHERE=(*expression*) | selects only the observations for which the expression is true. Example: see Program 2_1_1. |

# A.3 Some SAS Procedures

There is a lot of SAS procedures and only a small selection of them is used in the programs of this book. Here we present just three of them. They are examples for the usage and the capabilities of SAS.

## A.3.1 PROC MEANS

This procedure belongs to the Base SAS procedures. They are documented in the SAS Procedures Guide. PROC MEANS produces simple univariate descriptive statistics for numerical variables. There are some other procedures in Base SAS like UNIVARIATE, which have partly identical features.
The following selection of statements can be used with this procedure:

```
PROC MEANS ⟨options⟩;
    VAR variables;
    BY variables;
```

        CLASS *variables*;
        FREQ *variables*;
        OUTPUT OUT=*data set keyword=name(s)*;

The first statement PROC MEANS, which starts this procedure, is the only one necessary for its execution. There is a lot of possible options, for instance, DATA=*data set* for the selection of a data set to be processed; NOPRINT suppresses the standard output in the OUTPUT window (if you just want to create an interim output data set), MAXDE=*n* specifies the maximal number of decimals for printed results, and keywords invoke the computation of the following statistics:

| | |
|---|---|
| N | number of observations |
| MEAN | the mean $\bar{x}_n$ |
| STD | the standard deviation $s_n$ |
| MIN | the smallest value $x_{1:n}$ |
| MAX | the largest value $x_{n:n}$ |
| RANGE | the range $x_{n:n} - x_{1:n}$ |
| SUM | the sum $\sum_{i=1}^{n} x_i$ |
| VAR | the variance $s_n^2$ |
| USS | the uncorrected sum of squares $\sum_{i=1}^{n} x_i^2$ |
| CSS | the corrected sum of squares $\sum_{i=1}^{n} (x_i - \bar{x})^2$ |
| STDERR | the standard error of the mean $s_n/\sqrt{n}$ |
| CV | the coefficient of variation in percent $(s_n/\bar{x}_n) \cdot 100$ |
| MEDIAN | the median |
| P*n* | $n = 1, 5, 10, 25, 50, 75, 90, 95, 99$ $n\%$ quantile. |

Again this list is not complete. If no keyword is specified, PROC MEANS provides the number of observations, the mean, the standard deviation, the minimum and the maximum. If at least one of the keywords is used, only the specified statistics are given in the output.

Without an additional VAR statement, SAS computes the statistics for all numerical variables of the data set. If a VAR statement is specified, the computation is restricted to the specified variables.

Without an additional statement, SAS computes the statistics for the complete data set as one group; the statements BY and CLASS divide the data set in subgroups. Each combination of the specified BY or CLASS variables defines an own subgroup and the calculation is done separately for each subgroup. There are some differences in the possible options between these two statements; the BY statement, in particular, requires data sorted by the BY variables if no NOTSORTED option is specified.

When a FREQ statement appears with PROC MEANS, each observation in the input data set is assumed to represent *n* observations in the calculation of statistics, where $n$ is the value of the FREQ variable. If $n$ is not an integer, only the integer portion is used.

The OUTPUT statement requests that PROC MEANS output statistics are written to a new SAS data set. OUT=*data set* specifies the name of the output data set. By the keywords listed above one can select those statistics which are to appear in the output data set. In parentheses one can tell SAS for which variables these statistics are to be calculated. After an "=" one can coin names for the results. For example,

OUTPUT OUT=meandata MEAN(var1 var2)=mu_var1 mu_var

generates the new data set meandata containing the variables mu_var1 and mu_var2. The values of these variables are the mean of var1 and var2.

Examples are given in Programs 1_6_1 and 2_3_1.

## A.3.2 PROC GPLOT

This procedure is part of the GRAPH module. It generates plots in two and three dimensions. These can be simple scatterplots, overlay plots of more than two variables or bubble plots. In conjunction with a SYMBOL statement it is for instance possible to generate line plots, needle plots and box plots. Most plots in this book are produced by this procedure.

Some global statements like AXIS, LEGEND and SYMBOL are not part of this procedure, but they are crucial for the layout of these plots. Other global statements like TITLE or FOOTNOTE will also affect the graph.

The following statements can be used:

PROC GPLOT ⟨*options*⟩;
    BUBBLE *plot–request(s)* ⟨*/options*⟩;
    BUBBLE2 *plot–request(s)* ⟨*/options*⟩;
    PLOT *plot–request(s)* ⟨*/options*⟩;
    PLOT2 *plot–request(s)* ⟨*/options*⟩;

The PROC GPLOT statement initiates this procedure and, if necessary, specifies the data set that contains the plot data. If no data set is specified by a DATA=*data set* option, the most recently created one is used. Some other options are ANNOTATE=*data set* to annotate all graphs created by this procedure (see Program 3_2_1) and GOUT=*output catalog* to specify a SAS catalog where the graphics output could be stored for later usage, for instance in combination with other graphics (see Program 3_1_1).

The GPLOT procedure needs at least one additional BUBBLE or PLOT statement to create graphical output.

With the BUBBLE statement the procedure generates a plot of three variables, two of them being represented by values on the vertical ($y$) and horizontal ($x$) axes and the third one determines the size of the bubbles. A plot request is of the following form:

$$y\text{--}variable \; * \; x\text{--}variable = bubble\text{--}size\text{--}variable.$$

Multiple plot requests are separated by blanks.

There is a long list of options that can be used in the BUBBLE statement, here are only some of them:

| | |
|---|---|
| BCOLOR=*color* | specifies the bubble color. |
| BLABEL | labels the bubbles with the values of the third variable. |
| BSCALE=AREA\|RADIUS | specifies whether the scaling proportion is based on the area or the radius of the bubble, by default it is the area. |
| BSIZE=*multiples* | specifies an overall scaling factor for the bubble, by default the value is 5. |
| FRAME \|NOFRAME | specifies whether a frame is drawn around the axis area. The default in SAS version 8 is FRAME. |
| HAXIS=AXIS*n* | assigns an AXIS definition to the horizontal axis. |
| HREF=*values* | draws reference line(s) parallel to the vertical axis through the value(s) on the horizontal axis. |
| NAME=*entry-name* | specifies a name for the graph as a catalog entry. |
| VAXIS=AXIS*n* | assigns an AXIS definition to the horizontal axis. |

An example of PROC GPLOT with a BUBBLE statement is given in Program 4_2_3.

A frequently used statement is the PLOT statement. Typically, only two variables are displayed in a plot generated by this statement. They are specified in the following form:

$$y\text{--}variable \; * \; x\text{--}variable.$$

This can be expanded to the form:

$$y\text{--}variable \; * \; x\text{--}variable = n.$$

Here $n$ is the number of the SYMBOL definition which should by used. Another command line is similar to a BUBBLE statement:

$$y\text{--}variable \; * \; x\text{--}variable = third\text{--}variable.$$

But here the third variable is a classification variable with just a few different values. A legend is generated automatically showing the plot symbol and color for each value of the classification variable, optionally defined by SYMBOL statements.

More than one plot can be generated by either using two or more plot requests (as in the BUBBLE statement) or by specifying more than one $y$–variable and/or $x$–variable (set in parentheses and separated by a blank). The options

$$\text{FRAME} \,|\, \text{NOFRAME}, \; \text{HAXIS}{=}\text{AXIS}n, \; \text{HREF}{=}value(s),$$
$$\text{NAME}{=}entry\text{-}name, \; \text{VAXIS}{=}\text{AXIS}n, \; \text{FREF}{=}value(s)$$

are also applicable here. There are additional options, two of them should be mentioned here:

LEGEND | LEGEND=LEGEND*n*     controls a legend (if automatically generated) or specifies the legend to be used for this plot (defined in a LEGEND*n* statement).

OVERLAY     places all the plots generated by the PLOT statement on one set of axes. It does not produce a legend.

While the BUBBLE and the PLOT statements automatically plot the values of the dependent variable on a vertical axis on the left side of a plot, an additional BUBBLE2 or PLOT2 statement creates a second vertical axis on the right side. All options mentioned above for the BUBBLE and PLOT statements could be used here as well, except for the ones concerning the horizontal axis (HAXIS and HREF) and the NAME option.

The AXIS, LEGEND and SYMBOL statements are not part of PROC GPLOT, but they are global statements of the GRAPH module. AXIS can be used in combination with the BUBBLE and PLOT statements (by the HAXIS and VAXIS options), LEGEND (with the LEGEND option) and SYMBOL are only applicable in a PLOT statement.

AXISn options;

where $n$ is an integer between 1 and 99.

With the following selection of options one can define the appearance of the axis:

ORDER=*value list*     specifies the data values in the order they are to appear on the axis, useful especially for categorical variables.

OFFSET=$n_1, n_2\langle unit\rangle$;               specifies the distances from the origin
                                             of the first ($n_1$) and the last ($n_2$) tick
                                             mark, optionally with a unit like CM
                                             (centimeters), IN (inches) or PCT
                                             (percentage of graphics output area).

WIDTH=*thickness factor*                     specifies the thickness of the axis in
                                             form of a factor, default is 1.

LABEL=(*text argument(s)*) | NONE            defines the appearance or the text of
                                             an axis label, or both. Labels can be
                                             suppressed with NONE. Text is de-
                                             fined as a quoted string '*text*' and for-
                                             matted with text description parame-
                                             ters for the font (FONT=*font*) or the
                                             height (HEIGHT=$n$), separated by a
                                             blank.

VALUE=(*text argument(s)*) | NONE            defines the appearance or the text, or
                                             both of the major tick mark values.
                                             The major tick mark values on the
                                             axis can be suppressed (NONE) or
                                             defined by text arguments like FONT
                                             and COLOR.

LEGEND*n* options;

In a LEGEND statement one can define by the options how and where the
legend should appear. A selection of options is:

ACROSS=$n$                                   places the legend entries in rows $n$ en-
                                             tries wide.

DOWN=$n$                                     places the legend entries in columns $n$
                                             entries wide.

FRAME                                        draws a frame around the legend.

LABEL=*text argument(s)* | NONE              defines the appearance or the text of a
                                             legend label, or both. Like the axis label
                                             it can be suppressed with NONE or ex-
                                             plicitly defined with a string in quotes
                                             and attributes.

POSITION=(*position declaration*)            defines the position of the legend on
                                             the graph. Possible values are: IN-
                                             SIDE | OUTSIDE, BOTTOM | MID-
                                             DLE | TOP and LEFT | CENTER
                                             | RIGHT. The default is (BOTTOM
                                             CENTER OUTSIDE).

SYMBOL*n* options;

SYMBOL statements used in the GPLOT procedure control the plot symbols and lines and the interpolation methods.

COLOR=*color*
C=*color*
    specifies a color for all elements, unless a more explicit specification is included.

HEIGHT=*n*⟨*unit*⟩
H=*n*⟨*unit*⟩
    determines the height of the plot symbols in number of units, *n*.

INTERPOL=*option*
I=*option*
    determines whether and how succeeding points of the data set are interpolated (for example by splines or straight lines) or not. It can also create boxplots, a needle plot or a regression line. Some possible values are: BOXT for a boxplot with tops and bottoms of the whiskers, JOIN for a connection of the data points with straight lines, NEEDLE for vertical lines on a horizontal line at the value 0 on the vertical axis, RL for a linear regression line, SPLINE for a smooth interpolation with splines and STEPL for a step function with vertical lines, where the data point is displayed on the left side of the step.

LINE=*n*
L=*n*
    specifies the line type of the plot line with values from 1 to 46. Default is LINE=1, a solid line.

VALUE=*special symbol* | *text string* | NONE
V=*special symbol* | *text string* | NONE

defines the plot symbol for the data points. Default is V=PLUS, other symbols are DOT or STAR, but the symbol can also be an arbitrary character given in quotes. With NONE, no symbol will be plotted, which is especially useful when drawing a smooth curve.

WIDTH=*n*
W=*n*

specifies the thickness of interpolation lines, by default, WIDTH=1.

There are many examples of the PLOT statement in combination with especially different symbol definitions used in this book (see Programs 1_1_3, 1_5_2, 1_6_1 and 3_2_3).

## A.3.3 PROC IML

IML (Interactive Matrix Language) is a SAS module, which provides its own powerful programming language. Operators, functions and call routines are working directly on matrices.

The procedures starts with

   PROC IML;

possibly with an additional option. All following statements are IML code until you finish the procedure with

   QUIT;

A new matrix can be defined by

$$matrix = expression;$$

where *expression* can be an operation on already defined matrices or it can be the specification of the matrix elements like in $a = \{1\ 1\ 1,\ 1\ 0\ 2\}$. This code generates the matrix

$$a = \begin{pmatrix} 1 & 1 & 1 \\ 1 & 0 & 2 \end{pmatrix}.$$

Here are some IML operators working on matrices:

+   adds corresponding matrix elements (in the case of two matrices of the same dimension) or a scalar to each matrix element.

−   subtracts corresponding matrix elements (in the case of two matrices of the same dimension) or a scalar from each matrix element.

\*   performs usual matrix multiplication for matrices with appropriate dimensions.

\#   performs elementwise multiplication either of two matrices of equal dimensions or of a matrix and a scalar.

'   transposes a matrix: $a^T$.

\#\#   raises each element to a power, again either elementwise of two matrices of equal dimensions or of a matrix and a scalar.

The following IML functions are just a small selection of the possible ones:

| | |
|---|---|
| DIAG(*argument*) | creates a diagonal matrix; the *argument* can be a vector (the diagonal elements of the new matrix) or a square matrix. |
| I(*n*) | creates an identity matrix of dimension *n*. |
| INV(*matrix*) | computes the inverse matrix of a square non-singular matrix. |
| J(*n, m, x*) | creates a matrix with *n* rows, *m* columns and identical values equal to *x*. |
| NCOL(*matrix*) | returns the number of columns of a matrix. |
| NROW(*matrix*) | returns the number of rows of a matrix. |
| SQRT(*matrix*) | calculates the (positive) square root of each element of a matrix. |
| XMULT(*matrix1, matrix2*) | computes the matrix product of two matrices with suitable dimensions (like the operator \*, but with extended precision). |

Some matrix calculation can be done by calling subroutines with the CALL statements, for example:

| | |
|---|---|
| CALL EIGEN(*eigenval, eigenvec, matrix*) | computes eigenvalues and eigenvectors of a square matrix; *eigenval* and *eigenvec* are matrices to which the eigenvalues and eigenvectors are returned. |
| CALL SVD (*u, q, v, a*) | computes a singular value decomposition of the matrix *a*; *u*, *q* and *v* are the returned decomposition matrices: $a = u \operatorname{diag}(q) v^T$. |

Columns, rows or arbitrary elements of a matrix can be specified with subscripts:

$a[i,j]$  selects the elements in the $i$-th row and the $j$-th column of the matrix $a$.

$a[i]$  selects all elements of the $i$-th row of the matrix $a$.

$a[,j]$  selects all elements of the $j$-th column of the matrix $a$.

As in every programming language one can use control statements like DO loops or the IF – THEN / ELSE statement, for example

> DO *variable* = *start* TO *stop* ⟨BY *increment*⟩;
>    *statements*
> END;

or

> IF *expression* THEN *statement1*;
> ⟨ELSE *statement2*;⟩

One can print the values of matrices by the PRINT statement:

> PRINT *matrix*;

There are also statements to import data from data sets into a matrix or to export values of a matrix into a data set, for example:

> USE *data set*;
> READ ALL VAR *variable list* INTO *matrix*;

The first statement chooses a data set, the second one reads all observations (rows) and the variables of the *variable list* (columns) into the matrix.

> CREATE *data set* FROM *matrix* ⟨[COLNAME = *vector*] ⟩
> APPEND FROM *matrix*;

A new data set with the columns of the matrix as variables is created by the CREATE statement. To avoid the automatic variable names COL1, ... one can specify variable names by using a previously defined vector which contains the names of the variable. The APPEND statement writes each row of the matrix as an observation to the data set.

Especially in the framework of multivariate analysis, IML can be an enlargement of the capabilities of the standard SAS procedures.

Examples for the application of PROC IML are given in Programs 6_4_3, 7_2_2 and 7_3_2.

# Bibliography

Agresti, A. (1992). A survey of exact inference for contingency tables. *Statistical Science* **7**, 131–177.

Anderson–Sprecher, R. (1994). Model comparison and $R^2$. *The American Statistician* **48**, 113–116.

Andrews, D.F. and Herzberg, A.M. (1985). *Data.* Springer Series in Statistics, Springer, New York.

Balanda, K.P. and MacGillivray, H.L. (1988). Kurtosis: a critical review. The American Statistician **42**, 111–119.

Berberian, S.K. (1999). *Fundamentals of Real Analysis.* Universitext, Springer, New York.

Best, D.I. and Rayner, C.W. (1987). Welch's approximate solution for the Behrens–Fisher problem. *Technometrics* **29**, 205–220.

Cailliez, F. (1983). The analytical solution of the additive constant problem. *Psychometrika* **48**, 305–308.

Chapman, H. and Demeritt, D. (1936). *Elements of Forest Mensuration.* Williams Press, Nashville.

Cheng, B. and Titterington, D.M. (1994). Neural networks: a review from a statistical perspective (with discussion). *Stat. Science* **9**, 2–54.

Chernick, M.R. (1999). *Bootstrap Methods. A Practitioner's Guide.* Wiley Series in Probability and Statistics, Wiley, New York.

Cox, R.D. (1972). Regression models and lifetables (with discussion). *J. Roy. Statist. Soc. B* **34**, 187–220.

Dalal, S.R., Fowlkes, E.B., Hoadley, B. (1989). Risk analysis of the space shuttle: Pre–Challenger prediction of failure. *J. Amer. Statist. Assoc.* 84, 945–957.

Davies, L. and Gather, U. (1993). The identification of multiple outliers (with comments). *J. Amer. Statist. Assoc.* **88**, 782–801.

Dowdall, J.A. (1974). Women's attitudes toward employment and family roles. *Soc. Anal.* **35**, 251–262.

Efron, B. and Tibshirani, R.J. (1993). *An Introduction to the Bootstrap.* Monographs on Statistics and Applied Probability **57**, Chapman & Hall, New York, London.

Falk, M., Hüsler, J. and Reiss, R.-D. (1994). *Laws of Small Numbers: Extremes and Rare Events.* DMV Seminar **23**, Birkhäuser, Basel.

Fahrmeir, L. and Hamerle, A. (1984). *Multivariate statistische Verfahren.* De Gruyter, Berlin.

Fahrmeir, L. and Tutz, G. (1994). *Multivariate Statistical Modelling Based on Generalized Linear Models.* Springer Series in Statistics, Springer, New York.

Feller, W. (1968). *An Introduction to Probability Theory and its Application, Vol. I.* 3rd ed., Wiley, New York.

Fine, T.L. (1999). *Feedforward Neural Network Methodology.* Springer, New York.

Fisher, R.A. (1936). The use of multiple measurements in taxonomic problems. *Ann. Eugenics* **7** (part 2), 179–188.

Freund, R.J., Littell, R.C. and Spector, P.C. (1986). *SAS System for Linear Models.* SAS Institute Inc. Cary, NC.

Friendly, M. (1994). Mosaic displays for multi–way contingency tables. *J. Amer. Statist. Assoc.* **89**, 190–200.

Fristedt, B. and Gray, L. (1997). *A Modern Approach to Probability Theory.* Birkhäuser, Boston.

Gibbons, D.I., McDonald, G.C. and Gunst, R.F. (1987). The complementary use of regression diagnostics and robust estimators. *Naval Research Logistics* **34**, 109–131.

Haberman, S. (1978). *Analysis of Qualitative Data.* Academic Press, New York.

Hampson, R. and Walker, R. (1961). Vapor pressures of platinum, iridium, and rhodium. *J. Res. Nat. Bur. Stand.* **65** A, 289–295.

Hettmansperger, T.P. and Sheather, S.J. (1992). A cautionary note on the method of least median squares. *The American Statistician* **46**, 79–83.

Heuer, G. (1979). *Selbstmord bei Kindern und Jugendlichen.* Klett–Cotta, Stuttgart.

Huber, P.J. (1985). Projection Pursuit (with discussion). *Ann. Statist.* **13**, 435–525.

Huff, D. (1992). *How to Lie with Statistics.* Penguin Books, Harmondsworth, Middlesex.

Institut der deutschen Wirtschaft (1993). *Zahlen zur wirtschaftlichen Entwicklung der Bundesrepublik Deutschland.* Deutscher Instituts–Verlag, Köln.

Jobson, J.D. (1991). *Applied Multivariate Data Analysis. Volume I: Regression and Experimental Design.* Springer Texts in Statistics, Springer, New York.

Jobson, J.D. (1992). *Applied Multivariate Data Analysis. Volume II: Categorical and Multivariate Methods.* Springer Texts in Statistics, Springer, New York.

Jones, M.C. and Sibson, R. (1987). What is projection pursuit? *J. Roy. Statist. Soc.* A **150**, 1–36.

Jones, M.C., Marron, J.S. and Sheather, S.J. (1996). A brief survey of bandwidth selection for density estimation. *Amer. Statist. Assoc.* **91**, 401–407.

Jurečková, J. and Sen, P.K. (1996). *Robust Statistical Procedures. Asymptotics and Interrelations.* Wiley Series in Probability and Statistics, Wiley, New York.

Kaiser, H.F. (1958). The varimax criterion for analytic rotation in factor analysis. *Psychometrica* **23**, 187–200.

Khattree, R. and Naik, D.N. (1999). *Applied Multivariate Statistics with SAS Software.* 2nd ed., Wiley, New York.

Kruskal, J.B. (1964). Multidimensional scaling by optimizing goodness-of-fit to a nonmetric hypothesis. *Psychometrika.* **29**, 1–28, 115–129.

Kwak, J.H. and Hong, S. (1997). *Linear Algebra.* Birkhäuser, Boston.

Lang, S. (1987). *Linear Algebra.* 3rd ed., Springer, New York.

Läuter, H. and Pincus, R. (1989). *Mathematisch–Statistische Datenanalyse.* Oldenbourg Verlag, München–Wien.

Lehmann, E. (1975). *Nonparametrics: Statistical Methods based on Ranks.* Holden-Day, San Francisco.

Lehmann, E. (1993). The Fisher, Neyman–Pearson theories of testing hypothesis: one theory or two? *J. Amer. Statist. Assoc.* **88**, 1242–1249.

Li, X. and Morris, J.M. (1991). On measuring asymmetry and the reliability of the skewness measure. Statist. Probab. Letters **12**, 267–271.

Little, R.J.A. and Wu, M.M. (1991). Models for contingency tables with known margins when target and sampled populations differ. *J. Amer. Statist. Assoc.* **86**, 87–95.

Manteiga, W.G., Sánchez, J.M.P. and Romo, J. (1994). The bootstrap — a review. *Computational Statistics* **9**, 165–205.

Mardia, K.V., Kent, J.T. and Bibby, J.M. (1979). *Multivariate Analysis.* Academic Press, London.

McCulloch, W.S. and Pitts, W. (1943). A logical calculus of ideas immanent in nervous activity. *Bulletin of Mathematical Biophysics* **5**, 115–133.

Miller, R. (1981). *Simultaneous Statistical Inference.* Springer, New York.

Morton, D. et al. (1982). Lead absorption in children of employees in a lead–related industry. *American Journal of Epidemiology* **155**, 549–555.

Müller, D.W. and Sawitzki, G. (1991). Excess mass estimates and tests for multimodality. *J. Amer. Statist. Assoc.* **86**, 738–746.

National Center for Health Statistics (1982). *National Health and Nutrition Examination Survey, Public User Data Tape Documentation.* Catalog Number 5411, Hyattsville.

Natrella, M. (1963). *Experimental Statistics.* National Bureau of Standards Handbook **91**, Washington, D.C.

Pruscha, H. (1989). *Angewandte Methoden der Mathematischen Statistik.* Teubner, Stuttgart.

Rao, C.R. (1973). *Linear Statistical Inference and its Applications.* 2nd ed., Wiley, New York.

Reiss, R.–D. (1989). *Approximate Distributions of Order Statistics (With Applications to Nonparamatric Statistics).* Springer Series in Statistics, Springer, New York.

Reiss, R.–D. and Thomas, M. (2001). *Statistical Analysis of Extreme Values.* 2nd. ed., Birkhäuser, Basel.

Rice, J.A. (1995). *Mathematical Statistics and Data Analysis*. 2nd ed., Duxbury Press, Belmont.

Rohatgi, V.K. (1976). *An Introduction to Probability Theory and Mathematical Statistics*. Wiley, New York.

Rosenbaum, P.R. (1993). Hodges–Lehmann point estimates of treatment effect in observational studies. *J. Amer. Statist. Assoc.* **88**, 1250–1253.

Rousseuw, P.J. and Leroy, A. (1987). *Robust Regression and Outlier Detection*. Wiley, New York.

Sackrowitz, H. and Samuel–Cahn, E. (1999). *P* values as random variables — expected *p* values. *The American Statistician* **53**, 326–331.

SAS Institute Inc. (1990). *SAS/GRAPH Software, Reference, Version 6, First Edition*. SAS Institute Inc. Cary, NC.

SAS Institute Inc. (1992). *SAS Procedures Guide, Version 6, Third Edition*. SAS Institute Inc. Cary, NC.

Satterthwaite, F.W. (1946). An approximate distribution of estimates of variance components. *Biometrics Bulletin* **2**, 110–114.

Schiffman, S.S., Reynolds, M.L. and Young, F.W. (1981). *Introduction to Multidimensional Scaling*. Academic Press, New York.

Schervish, M.J., (1996). *P* values: What they are and what they are not. *The American Statistician*. **50**, 203–206.

Sen, P.K. and Singer, J.M. (1993). *Large Sample Methods in Statistics. An Introduction with Applications*. Chapman & Hall, New York.

Serfling, R.J. (1980). *Approximation Theorems of Mathematical Statistics*. Wiley, New York.

Simonoff, J.S. (1996). *Smoothing Methods in Statistics*. Springer Series in Statistics, Springer, New York.

Smith, K.T. (1983). *Primer of Modern Analysis*. Springer, New York.

Steel, R.G.D. and Torrie, J.H. (1980). *Principles and Procedures of Statistics*. 2nd ed., McGraw–Hill, New York.

Stone, C.J. (1996). *A Course in Probability and Statistics*. Duxbury Press, Belmont.

Takane, Y., Young, F.W. and de Leeuw, J. (1977). Nonmetric individual differences multidimensional scaling: An alternating least squares method with optimal scaling features. *Psychometrika* **42**, 7–67.

Thornes, B. and Collard, J. (1979). *Who divorces?* Routledge & Kegan, London.

Tukey, J.W. (1949). One degree of freedom for non–additivity. *Biometrics* **5**, 232–242.

Tukey, J.W. (1977). *Exploratory Data Analysis*. Addison–Wesley, Reading.

Welch, B.L. (1947). The generalization of "Student's" problem when several different population variances are involved. *Biometrika* **34**, 28–35.

White, J., Riethof, M. and Kushnir, I. (1960). Estimation of microcrystalline wax in beeswax. *J. Assoc. Offic. Anal. Chem.* **43**, 781–790.

# Data Index

The data sets used in this book can be downloaded as raw data from http://statistics-with-sas.ku-eichstaett.de/ in ASCII format as well as in SAS system formats, see the above address for more information.

| Name of data set | File name | Description |
|---|---|---|
| Air Pollution Data | air.dat | Example 3.3.4 |
| Air Pollution II Data | polution.dat | Example 7.1.1 |
| Beeswax Data | beeswax.dat | Example 1.1.11 |
| Chestnut Data | chestnut.dat | Example 3.2.1 |
| CNS Data | cns.dat | Example 1.1.3 |
| Claim Size Data | claim.dat | Example 4.1.9 |
| Crystal Data | crystal.dat | Example 2.3.1 |
| Course Data | course.dat | Example 5.1.1 |
| Economy Data | economy.dat | Example 3.3.1 |
| Geo Data | geo.dat | Example 3.2.6 |
| Grounds for Divorce Data | grounds.dat | Example 7.2.7 |
| Ice Data | ice.dat | Example 2.4.4 |
| Lead Data | lead.dat | Example 2.3.8 |
| Mathematics Data | math.dat | Exercise 18, Chapter 8 |
| O–Ring Data | oring.dat | Example 4.2.6 |
| pH Data | ph.dat | Exercise 1, Chapter 1 |
| Platinum Data | platinum.dat | Example 1.2.2 |
| Random Data | random.dat | Example 7.5.1 |
| $SO_2$ Data | so2.dat | Example 5.1.4 |
| Suicide Data | suicide.dat | Example 4.1.11 |
| Sunspot Data | sunspot.dat | Example 1.5.1 |

The following data sets are not stored in separate files, instead they are contained in the corresponding program and read by the CARDS statement:

| Divorce Data | | Example 4.2.1 |
|---|---|---|
| Drug Data | | Figure 4.1.4 |
| Habitude Data | | Example 7.3.4 |
| Hair Color Data | | Example 4.1.13 |
| NHANES II Data | | Example 4.1.2 |
| Wheat Data | | Example 5.2.4 |

The following data sets are listed in the text only:

| Ethno Data | | Exercise 13, Chapter 4 |
|---|---|---|
| Yoga Data | | Exercise 25, Chapter 2. |

# Index

# STATISTICS FOR INDUSTRY AND TECHNOLOGY

Publications in the series will contain statistical information that is accessible to an interdisciplinary audience: carefully organized authoritative presentations; numerous illustrative examples based on current practice; reliable methods; realistic data sets; and discussions of select new emerging methods and their application potential.

Reiss, R.D. / Thomas, M., Universität-Gesamthochschule
Siegen, Germany

## Statistical Analysis of Extreme Values

from Insurance, Finance, Hydrology
and Other Fields
Second Edition
2001. 462 pages. Softcover. CD-ROM included
€ 58.50 / sFr. 88.–
ISBN 3-7643-6487-4

The statistical analysis of extreme data is important for various
disciplines, including hydrology, insurance, finance, engineering and
environmental sciences. This book provides a self-contained introduction
to the parametric modeling, exploratory analysis and statistical inference
for extreme values.

Besides numerous data-based examples, the book contains special
chapters about flood frequency analysis (coauthored by J.R.M. Hosking),
insurance (coauthored by M. Radtke) and finance (coauthored by C.G. de
Vries and S. Caserta). In addition, five longer case studies are included
that replace those presented in the first edition.

The assessment of the adequacy of the parametric modeling and the
statistical inference is facilitated by the included statistical software
Academic Xtremes, an interactive menu-driven system which runs under
Windows 95, 98, 2000, NT. The applicability of the system is enhanced
by the integrated programming language StatPascal.

It is the declared aim of the second extended edition to enforce the
characteristic of the book of providing a broad statistical background.
The new highlights, elaborated on about 160 pages, include

- the statistical modeling of tails in conjunction with the global modeling
  of distributions with special emphasis laid on heavy-tailed distributions;
- the Bayesian methodology with applications to regional flood
  frequency analysis and credibility estimation in reinsurance business;
- a thorough treatment of the phenomenon of penultimate distributions;
- a section about conditional extremes;
- an extension of the chapter about multivariate extreme value models,
  especially for the Gumbel-McFadden model with an application to the
  theory of economic choice behavior.

\* € price is net price

For orders originating from all over
the world except USA and Canada:
**Birkhäuser Verlag AG**
c/o Springer GmbH & Co
Haberstrasse 7
D-69126 Heidelberg
Fax: ++49 / 6221 / 345 229
e-mail: birkhauser@springer.de

For orders originating in the USA
and Canada:
Birkhäuser
333 Meadowland Parkway
USA-Secaucus
NJ 07094-2491
Fax: +1 201 348 4505
e-mail: orders@birkhauser.com

**http://www.birkhauser.ch**

Probability & Statistics with Birkhäuser

**Good, P.I.**, Information research, Huntington Beach, USA

# Resampling Methods
## A Practical Guide to Data Analysis
Second Edition
2002. 250 pages. Hardcover
€ 98.13 / sFr. 158.–
ISBN 0-8176-4243-9

The goal of this book is to introduce statistical methodology-estimation, hypothesis, testing and classification-to a wide applied audience through resampling from existing data via the bootstrap, and estimation or cross-validation methods. The book provides an accessible introduction and practical guide to the power, simplicity and veritability of the bootstrap, cross-validation and permutation tests. Industrial statistical consultants, professionals and researchers will find the book's methods and software imimediately helpful.

This second edition is a practical guide to data analysis using the bootstrap, cross-validation, and permutation tests. It is an essential resource for industrial statisticians, statistical consultants, and research professionals in science, engineering, and technology.

Only requiring minimal mathematics beyond algebra, it provides a table-free introduction to data analysis  utilizing numerous exercizes, practical data sets, and freely available statistical shareware.

Topics and features:
Thoroughly revised text features more practical examples plus an additional chapter devoted to regression and data mining techniques and their limitations

- Uses resampling approach to introduction statistics
- A Practical presentation that covers all three sampling methods – bootstrap, density-estimation, and permutations
- Includes systematic guide to help one select correct procedure for a particular application
- Detailed coverage of all three statistical methodologies – classification, estimation, and hypothesis testing
- Suitable for classroom use and individual, self-study purposes
- Numerous practical examples using popular computer programs such as SAS, Stata, and StatXact

\* € price is net price

For orders originating from all over the world except USA and Canada:
**Birkhäuser Verlag AG**
c/o Springer GmbH & Co
Haberstrasse 7
D-69126 Heidelberg
Fax: ++49 / 6221 / 345 229
e-mail: birkhauser@springer.de

For orders originating in the USA and Canada:
**Birkhäuser**
333 Meadowland Parkway
USA-Secaucus
NJ 07094-2491
Fax: +1 201 348 4505
e-mail: orders@birkhauser.com

http://www.birkhauser.ch

*Probability & Statistics with Birkhäuser*